METHODS IN MOLECULAR BIOLOGY™

Series Editor
John M. Walker
School of Life Sciences
University of Hertfordshire
Hatfield, Hertfordshire, AL10 9AB, UK

For further volumes:
http://www.springer.com/series/7651

Microglia

Methods and Protocols

Edited by

Bertrand Joseph

Department of Oncology-Pathology, Karolinska Institutet, Cancer Centrum Karolinska, Stockholm, Sweden

José Luis Venero

Departamento de Bioquímica y Biología Molecular, Facultad de Farmacia, Universidad de Sevilla, Sevilla, Spain; Instituto de Biomedicina de Sevilla, Hospital Universitario, Sevilla, Spain

Humana Press

Editors
Bertrand Joseph
Department of Oncology-Pathology
Karolinska Institutet
Stockholm, Sweden

José Luis Venero
Departamento de Bioquímica
y Biología Molecular
Facultad de Farmacia
Universidad de Sevilla
Sevilla, Spain
Instituto de Biomedicina de Sevilla
Hospital Universitario,Sevilla, Spain

ISSN 1064-3745 ISSN 1940-6029 (electronic)
ISBN 978-1-4939-6051-4 ISBN 978-1-62703-520-0 (eBook)
DOI 10.1007/978-1-62703-520-0
Springer New York Heidelberg Dordrecht London

Printed on acid-free paper

Humana Press is a brand of Springer
Springer is part of Springer Science+Business Media (www.springer.com)

Preface

Microglia are resident myeloid and immune effector cells of the central nervous system. Pío del Río-Hortega, a disciple of Santiago Ramón y Cajal, named these cells "microglia" in 1919 [1]. The cells had been previously described in the 1880s by Franz Nissl and William Ford Robertson, but Pío del Río-Hortega conducted the first systematic studies on this cell type and remarkably many of his observations are still valid [2]. He is accordingly considered as the "Father of Microglia."

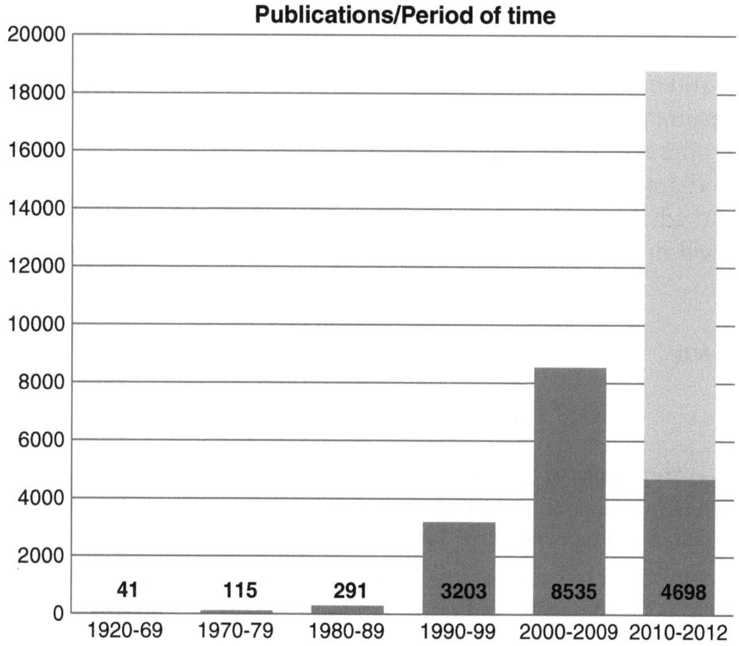

Considering this historical note, it becomes evident that for a long period of time little improvement was made in our knowledge of microglia. Indeed, this field of research remained rather confidential for more than half a century. However, ever since the 1990s there is a clear regain of interest for microglia as evidenced by the constant increase at an almost exponential rate in the number of publications. In fact, there are none less than 8,000 publications including the term microglia during the past decade and if one considers the first quarter of the current decade one may expect over 18,000 publications on the subject at the end of this decade.

The reasons for this recent and considerable interest for microglia are certainly multiple. First and foremost, key discoveries concerning the different biological functions of microglia in health and disease have attracted scientists from various fields. Indeed, microglia act as sentinels of infection and injury, and participate in both innate and adaptive immune responses in the central nervous system. Microglia can also be dysregulated in the context of neurodegenerative disease and cancer, and thereby contribute to disease severity. Microglia were as well as recently shown to play roles in normal brain development, adult neuronal plasticity, and circuit function. A second factor which has undoubtedly contributed to the expansion of this field of research is the development of model systems and methods to investigate microglia biological functions both in vitro and in vivo.

In light of the interest for microglia, the aim of *Microglia: Methods and Protocols* is to provide a selection of the key cellular, molecular, and biochemical techniques that are used in studying the many and varied functions of this fascinating cell. The chapters of *Microglia: Methods and Protocols* are written by experts who have hands-on experience with the particular method. They provide a comprehensive step-by-step guide to many techniques for microglia cell culture, for studying microglia activation and functions, as well as their interaction with other cell types, both in vitro and in vivo. These methods and protocols should provide a useful resource for cell biologists, molecular biologists, immunologists, oncologists, and neuroscientists.

This book is dedicated to the memory of Dr. Laia Acarin (1970–2011), Professor at the Medical School and member of the Institute of Neuroscience at the Autonomous University of Barcelona, Spain. She was a brilliant scientist, teacher, and mentor and made prominent contributions to the field of microglia.

We wish to thank all the authors for their excellent contributions and Prof. John M. Walker for sound advice and assistance throughout the editorial process. It is our sincerest hope that you will find *Microglia: Methods and Protocols* a useful lab companion for years to come.

Stockholm, Sweden — ***Bertrand Joseph***
Sevilla, Spain — ***José Luis Venero***

1. Del Río-Hortega P (1919)El 'tercer elemento' de los centros nerviosos. I. La microglia en estado normal. II. Intervencíon de la microglia en los procesos patológicos. III. Naturaleza probable de la microglia. Bol de la Soc esp de biol 9:69–129
2. Del Río-Hortega P (1932) Microglia. In: Cytology & cellular pathology of the nervous system. Paul B. Hoeber, Inc., New York, p 483–534

Contents

Contributors

BEATRIZ ALMOLDA • *Universitat Autònoma de Barcelona, Barcelona, Spain*
JACK P. ANTEL • *Neuroimmunology Unit, Montréal Neurological Institute, McGill University, Montreal, Canada*
ALICIA A. BABCOCK • *Institute of Molecular Medicine, University of Southern Denmark, Odense, Denmark*
ANIRBAN BASU • *National Brain Research Centre, Manesar, India*
ALAIN BESSIS • *Ecole Normale Supérieure, Paris, France*
MANON BLAIN • *McGill University, Montreal, Canada*
GUY C. BROWN • *University of Cambridge, Cambridge, UK*
MIGUEL A. BURGUILLOS • *Karolinska Institutet, Stockholm, Sweden*
ASTRID E. CARDONA • *Department of Biology and South Texas Center for Emerging Infectious Diseases, The University of Texas at San Antonio, San Antonio, TX, USA*
SANDRA M. CARDONA • *The University of Texas at San Antonio, San Antonio, TX, USA*
BERNARDO CASTELLANO • *Universitat Autònoma de Barcelona, Bellaterra, Spain*
SHIH-HENG CHEN • *National Institute of Environmental Health Sciences, Research Triangle Park, NC, USA*
BETTINA CLAUSEN • *University of Southern Denmark, Odense, Denmark*
SABRINA COLASSE • *Ecole Normale Supérieure, Paris, France*
FERNANDO G. CORREA • *Instituto Cajal, Consejo Superior de Investigaciones Científicas, Madrid, Spain; Universidad de Buenos Aires-Consejo Nacional de Investigaciones Científicas y Técnicas, Buenos Aires, Argentina*
ROCÍO M. DE PABLOS • *Universidad de Sevilla, Sevilla, Spain; Instituto de Biomedicina de Sevilla, Sevilla, Spain*
TOMAS DEIERBORG • *Lund University, Lund, Sweden*
MIKE DRAGUNOW • *The University of Auckland, Auckland, New Zealand*
BRYCE A. DURAFOURT • *McGill University, Montreal, Canada*
CLAUDIA EDER • *St. Georges', University of London, London, UK*
ANA M. ESPINOSA-OLIVA • *Universidad de Sevilla, Sevilla, Spain; Instituto de Biomedicina de Sevilla, Sevilla, Spain*
RICHARD L.M. FAULL • *The University of Auckland, Auckland, New Zealand*
CHRISTINA FENGER • *University of Southern Denmark, Odense, Denmark*
BENTE FINSEN • *Neurobiology Research, Institute of Molecular Medicine, University of Southern Denmark, Odense, Denmark*
SADANAND M. GAIKWAD • *University of Bonn, Bonn, Germany*
JENNY A. GARCIA • *The University of Texas at San Antonio, San Antonio, TX, USA*
HANNAH M. GIBBONS • *The University of Auckland, Auckland, New Zealand*
BERTA GONZÁLEZ • *Universitat Autònoma de Barcelona, Barcelona, Spain*
CARMEN GUAZA • *Instituto Cajal, Consejo Superior de Investigaciones Científicas, Madrid, Spain*
ASHLEY S. HARMS • *Emory University School of Medicine, Atlanta, GA, USA*

Michael T. Heneka • *Clinical Neuroscience Unit, Department of Neurology, University of Bonn, Bonn, Germany*
Miriam Hernangómez • *Instituto Cajal, Consejo Superior de Investigaciones Científicas, Madrid, Spain*
Antonio J. Herrera • *Facultad de Farmacia, Departamento de Bioquímica y Biología Molecular, Universidad de Sevilla, Sevilla, Spain; Instituto de Biomedicina de Sevilla, Sevilla, Spain*
Jau-Shyong Hong • *National Institute of Environmental Health Sciences, Research Triangle Park, NC, USA*
Kazuhide Inoue • *Kyushu University, Fukuoka, Japan*
Bertrand Joseph • *Department of Oncology-Pathology, Karolinska Institutet, Cancer Centrum Karolinska, Stockholm, Sweden*
Deepak Kumar Kaushik • *Department of cellular and Molecular Neurosciences, National Brain Research Centre, Manesar, India*
Edel Kavanagh • *Karolinska Institutet, Stockholm, Sweden*
Satoru Kondo • *Department of Molecular Physiology, Kyushu University, Fukuoka, Japan*
Jae-Kyung Lee • *Emory University School of Medicine, Atlanta, GA, USA*
Claire Lill • *Department of Anatomy with Radiology, The University of Auckland, Auckland, New Zealand*
Takahiro Masuda • *Kyushu University, Fukuoda, Japan*
Craig S. Moore • *McGill University, Montreal, QC, Canada*
Urte Neniskyte • *University of Cambridge, Cambridge, UK*
Shigeo Okabe • *University of Tokyo, Tokyo, Japan*
Esteban A. Oyarzabal • *National Institute of Environmental Health Sciences, Research Triangle Park, NC, USA*
Lorena Pont-Lezica • *Ecole Normale Supérieure, Paris, France*
Kandiyil Prajeeth Chittappen • *Hannover Medical School, Hannover, Germany*
Refik Pul • *Hannover Medical School, Hannover, Germany*
Johanna Rodhe • *Karolinska Institutet, Stockholm, Sweden*
Manuel Sarmiento • *Department of Oncology, CR-UK/MRC Gray Institute for Radiation Oncology and Biology, University of Oxford, Oxford, UK*
Tom Schilling • *St. George's, University of London, London, UK*
Amy M. Smith • *The University of Auckland, Auckland, New Zealand*
Martin Stangel • *Hannover Medical School, Hannover, Germany*
Malú G. Tansey • *Emory University School of Medicine, Atlanta, GA, USA*
Hidetoshi Tozaki-Saitoh • *Kyushu University, Fukuoda, Japan*
Makoto Tsuda • *Kyushu University, Fukuoda, Japan*
José Luis Venero • *Departamento de Bioquímica y Biología Molecular, Facultad de Farmacia, Universidad de Sevilla, Sevilla, Spain; Instituto de Biomedicina de Sevilla, Hospital Universitario, Sevilla, Spain*
Nàdia Villacampa • *Universitat Autònoma de Barcelona, Barcelona, Spain*
Martin Wirenfeldt • *University of Southern Denmark, Odense, Denmark*

Part I

Overview of Microglia Biology

Chapter 1

A Brief Overview of Multitalented Microglia

Bertrand Joseph and José Luis Venero

Abstract

Microglia are the resident immune cells of the central nervous system, and accumulating data demonstrates a vast array of tasks in the healthy and injured brain. Microglia participate in both innate and adaptive immune responses. These cells contribute to the brain homeostasis, including the regulation of cell death, synapse elimination, neurogenesis, and neuronal surveillance. However, microglia can also become activated and/or deregulated in the context of neurodegenerative diseases, brain injuries, and cancer and thereby contribute to disease severity. As a consequence of these developments, microglia have attracted substantial attention on themselves.

Key words Immune cells, Central nervous system, Microglia, Brain homeostasis, Cell death

1 Overview of the Multitalented Microglia

Microglia are key glial cell elements of the central nervous system (CNS) and are considered the major immunocompetent cells in the brain. Consequently, microglia trigger signaling cascades well established in the immune system involving chemokines and cytokines and their receptor systems. In keeping with this view, microglia share many features of monocytes, and they are constantly scavenging for damaged neurons, plaques, and infectious agents using their multiple surface receptors [1, 2].

Microglia were first described by Rio-Hortega, a disciple of Ramón y Cajal, in 1919 in a visionary article [3] in which he anticipated key features of microglia cells. For instance, Rio-Hortega deduced that (1) microglia enter the brain during early development; (2) they acquire a branched, ramified morphology in the mature brain; (3) they undergo a morphological transformation into an amoeboid-shape morphology in response to different types of injury; and (4) they have the capacity to migrate, proliferate, and phagocytose. As Rio-Hortega predicted, microglia populate the mammalian CNS in early embryonic development. By adulthood,

Bertrand Joseph and José Luis Venero (eds.), *Microglia: Methods and Protocols*, Methods in Molecular Biology, vol. 1041,
DOI 10.1007/978-1-62703-520-0_1,

microglia are found in all regions of the brain and spinal cord and comprise 10–15 % of all CNS cells. In addition to being sensitive to changes in their environment, each microglial cell also regularly physically surveys its domain. Microglia are believed to derive from monocytes that invade the developing CNS and persist over the adult life as resident macrophages [4, 5]. An elegant study using fate-mapping analysis confirmed that these glial cells derive from primitive myeloid progenitors that arise before embryonic day 8 and that postnatal hematopoietic progenitors do not contribute to microglia homeostasis in the adult brain [6].

Recent studies using in vivo two-photon imaging in undamaged normal brain reveal that microglial cells are highly active in their presumed resting state, continually surveying their microenvironment with extremely motile processes and protrusions [7, 8]. In the healthy brain, microglia habitually interact with neuronal and nonneuronal elements, both structurally and functionally, including phagocytosis of synaptic structures during postnatal development, phagocytosis of newborn neurons during adult neurogenesis, and active remodeling of the perisynaptic environment and release of soluble factors in the mature and aging brain [9]. Thus, microglial cells fulfil an astonishing variety of tasks within the CNS; they are multitalented cells demanding multidisciplinary approaches and methodologies to get further insights into their multiple tasks.

When microglia are "activated," they take on an amoeboid shape with phagocytic activity (again anticipated by del Rio-Hortega [3]), and they increase their gene expression leading to the production of numerous potentially neurotoxic mediators, such as proteases and proinflammatory cytokines. Another group of potentially neurotoxic mediators are reactive oxygen species (ROS) and NO. ROS, including superoxide, hydroxyl radicals, and hydrogen peroxide, are highly reactive molecules, involved in various signal transduction cascades [10]. ROS form little threat in low concentrations as cells possess various defense and repair mechanisms for minor oxidative stress. However, during inflammation, they are released in high concentrations by the oxidative burst of activated microglia. High ROS concentrations overrule cellular defense mechanisms and cause oxidative damage to proteins, lipids, and nucleic acids with subsequent risk for neuronal populations.

Activated microglia also release various neurotrophic factors and cytokines that can modify neuronal circuits [11, 12]. These mediators are important in the normal functions of microglia, and their production is usually decreased once their task is complete. During recent years, researchers have tended to differentiate between acute inflammation and chronic inflammation and their effects on neurons. Neuroinflammation is a complex innate response of neural tissue against the harmful effects of diverse stimuli within the central nervous system (CNS). Infections, trauma, stroke, toxins, and other stimuli are capable of producing an acute neuroinflammatory response within the CNS with the activation of resident immune cells (microglia, astrocytes) and the release of inflammatory mediators such as cytokines and chemokines. Neurodegenerative disorders, including Alzheimer's disease (AD), Parkinson's disease (PD), Huntington's disease (HD), multiple sclerosis (MS), and amyotrophic lateral sclerosis (ALS), are associated with chronic neuroinflammation [10]. Chronic neuroinflammation can be seen as a long-lasting and often self-perpetuating neuroinflammatory response. Chronic neuroinflammation involves the sustained activation of microglia, and consequent release of proinflammatory mediators, but also an increased oxidative and nitrosative stress which are both detrimental for the neuronal cell population. Increasing evidences indicate that microglia can become chronically activated in response to dying/damaged neurons, fuelling a self-renewing cycle of microglial activation, causing a self-propelling cycle of neuron death. This repeating cycle of neurotoxic microglial activation in response to neuron injury is commonly referred to as reactive microgliosis, which is a proposed mechanism of chronic neuronal loss across diverse neurodegenerative diseases [13]. However, not all immune responses in the CNS are detrimental, and in many cases, they actually aid repair and regeneration [14]. For example, microglia clear debris after myelin damage and when this is impeded, delayed regeneration occurs [15]. Immune activation is also crucial to limit neurotropic viral infections, and immune cells remove necrotic cells following ischemia [16]. Consequently, when studying the roles of microglia in different scenarios, it is very important to define the paradigm of microglia/macrophage activation [17]. Microglia and macrophages can be classified into at least two subsets with distinct molecular phenotypes and effector functions depending on the activation pathway. The "classically activated" proinflammatory M1 macrophages, activated by LPS and by the proinflammatory cytokine IFN-γ, produce high amounts of oxidative metabolites, proteases, and proinflammatory cytokines. They play a central role in the host defense against pathogens and tumor cells, but they can also damage healthy cells such as neurons and glial cells. In contrast, M2 macrophages or "alternatively activated," induced by IL-4 and IL-13, exhibit anti-inflammatory properties and promote tissue remodeling/repair and angiogenesis [17, 18]. Microglial cells are attracted toward glioma in large numbers—glioma tissue consists of as much as 30 % microglial cells—and microglia density in gliomas positively correlates with malignancy, invasiveness, and grading of tumors. Soluble factors released from glioma stimulate microglial Toll-like receptors, resulting in microglia activation and infiltration of the tumor. Thereafter, tumor cells shut down their immune properties and stimulate the microglia to release tumor growth-promoting factors. In fact, microglia release many factors, including extracellular matrix proteases and cytokines, which directly or indirectly may influence tumor invasiveness and growth [19–21]. As further evidence of their essential role in glioma biology, their removal has been shown in vivo to inhibit glioma invasiveness.

Another major flow of microglia research relies on knowledge of the impressive array of multiple receptors located on these cells. Among others, they include α-amino-3-hydroxy-5-methyl-4-isoxazolepropionic acid (AMPA) receptors to respond to glutamate [22], P2 purinoceptors to respond to ATP and other nucleotides [23, 24], cytokine receptors to respond to tumor necrosis factor (TNF)-α and interleukin (IL)-1β [24], and receptors for the major histocompatibility complex (MHC) class antigens and chemokines [25–27]. Activation of these receptors induces signaling cascades in microglia that have been shown to be involved in microglia chemotaxis, activation, and phagocytosis. Furthermore, microglial P2XRs and P2Yrs receptors play a key role in pain signaling in the spinal cord under pathological conditions, such as following peripheral nerve injury (neuropathic pain) [28]. Further, microglia express pattern recognition receptors (PRRs), which act as sensor for inflammatory triggers. The inflammatory response can be triggered by the recognition of several microbial signals, through conserved molecular pattern known as pathogen-associated molecular patterns (PAMPs) [29] or intracellular molecules, normally found in the cytosol or in the nucleus and are released from the cell after tissue damage known as danger-associated molecular patterns (DAMPs) [30]. PAMPs and DAMPs are detected by members of Toll-like receptors (TLRs) family of PRR [31]. The nucleotide oligomerization domain (NOD)-like receptor (NLR) family is a large family of cytosolic PRRs, specialized in the recognition of intracellular signals [32]. Once activated by the inflammatory stimuli, the PRRs (TLRs and NLRs) participate in the formation of multiprotein oligomers known as inflammasomes, which orchestrates the activation of caspase-1, which in turn contributes to cytokine maturation and the inflammatory responses [33]. We recently described a novel and unexpected role for so-called apoptotic caspases in the activation of microglia and associated neurotoxicity [34]. Activation of TLRs with proinflammogens leads to the orderly activation of caspase-8 and caspase-3. Caspase-3 activates the NF-κB pathways through processing and activation of protein kinase C-δ. This finding may be relevant in the onset of chronic neurodegenerative diseases, mainly PD and AD [35]. In fact, several intracellular signaling pathways contribute to microglial activation. Activation of most TLRs results in the recruitment of the adaptor protein myeloid differentiation factor-88 (MyD88), serine/threonine kinase IL-1R-associated kinase (IRAK), tumor necrosis factor (TNF) receptor-associated factor (TRAF) adaptor protein TRAF6, NF-kB-inducing kinase (NIK), and IkB kinase (IKK), thereby leading to the nuclear translocation of NF-kB [29]. However, other alternative adaptor molecules including TIR-associated protein (TIRAP)/MyD88-adaptor-like (MAL), TIR-domain-containing adaptor protein-inducing IFN-β (TRIF)/TIR-domain-containing molecule 1 (TICAM1), and TRIF-related adaptor molecule (TRAM) exist, which transduce signals from TLRs via an MyD88-independent pathway [29]. In canonical interferon (IFN)-γ-Janus kinase (Jak)-signal transducer and activator of transcription 1 (STAT1) signaling (reviewed in ref. [36]), the binding of IFN-γ to its receptor leads to activation of receptor-associated Jak1 and Jak2 and phosphorylation of a receptor tyrosine residue working as a docking site for STAT1. STAT1 is then activated by phosphorylation and translocates to the nucleus where it stimulates transcription of STAT1 target genes. Another way for IFN-γ to activate macrophages is by enhancing macrophage responsiveness to other inflammatory stimuli, such as TLR ligands and TNF. This mechanism is known by the name of "priming." By this "priming" of IFN-γ, the TLR-dependent expression of several proinflammatory cytokines and chemokines is greatly augmented [37].

The clarification of the microglia biological functions in the healthy brain, the discovery of their significant contribution to various CNS diseases, ranging from neurodegenerative disorders, stroke, brain injuries, to cancer, as well as the advances in the understanding of the signaling pathways controlling their activation and functions have collectively contributed to the literal "explosion of interest" for these cells which used to be rather ignored. As a result, methods, protocols, and tools have been developed to study the multitalented microglia.

References

1. Perry VH, Gordon S (1988) Macrophages and microglia in the nervous system. Trends Neurosci 11(6):273–277
2. Kreutzberg GW (1996) Microglia: a sensor for pathological events in the CNS. Trends Neurosci 19(8):312–318
3. Rio-Hortega Pd (1919) El "tercer elemento" de los centros nerviosos. Poder fagocitario y movilidad de la microglia. Bol de la Soc espanola de Biol. Ano ix:154–166
4. Alliot F, Godin I, Pessac B (1999) Microglia derive from progenitors, originating from the yolk sac, and which proliferate in the brain. Brain Res Dev Brain Res 117(2): 145–152
5. Gomes-Leal W (2012) Microglial physiopathology: how to explain the dual role of microglia after acute neural disorders? Brain Behav 2(3):345–356
6. Ginhoux F, Greter M, Leboeuf M, Nandi S, See P, Gokhan S, Mehler MF, Conway SJ, Ng LG, Stanley ER, Samokhvalov IM, Merad M (2010) Fate mapping analysis reveals that adult microglia derive from primitive macrophages. Science 330(6005):841–845
7. Davalos D, Grutzendler J, Yang G, Kim JV, Zuo Y, Jung S, Littman DR, Dustin ML, Gan WB (2005) ATP mediates rapid microglial response to local brain injury in vivo. Nat Neurosci 8(6):752–758
8. Nimmerjahn A, Kirchhoff F, Helmchen F (2005) Resting microglial cells are highly dynamic surveillants of brain parenchyma in vivo. Science 308(5726):1314–1318
9. Tremblay ME, Stevens B, Sierra A, Wake H, Bessis A, Nimmerjahn A (2011) The role of microglia in the healthy brain. J Neurosci 31(45):16064–16069

10. Venero JL, Burguillos MA, Brundin P, Joseph B (2011) The executioners sing a new song: killer caspases activate microglia. Cell Death Differ 18(11):1679–1691
11. Merrill JE (1992) Tumor necrosis factor alpha, interleukin 1 and related cytokines in brain development: normal and pathological. Dev Neurosci 14(1):1–10
12. Biber K, Neumann H, Inoue K, Boddeke HW (2007) Neuronal 'On' and 'Off' signals control microglia. Trends Neurosci 30(11): 596–602
13. Block ML, Zecca L, Hong JS (2007) Microglia-mediated neurotoxicity: uncovering the molecular mechanisms. Nat Rev Neurosci 8(1):57–69
14. Aguzzi A, Barres BA, Bennett ML (2013) Microglia: scapegoat, saboteur, or something else? Science 339(6116):156–161
15. Neumann H, Kotter MR, Franklin RJ (2009) Debris clearance by microglia: an essential link between degeneration and regeneration. Brain 132(Pt 2):288–295
16. Amor S, Puentes F, Baker D, van der Valk P (2010) Inflammation in neurodegenerative diseases. Immunology 129(2):154–169
17. Gordon S, Martinez FO (2010) Alternative activation of macrophages: mechanism and functions. Immunity 32(5):593–604
18. Czeh M, Gressens P, Kaindl AM (2011) The yin and yang of microglia. Dev Neurosci 33(3–4):199–209
19. Badie B, Schartner J (2001) Role of microglia in glioma biology. Microsc Res Tech 54(2): 106–113
20. Markovic DS, Vinnakota K, Chirasani S, Synowitz M, Raguet H, Stock K, Sliwa M, Lehmann S, Kalin R, van Rooijen N, Holmbeck K, Heppner FL, Kiwit J, Matyash V, Lehnardt S, Kaminska B et al (2009) Gliomas induce and exploit microglial MT1-MMP expression for tumor expansion. Proc Natl Acad Sci USA 106(30):12530–12535
21. Saijo K, Glass CK (2011) Microglial cell origin and phenotypes in health and disease. Nat Rev Immunol 11(11):775–787
22. Noda M, Nakanishi H, Nabekura J, Akaike N (2000) AMPA-kainate subtypes of glutamate receptor in rat cerebral microglia. J Neurosci 20(1):251–258
23. Tsuda M, Shigemoto-Mogami Y, Koizumi S, Mizokoshi A, Kohsaka S, Salter MW, Inoue K (2003) P2X4 receptors induced in spinal microglia gate tactile allodynia after nerve injury. Nature 424(6950):778–783
24. Koizumi S, Shigemoto-Mogami Y, Nasu-Tada K, Shinozaki Y, Ohsawa K, Tsuda M, Joshi BV, Jacobson KA, Kohsaka S, Inoue K (2007) UDP acting at P2Y6 receptors is a mediator of microglial phagocytosis. Nature 446(7139):1091–1095
25. Finsen BR, Jorgensen MB, Diemer NH, Zimmer J (1993) Microglial MHC antigen expression after ischemic and kainic acid lesions of the adult rat hippocampus. Glia 7(1):41–49
26. Bessis A, Bechade C, Bernard D, Roumier A (2007) Microglial control of neuronal death and synaptic properties. Glia 55(3):233–238
27. Cullheim S, Thams S (2007) The microglial networks of the brain and their role in neuronal network plasticity after lesion. Brain Res Rev 55(1):89–96
28. Tsuda M, Tozaki-Saitoh H, Inoue K (2012) Purinergic system, microglia and neuropathic pain. Curr Opin Pharmacol 12(1):74–79
29. Akira S, Uematsu S, Takeuchi O (2006) Pathogen recognition and innate immunity. Cell 124(4):783–801
30. Beg AA (2002) Endogenous ligands of Toll-like receptors: implications for regulating inflammatory and immune responses. Trends Immunol 23(11):509–512
31. Akira S, Takeda K (2004) Functions of toll-like receptors: lessons from KO mice. C R Biol 327(6):581–589
32. Strober W, Murray PJ, Kitani A, Watanabe T (2006) Signalling pathways and molecular interactions of NOD1 and NOD2. Nat Rev Immunol 6(1):9–20
33. Yazdi AS, Guarda G, D'Ombrain MC, Drexler SK (2010) Inflammatory caspases in innate immunity and inflammation. J Innate Immun 2(3):228–237
34. Venero JL, Burguillos MA, Joseph B (2013) Caspases playing in the field of neuroinflammation: old and new players. Dev Neurosci. doi:10.1159/000346155
35. Burguillos MA, Deierborg T, Kavanagh E, Persson A, Hajji N, Garcia-Quintanilla A, Cano J, Brundin P, Englund E, Venero JL, Joseph B (2011) Caspase signalling controls microglia activation and neurotoxicity. Nature 472(7343):319–324
36. Stark GR (2007) How cells respond to interferons revisited: from early history to current complexity. Cytokine Growth Factor Rev 18(5–6):419–423
37. Hu X, Paik PK, Chen J, Yarilina A, Kockeritz L, Lu TT, Woodgett JR, Ivashkiv LB (2006) IFN-gamma suppresses IL-10 production and synergizes with TLR2 by regulating GSK3 and CREB/AP-1 proteins. Immunity 24(5): 563–574

Part II

Isolation and Culture of Microglia

Chapter 2

Cell Culturing of Human and Murine Microglia Cell Lines

Johanna Rodhe

Abstract

Despite the fact that microglia cells were first described almost a century ago, microglia-derived immortalized cell lines have only been established in the last two decades. One should be aware of their limitations but also of their advantages. Cell lines offer a potentially powerful tool to investigate some functional aspects of microglia. Cell culturing of human and murine microglia cell lines will be described in this chapter. It includes a presentation of equipment needed, cell culture medium and supplements, cell culture monitoring, and a protocol describing the steps for subculturing of microglia cell lines.

Key words Microglia cell line, BV2, CHME5, Cell culture, Medium, Serum

1 Introduction

The concept of microglia cells was first described by Rio-Hortega in 1919 in studies of tissue sections of the brain [1]. Even though isolation of primary microglia cells for in vitro cultures was already described in 1930, the first immortalized cell lines from rodents were not established until the late 1980s when microglia cells from primary cultures of mouse brain were infected with myc oncogenes [2, 3]. Several rodent cell lines were established in the following years, before microglia cells could be isolated from human embryonic spinal cord and cortex and immortalized to form the first human microglia cell line in 1995 [4].

Microglia cell lines are derived from primary cultures of microglia cells from the brain or spinal cord, which are then immortalized by, e.g., infection by a retroviral vector carrying a v-raf, v-myc, or c-myc oncogene, the simian virus 40 (SV40) large T antigen, or spontaneously immortalized in culture (*see* Table 1) [2–12].

Immortalized microglia cells can express several typical microglia/macrophage markers, like CD11b/Mac1, CD68, FcR, MHC class I, MHC class II, B4, and Mac2 [13]. They should keep some of

Bertrand Joseph and José Luis Venero (eds.), *Microglia: Methods and Protocols*, Methods in Molecular Biology, vol. 1041, DOI 10.1007/978-1-62703-520-0_2, © Springer Science+Business Media New York 2013

Table 1
Table showing some of the existing murine and human microglia cell lines

Cell line	Species	Immortalization method	Year	Authors
N9, N13	Mouse	v-myc oncogene	1989	Righi et al. [2]
BV2 (BV-2)	Mouse	v-raf/v-myc oncogenes	1990	Blasi et al. [3]
EOC	Mouse	Spontaneous	1995	Walker et al. [5]
C8-B4	Mouse	Spontaneous	1996	Alliot et al. [6]
MG5	Mouse	p53-deficient mice	1997	Ohsawa et al. [7]
MG6 (MG6-1)	Mouse	c-myc oncogene	2005	Takenouchi et al. [8]
MLS-9	Rat		1998	Zhou et al. [9]
Ra2	Rat	Spontaneous	1998	Sawada et al. [10]
HAPI	Rat	Spontaneous	2001	Cheepsunthorn et al. [11]
CHME	Human	SV40 large T antigen	1995	Janabi et al. [4]
HMO6	Human	v-myc oncogene	2001	Nagai et al. [12]

the important characteristics of microglial cells, like the ability of phagocytosis and secretion of cytokines upon activation, e.g. IL-6 and TNF-α [2, 11, 12].

A comparison with primary microglia culture shows that there are both advantages as well as limitations when using continuous cell lines. A primary cell culture is time-consuming to prepare and gives a limited number of cells per animal, which can only be used for a small number of passages. The established cell lines on the other hand can be thawed and frozen when needed and give a virtually unlimited number of cells, which can be used for numerous of experiments over many passages. Cell lines also have the advantage of being characterized and are generally at the same stage of maturation. They are a more homogenous population of cells derived from one clone, which gives less variation in experiments and makes it easier to compare results between different labs. This can also be seen as a disadvantage; since the cell lines are selected from the expansion of one clone, there is a possibility that the cell line has lost important characteristics or markers from the population in primary culture. Continuous cell lines are also rapidly dividing and have therefore the ability to differentiate and change characteristics over time, which can affect the outcome of long-term experiments. Since there are differences between the cell lines in expression of antigens and in response to different stimuli, it is important to select the cell line which has the characteristics needed to study the question in focus.

The access to human samples for isolating microglia from the brain and spinal cord is very limited in both availability and yield, which makes it a great advantage of having continuous cell lines. Microglia cell lines are adherent cells that grow in monolayer in cell culture. The cells are grown in a H_2O-saturated atmosphere with 5 % CO_2 at 37 °C. The following section describes a routine protocol for subculturing of common microglia cell lines, such as BV2 and CHME5.

2 Material

1. Dulbecco's Modified Eagle Medium (DMEM: high glucose with glutamax/glutamine), supplemented with 10 % heat-inactivated Fetal Bovine Serum (FBS), penicillin (100 U/ml), and streptomycin (100 µg/ml) (*see* **Notes 1** and **2**).
2. 0,05 % trypsin–EDTA.
3. Phosphate-buffered saline (PBS) pH 7.4.
4. Cell culture flasks.
5. 5 and 10 ml pipettes.
6. Falcon tubes.

3 Methods

3.1 Preparation

1. Warm cell culture medium and trypsin bottle to 37 °C.
2. Prepare working area and bring out all equipment needed. All cell culture work should be performed in laminar flow hood using sterile techniques, and all equipment and solutions used for cell culture must be sterile.
3. Check the cells under an inverted microscope to ensure healthy cells (*see* **Note 3**), mainly attached to bottom of flask and expanding at normal growth rate (*see* **Note 4**). Observe cell culture to ensure absence of contamination (*see* **Note 5**) and look out for color changes in the culture medium.

3.2 Subculturing

1. Remove and discard the medium from the cell flask.
2. Wash the cells with 5 ml room temperature PBS, tip the flask gently to rinse off remaining FBS from the culture media, and then remove the PBS.
3. Add 2–3 ml of trypsin–EDTA (ca. 1 ml per 25 cm^2 of surface area), rock the flask gently for complete coverage of the cell layer, and leave in an incubator at 37 °C for 1–5 min for the cells to detach. Avoid prolonged periods in trypsin by checking the cells oft during the trypsinization. If cells are slow to detach, tap gently on the bottle for the cells to lift off.

4. As soon as the cells have detached, add 5–10 ml of complete medium (2–3 times the amount of trypsin added) to the cells and transfer cell suspension to a Falcon tube.
5. Spin down the cells in a centrifuge at $200 \times g$ for 5 min.
6. Remove and discard the supernatant and resuspend the cell pellet in new medium.
7. Add desired amount of cell suspension into a new culture flask and add complete culture medium (*see* **Notes 6** and 7).
8. Return cell flask to incubator.

4 Notes

1. RPMI 1640 medium can also be used for culturing of microglia cell lines [2, 3] and can be used to replace DMEM if preferred. Another option for some microglia cell lines is to use MEM as culture medium [9, 10]. Although most cell lines are cultured in regular culture medium, some cell lines do require special supplements to proliferate, like GM–CSF or insulin [10]. Consult original publications of the cell lines for information about eventual need of specific supplements.
2. For experiments with microglia cells, the medium used for the treatment of cells can be supplemented with a lower amount of FBS than in continuous cell culture, e.g., when activating cells using the proinflammatory stimulant lipopolysaccharide (LPS). Medium containing 5 % FBS can be used for experiments [11, 14] or even serum-free medium during the treatments [15].

   The amount of serum that is used, the source, batch, and storage of serum (e.g., repeated freeze and thaw cycles) can be a source of diversity between experiments both within and between labs. Keeping the same standards within experiments is important for consistency in results. The choice of what amount of serum to use depends on duration of experiments and possible effect of components in the serum. In cell culture, serum is the source of nutrients for the cells, but in vivo the cells only come in contact with complete serum in pathological conditions when breakdown of blood–brain barrier exposes microglia to several serum components that can cause microglia activation [16, 17]. This is something to keep in mind for experiments using compounds to stimulate activation of microglia cells, since the serum exposure can give a higher baseline of microglia activation. An example of this is the LPS-binding protein (LBP), which can be found in serum. The binding of LBP to LPS increases its affinity for binding to microglia surface receptors, which results in an increased sensitivity to LPS and a higher degree of microglia activation [18].

Another source of diversity here can also be the choice of LPS. The source, batch, storage conditions, and methods of application of compound affect the sensitivity of microglia cells to LPS.

3. A good habit is to look at the cells under the microscope at a regular basis. Get to know how the cell line looks under normal conditions to be able to observe any changes in the culture and recognize changes in morphology and signs of activated cells when evaluating effect of experiments. Also note that healthy cells should have light refracting in a ring around the cells in the inverted light microscope.
4. Many microglia cell lines are sensitive to being over-confluent and should be subcultured at ca. 80–90 % confluency. Subculturing should be done every 2–3 days depending on growth speed of cell line.
5. Always look out for traces of contamination in the cell culture. Even very low levels of bacterial contamination can activate microglia cells.
6. Cells are often split in a ratio of 1:4–1:10, depending on growth rate of the cell line. Add ca. 5–10 ml of complete DMEM per 25 cm^2 of surface area of the flask.
7. It is good to keep track of the number of passages for the cell line, e.g., write passage number on the culture bottle before returning it to the incubator. Keep the number of passages relatively low to avoid selection pressure and diversion from the parental cell line. The maximum number of passages can differ between cell lines. Never let the amount of passages exceed the number where you have to question if the characteristics that are studied are compromised.

References

1. Rio Hortega PD (1919) El tercer elemento de los centros nerviosos. Bol Soc Esp Biol 9:69–129
2. Righi M, Mori L, De Libero G, Sironi M, Biondi A, Mantovani A, Donini SD, Ricciardi-Castagnoli P (1989) Monokine production by microglial cell clones. Eur J Immunol 19(8):1443–1448
3. Blasi E, Barluzzi R, Bocchini V, Mazzolla R, Bistoni F (1990) Immortalization of murine microglial cells by a v-raf/v-myc carrying retrovirus. J Neuroimmunol 27(2–3):229–237
4. Janabi N, Peudenier S, Héron B, Ng KH, Tardieu M (1995) Establishment of human microglial cell lines after transfection of primary cultures of embryonic microglial cells with the SV40 large T antigen. Neurosci Lett 195(2):105–108
5. Walker WS, Gatewood J, Olivas E, Askew D, Havenith CE (1995) Mouse microglial cell lines differing in constitutive and interferon-gamma-inducible antigen-presenting activities for naive and memory CD4+ and CD8+ T cells. J Neuroimmunol 63(2):163–174
6. Alliot F, Marty MC, Cambier D, Pessac B (1996) A spontaneously immortalized mouse microglial cell line expressing CD4. Brain Res Dev Brain Res 95(1):140–143
7. Ohsawa K, Imai Y, Nakajima K, Kohsaka S (1997) Generation and characterization of a microglial cell line, MG5, derived from a p53-deficient mouse. Glia 21(3):285–298

8. Takenouchi T, Ogihara K, Sato M, Kitani H (2005) Inhibitory effects of U73122 and U73343 on Ca2+ influx and pore formation induced by the activation of P2X7 nucleotide receptors in mouse microglial cell line. Biochim Biophys Acta 1726(2):177–186
9. Zhou W, Cayabyab FS, Pennefather PS, Schlichter LC, DeCoursey TE (1998) HERG-like K+ channels in microglia. J Gen Physiol 111(6):781–794
10. Sawada M, Imai F, Suzuki H, Hayakawa M, Kanno T, Nagatsu T (1998) Brain-specific gene expression by immortalized microglial cell-mediated gene transfer in the mammalian brain. FEBS Lett 433(1–2):37–40
11. Cheepsunthorn P, Radov L, Menzies S, Reid J, Connor JR (2001) Characterization of a novel brain-derived microglial cell line isolated from neonatal rat brain. Glia 35(1):53–62
12. Nagai A, Nakagawa E, Hatori K, Choi HB, McLarnon JG, Lee MA, Kim SU (2001) Generation and characterization of immortalized human microglial cell lines: expression of cytokines and chemokines. Neurobiol Dis 8(6):1057–1068
13. De Vries GH, Boullerne AI (2010) Glial cell lines: an overview. Neurochem Res 35(12): 1978–2000
14. Burguillos MA, Deierborg T, Kavanagh E, Persson A, Hajji N, Garcia-Quintanilla A, Cano J, Brundin P, Englund E, Venero JL, Joseph B (2011) Caspase signalling controls microglia activation and neurotoxicity. Nature 472(7343):319–324
15. Pocivavsek A, Burns MP, Rebeck GW (2009) Low-density lipoprotein receptors regulate microglial inflammation through c-Jun N-terminal kinase. Glia 57(4):444–453
16. Möller T, Weinstein JR, Hanisch UK (2006) Activation of microglial cells by thrombin: past, present, and future. Semin Thromb Hemost 32(Suppl 1):69–76
17. Ransohoff RM, Perry VH (2009) Microglial physiology: unique stimuli, specialized responses. Annu Rev Immunol 27:119–145
18. Schumann RR, Leong SR, Flaggs GW, Gray PW, Wright SD, Mathison JC, Tobias PS, Ulevitch RJ (1990) Structure and function of lipopolysaccharide binding protein. Science 249(4975):1429–1431

Chapter 3

Microglia Isolation from Adult Mouse Brain

Jae-Kyung Lee and Malú G. Tansey

Abstract

Although microglia isolation from embryonic or postnatal mouse brain is possible using a number of different protocols, microglia isolation from adult brain is more challenging and often results in low yields. Here, we describe a protocol to isolate intact microglia from adult mouse brain for functional assays, immunocytochemistry, and/or flow cytometry analysis. This protocol involves enzymatic dissociation in medium supplemented with dispase II, papain, and DNase I followed by mechanical dissociation. Cell separation is achieved via percoll gradients of various densities. Microglia isolated using this protocol is suitable for flow cytometry analysis, RNA isolation for gene expression by real-time PCR or microarrays, and for functional assays including cytokine production, chemotaxis, and phagocytosis.

Key words Microglia, CD11b, Inflammatory gene expression, Flow cytometry, Immunocytochemistry

1 Introduction

Microglia, the myeloid-derived resident macrophages of the brain, play the primary role of immune surveillance and respond to environmental stress and immunological challenges [1, 2]. Initial physical or pathogenic events in the CNS can trigger microglial expansion through recruitment of peripheral macrophages to the CNS, as a result of increased permeability of the BBB, differentiation from progenitor cells, or proliferation of residual microglia [3]. Acute activation of microglia often results in secretion of neurotrophic factors such as glial-derived neurotrophic factor family ligands (GFLs) that limit tissue injury by protecting vulnerable populations of neurons and aiding in repair processes [4, 5]. However, activated microglia can also overproduce prostaglandins, chemokines, cytokines, and reactive oxygen and nitrogen species (ROS and RNS) including nitric oxide (NO), which can have a deleterious effect on neuronal survival by enhancing oxidative stress and activating cell death pathways [6]. If glial activation persists for extended periods, a cycle of chronic neuroinflammation ensues and contributes to tissue damage. This protocol describes a

Bertrand Joseph and José Luis Venero (eds.), *Microglia: Methods and Protocols*, Methods in Molecular Biology, vol. 1041, DOI 10.1007/978-1-62703-520-0_3, © Springer Science+Business Media New York 2013

stepwise method (modified from one previously described [7]) to isolate intact microglia from adult mouse brain. Microglia isolated using this method are suitable for in vitro studies aimed at understanding the mechanisms that regulate the balance of neuroprotective and neurotoxic microglial activities.

2 Materials

2.1 Reagents

1. PBS-perfused brains from adult mice (age >8 weeks).
2. Hank's balanced salt solution (HBSS) without calcium and magnesium.
3. 10× HBSS.
4. DNase I (final concentration; 20 U/mL, Invitrogen).
5. Dispase II (final concentration; 1.2 U/mL, Roche).
6. Papain (1 mg/mL, Sigma-Aldrich).
7. DMEM/F12 medium.
8. 100× penicillin/streptomycin.
9. Percoll.

3 Media and Solutions

1. Dissection medium (50 mL).

   49.5 mL of HBSS plus 0.5 mL of penicillin/streptomycin (100×).

   Sterile filter all solutions after mixing/before use.
2. Glucose.

   Prepare 0.45 g/mL in PBS (100×).

   Add 500 μL in 50 mL reagents to make final concentration of 4,500 mg/L.
3. Serum-free medium (50 mL).

   49 mL of DMEM/F12 plus 0.5 mL of penicillin/streptomycin (100×) and 0.5 mL of glucose (100×).

   Sterile filter all solutions after mixing/before use.
4. Neutralization medium (50 mL).

   44 mL of DMEM/F12 plus 0.5 mL of penicillin/streptomycin (100×), 0.5 mL of glucose (100×), and 5 mL of heat-inactivated FBS (or FCS).

   Sterile filter all solutions after mixing/before use.

5. Culture medium (50 mL).

   44.5 mL of DMEM/F12 plus 0.5 mL of penicillin/streptomycin (100×) and 5 mL of heat-inactivated FBS (or FCS).

   Sterile filter all solutions after mixing/before use.

6. Dispase II

   Dissolve the non-sterile lyophilized enzyme in HEPES-buffered saline (50 mM HEPES/KOH pH 7.4, 150 mM NaCl) (10 mg/mL). Dilute further with the culture medium to be used for the isolated cells at the final concentration of 2.4 U/mL. Concentrations higher than 2.4 U/mL are not recommended. Sterilize through a 0.22 μm filter membrane.

7. Dissociation medium.

   Prepare dispase, DNase, and papain (DDP) solution.

   - 0.028 g papain (final concentration 1 mg/mL).
   - 14 mL of DMEM/F12.
   - 14 mL premade dispase II (final concentration 1.2 U/mL).

   Aliquot in −20 °C.

   Add DNase I (final concentration 20 U/mL) right before use.

8. Percoll preparation.

   Prepare stock isotonic percoll (SIP); mix nine parts of percoll with one part 10× HBSS.

   70 % percoll: 7.0 mL of SIP plus 2.0 mL of 1× HBSS.

   37 % percoll: 3.7 mL of SIP plus 5.3 mL of 1× HBSS.

   30 % percoll: 3.0 mL of SIP plus 6.0 mL of 1× HBSS.

4 Methods

1. Place the freshly perfused adult brain in 2 mL serum-free media into the 35 mm dish and mince the freshly perfused adult brain as finely as possible using the 15 T scalpel blade. Make sure the amount of serum-free medium is equivalent to the amount of dissociation medium.
2. Transfer minced brain to 15 mL tube containing 3 mL of dissociation medium.
3. Gently rock (or invert the tube every 5 min) cell suspension in tissue culture incubator for 20 min.
4. Neutralize the enzymes in dissociation media by adding 5 mL of neutralization medium.
5. Centrifuge 5 min at 250 ×*g* in room temperature.

6. Aspirate the medium slowly (be extremely cautious not to disturb the pellet as it can easily be aspirated by the aspirating pipette).
7. Resuspend the pellet in 5 mL of serum-free medium. Repeat **steps 5** and **6**.
8. Add 3 mL DMEM/F12 and pipette up and down with polished large size-hole Pasteur pipette against bottom of tube until large clumps of tissue are broken up. Let it sit for 1 min until large clumps settle. Transfer upper part (containing dissociated cells) into new 15 mL conical tube and keep it on ice.
9. Add 3 mL of DMEM/F12 and pipette up and down with polished Pasteur pipette (medium-size hole) against bottom of tube until large clumps of tissue are broken up. Let it sit for 1–2 min until large clumps settle. Transfer upper part into collecting tube with previous cell suspension and keep it on ice.
10. Add 2 mL DMEM/F12 and pipette up and down with polished Pasteur pipette (small-size hole) to break clumps. Let it sit for 1 min until large clumps settle.
11. Combine upper part (containing dissociated cells) into previous cell suspension.
12. Wet 40 μm cell strainer with 2 mL of DMEM/F12 and then filter cell suspension through cell strainer.
13. Spin at 250 × *g* for 4 min.
14. Resuspend cells in 5 mL DMEM/F12, spin 250 × *g* for 4 min, and remove supernatant.
15. Resuspend cell pellet in 4 mL per brain of 37 % SIP.
16. Transfer 4 mL of the 37 % SIP (from **step 15**) to 15 mL conical tubes and slowly underlay 4 mL of 70 % percoll (*see* **Note 1**). Then on top of the 37 % layer, slowly pipette 4 mL of 30 % percoll, followed by 2 mL of HBSS (Fig. 1).
17. Centrifuge gradient 40 min at 300 × *g* (18 °C) with no brake. ***Important!! Make sure centrifuge will stop with no brake so that the interphase is not disturbed.
18. Using a transfer pipette, gently remove layer of debris and collect 2.0–2.5 mL of the 70–37 % interphase into a clean 15 mL conical tube.
19. Add 6 mL of HBSS for each 2 mL of interphase volume collected to ensure the percoll containing the interphase is diluted about three times.
20. Centrifuge 7 min at 500 × *g* at 4 °C.

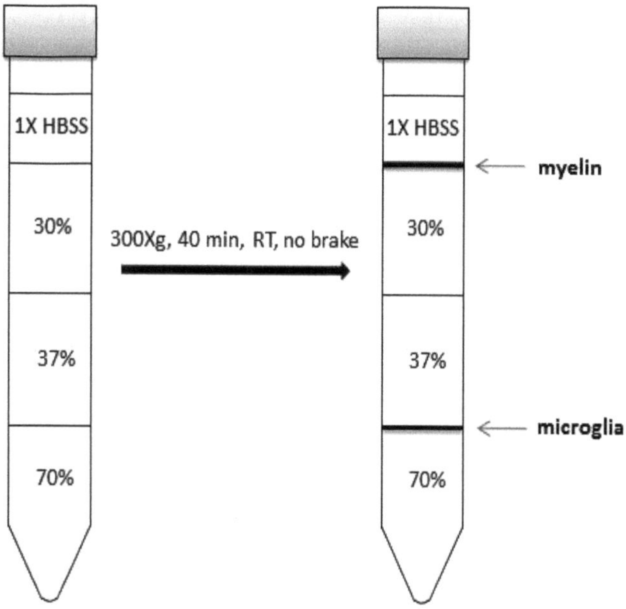

Fig. 1 Schematic of the percoll gradient setup for isolation of adult mouse microglia

21. Resuspend pellet in 500 μL of HBSS and transfer to small 0.6 mL or 1.5 mL tubes and wash three times in a volume of 500 μL, using a micro-centrifuge at 800 × *g* at 4 °C.
22. Count cells using a hemocytometer and seed cells on the culture plate for immunocytochemistry or functional assays (*see* **Notes 2–4**).

5 Notes

1. Percoll should be kept at room temperature.
2. Typical yield is from 300,000 to 500,000 microglial cells per mouse brain (*see* Fig. 2).
3. All the procedures need to be performed under the hood and reagents needs to be filtered to prevent contamination.
4. Cells can be maintained in the culture media for functional analysis (*see* Fig. 3).

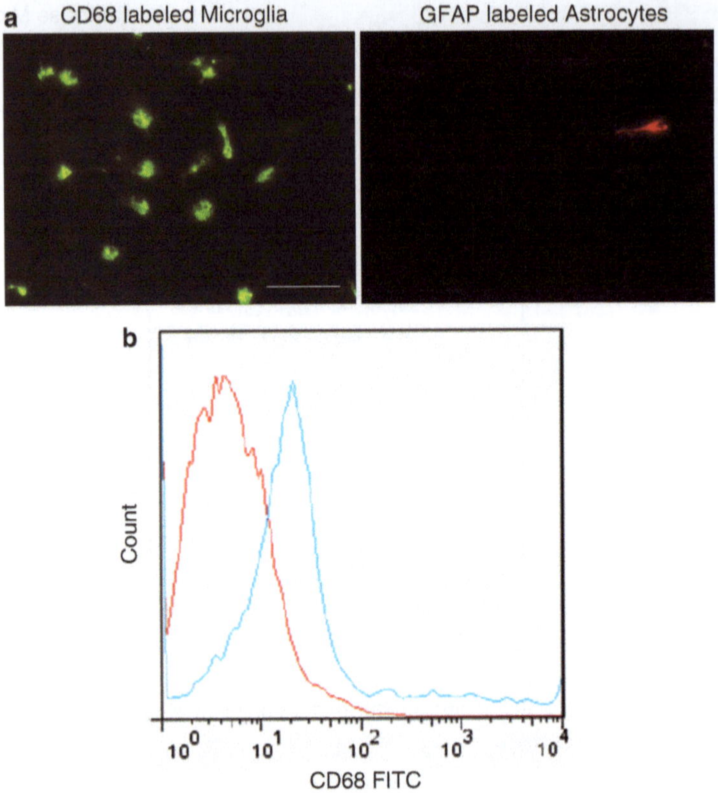

Fig. 2 Purity of isolated adult mouse microglia population. (**a**) Primary microglia cells from C57BL/6 mice were plated into a 4-well chamber plate at the density of 8,000 cells/well in growth medium. Twenty-four hours after seeding, cells were fixed with 4 % paraformaldehyde for 15 min. Cells were then stained with CD68 (as microglia marker, *green*) and GFAP (as astrocyte marker, *red*) antibodies followed by incubating with Alexa fluorophore-conjugated secondary antibody (Invitrogen). (**b**) Primary microglia cells were labeled with fluorescently conjugated CD68-FITC antibodies (*green*), and fixed and analyzed by flow cytometry. Unstained microglia (*red*) were used as negative controls. Histogram plot indicates the presence of a single-labeled population. Scale bar represents 100 μm

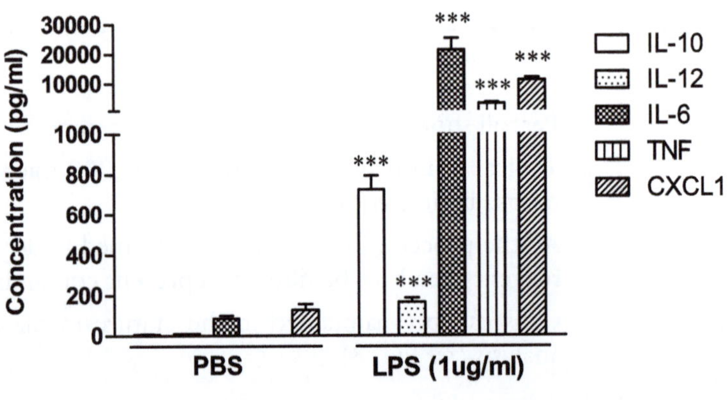

Fig. 3 Isolated adult mouse microglia secrete cytokines upon stimulation with LPS. Primary microglia cells were plated at a density of 3,000 cells per well in 96-well plate. Cells were treated with 1 μg/mL of LPS for 18 h. Conditioned media were collected and analyzed for inflammatory factor production by multiplexed immunoassay (Meso Scale Discovery). Student *t*-test was used for statistical analysis. The *asterisks* denote significant differences between PBS and LPS treatment at $p < 0.001$

Acknowledgements

This work was supported by grant 1R01 NS072467 (MGT) from NIH/NINDS.

References

1. Puntambekar SS, Doose JM, Carson MJ (2008) Microglia: a CNS-specific tissue macrophage. In: Lane TE, Carson M, Bergmann C, Wyss-Coray T (eds) Central nervous system diseases and inflammation. Springer, New York, pp 1–12
2. Tansey MG, Wyss-Coray T (2008) Cytokines in CNS inflammation and disease. In: Lane TEC, Carson M, Bergmann C, Wyss-Coray T (eds) Central nervous system diseases and inflammation. Springer, New York, pp 59–106
3. Bjorkqvist M, Wild EJ, Thiele J et al (2008) A novel pathogenic pathway of immune activation detectable before clinical onset in Huntington's disease. J Exp Med 205(8):1869–1877
4. Wyss-Coray T, Mucke L (2002) Inflammation in neurodegenerative disease—a double-edged sword. Neuron 35(3):419–432
5. Marchetti B, Abbracchio MP (2005) To be or not to be (inflamed) – is that the question in anti-inflammatory drug therapy of neurodegenerative disorders? Trends Pharmacol Sci 26(10):517–525
6. McGeer PL, McGeer EG (2004) Inflammation and neurodegeneration in Parkinson's disease. Parkinsonism Relat Disord 10(Suppl 1):S3–S7
7. Cardona AE, Huang D, Sasse ME et al (2006) Isolation of murine microglial cells for RNA analysis or flow cytometry. Nat Protoc 1(4): 1947–1951

Chapter 4

Preparation of Primary Microglia Cultures from Postnatal Mouse and Rat Brains

Tomas Deierborg

Abstract

Microglia are the inflammatory cells of the brain and are activated in neuropathological conditions. To study the biology of microglia, these cells can be isolated from the brain and analyzed in terms of pro- and anti-inflammatory cytokine production, involvement of intracellular signaling pathways upon inflammatory stimuli, phagocytosis, and several more biological aspects to understand their role in the brain. In this book chapter, I will discuss microglial cells and describe how these cells can be cultured from a postnatal mouse and rat brain.

Key words Microglia, Inflammation, Cytokine, Postnatal, Cell culture

1 Introduction

Microglial cells are the immune cells of CNS and play an important role not only in inflammation processes in neuropathological conditions but also to maintain normal homeostasis under physiological conditions [1].

In the situation of neurodegeneration and acute brain injuries, microglia are proliferating, adapting a hypertrophic morphology, and migrate towards an injury.

Microglia is a phagocytizing cell type and can typically be found at the site of injury where the cells take care of cell debris. Microglia in the brain have two origins: one is the parenchymal origin meaning that the cells come from precursors in the brain parenchyma and the second origin is from blood-borne monocytes. These two different origins cannot be distinguished at an immunohistological level, but the different populations can be distinguished in terms of expression levels of CD11b, where infiltrating cells from the blood have a high expression of CD11b [2].

Microglia respond quickly to a multitude of stimulus, which leads to a production of cytokines, predominantly proinflammatory

Bertrand Joseph and José Luis Venero (eds.), *Microglia: Methods and Protocols*, Methods in Molecular Biology, vol. 1041,
DOI 10.1007/978-1-62703-520-0_4,

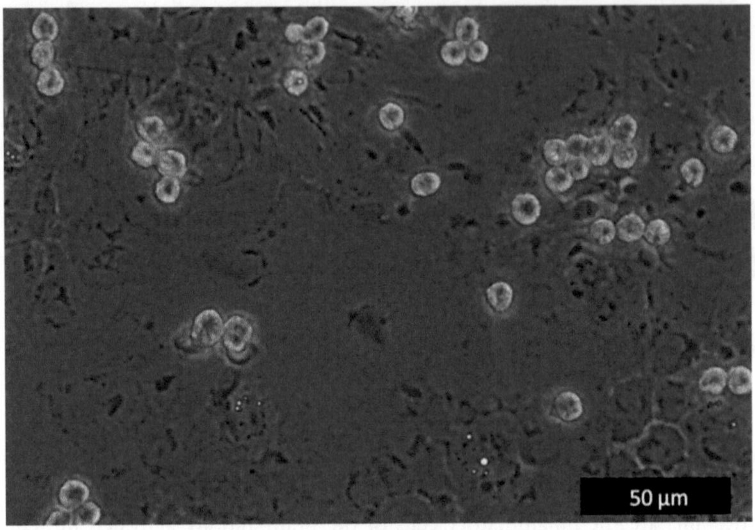

Fig. 1 Mixed culture of glial cells from postnatal mouse brains showing clearly the microglial cells as small round cells on top of the astrocytic monolayer. Scale bar, 50 μm

cytokines (e.g., IL-1β, TNF-α) that are initiating and maintaining the inflammatory response but also anti-inflammatory cytokines (e.g., IL-10 and IL-4) that are important in the regulation of the inflammatory cascade. Hence, determining the role of microglia in neuroinflammatory conditions and their role in inflammation propagation and attenuation is important for understanding and treating pathological conditions in the brain. Studying primary cultures of microglia from rats and mice provides important and reproducible methods to unravel the role of microglia in brain pathologies.

Isolation and culturing of microglia from the postnatal rodent brain is a well-characterized and reliable method to study microglia [3]. Though, it should be emphasized that microglia in culture are cells that have been altered by the isolation and culturing condition and might therefore not resemble the normal environment in the brain. The preparation and isolation protocol is straightforward where a mixed population of glial cells is isolated and grown together before the microglia is isolated after a couple of weeks. The small amoeboid microglial cells are growing on top of a monolayer of astrocytes where the microglia typically appears as clusters or colonies suggesting a clonal expansion of precursor cells or individual cells (*see* Fig. 1). Earlier studies have characterized this distinct phenotype of amoeboid microglial cells. Microglial cells are similar to macrophages and show ability to secrete cytokines, produce reactive oxygen species, phagocytize, and proliferate. The ramified microglia population on the other hand has a reduced proliferation rate and ability of phagocytosis. When cultures are maintained for longer time periods, the ramified microglial cells will integrate into the astrocytic layer [4, 5]. Interestingly, increasing intracellular calcium or cAMP will make the cells convert back to a round amoeboid morphology [4] showing, at least in vitro, that the two distinct phenotypes can convert from one to another. Microglia has an amoeboid morphology during embryogenesis that is reduced postnatally and increases following brain injuries.

It is suggested that brain-derived microglia is different to bone marrow-derived brain macrophages [6] in terms of inflammatory profile and morphology and that microglial cells respond to granulocyte/macrophage colony-stimulating factor (GM-CSF) unlike bone marrow-derived monocytes or tissue macrophages [7]. In mixed glial cultures, which will be described later in terms of how to harvest microglia, the astrocytes are important for the proliferation of microglia, and astrocytes also have an effect on the ramification of the microglia [7, 8]. Astrocytes produce trophic factors, e.g., macrophage colony-stimulating factor (M-CSF) and GM-CSF, that will regulate microglial cells [9, 10]. Since these factors will have the ability to change the phenotypes of the microglia, i.e., inducing ramification [11], augments phagocytic ability, changing cytokine production, and antigen-presenting ability [9, 12, 13], there are reasons to avoid these mitogens in the culturing procedures. It has been suggested that microglia from adult mice that has been treated with GM-CSF can acquire a dendritic cell phenotype, whereas the presence of the growth factor M-CSF can convert the microglia into a proliferative phenotype [14]. Also, the rat microglia after treatment with GM-CSF have a reduced period of proliferation which corresponds to shortening of telomeres [15].

In this book chapter I will describe a simple and reproducible method to isolate brain cells, culturing mixed glial cultures without additional mitogens, only using the supportive function of astrocytes, and harvest pure microglial cells that could be seeded and cultured for detailed cellular experiments.

2 Materials

The material needed to dissect out and culture microglial cells from mouse pups postnatal day 1–4.

1. Dissection microscope.
2. Ice-cold Hank's Balanced Salt Solution without bivalent ions (HBSS).
3. Freezing block.
4. Petri dishes.
5. Two pairs of fine forceps.
6. Fine scissors.
7. Trypsin (0.1 %).

8. DNase (0.05 %).
9. DMEM + GlutaMAX medium containing 4.5 g/l glucose without sodium pyruvate.
10. Fetal Bovine Serum (FBS).
11. Penicillin/streptomycin (100 U/ml and 0.1 mg/ml, respectively).
12. Eppendorf tubes.
13. Falcon tubes (15 or 50 ml).
14. T75 culture flasks.
15. Incubator.
16. Centrifuge.
17. Sterile dissections hood/bench.
18. Pipette 1 ml and pipette tips.

3 Methods

The isolation of tissue from mouse and rats is similar. I will describe the typical isolation of tissue from the mouse brain, but the same protocol can be used for the rat brain. Microglia is derived from mouse pups postnatal days 1–4.

1. Mouse pups are decapitated and heads are put in ice-cold HBSS.

   For the whole procedure, brains and dissected out tissue are kept in ice-cold HBSS on a cold freezing block or ice in the dissection hood.
2. The skull is cut open with a scissors along the sagittal fissure, and the bone and cartilage of the left and right hemisphere are removed or dragged to the side.
3. After the bone and cartilage have been removed, the cortical tissue can be removed in one operation by going down with a forceps sagittally around one hemisphere, squeezing the forceps while pulling upwards. By this maneuver, one hemisphere including cortex, striatum, and hippocampus can be isolated.
4. The hippocampus and striatum are removed from the cortex as well as the choroid plexus using a dissection microscope. The dura mater surrounding the cortex can be removed as much as reasonable possible. Even though the removal of dura mater is not essential for the microglia culture, this membrane can damage/remove the astrocytic monolayer when microglia is removed by smacking the culture flask.
5. The cortex tissue is cut up in pieces with scissors, thereafter trypsin (0.1 %) and DNase (0.05 %) are added to the HBSS (without bivalent ions) and incubated in an Eppendorf tube for 15–20 min at 37 °C.

6. The tissue is mechanically dissociated by using a 1 ml pipette tip by aspirating the tissue up and down 5–10 times until a cell suspension is obtained. Avoid air bubbles to minimize the damage to the cells.
7. The cell suspension is then spun down at 400 × *g* for 4 min in a Falcon tube.
8. After centrifugation the supernatant is removed and the pellet, containing the cells, is taken up from the tube and put into a T75 flask with 10 ml of DMEM + GlutaMAX medium containing 4.5 g/l glucose without sodium pyruvate (Gibco by Life Technologies) supplemented with 10 % of Fetal Bovine Serum (FBS) and penicillin/streptomycin (100 U/ml and 0,1 mg/ml, respectively). Cultures are maintained at 37 °C and 5 % CO_2.
9. Normally, the cortices from three mouse pups should be cultured in one T75 flask, and it is not beneficial for the cultures to have too low concentration of cells, which will compromise the yield of microglia in the end. For rat, the typical cell concentration per T75 flask should be 10 million cells.
10. Optionally, a cell strainer, 70 μm mesh, can be used to get rid of bigger tissue/cell aggregates. Even though some protocols suggest changing of medium after 1 day, it is often better to wait around 2–4 days before changing medium to let the cells attached to the bottom of the culture flask.
11. The medium can then be changed two times a week, by exchanging about 60 % of the culture medium, or once a week and then changing about 90 % of the culture medium.
12. After around 2–3 weeks of culture, the microglia can be harvested. The microglial cells will appear as small round cells on top of the monolayer of astrocytes, typically appearing in clusters (Fig. 1).
13. The microglial cells can easily be removed from the T75 flasks by smacking the cultures flask 10–20 times, which will let the round microglial cells let go from the astrocytic monolayer. Too vigorous smacking can result in that the astrocytic monolayer detaches from the bottom of the culture flask. Smacking the flasks and continuous looking in the microscope that satisfactory number of microglial cells have detached are recommended. Even though several protocols describe the use of a rotating shaker that could be used for up to 1 h to isolate the microglia from the flask, smacking the flask works equally good. All culture medium that now contains microglia should be removed from the T75 flask.
14. The cell-containing medium should be centrifuged at 400 × *g* for 4 min and the supernatant removed to get the pellet with the microglial cells. Around 95 % of microglial purity can be

expected using this approach. The experiments can be pursued directly with the harvested microglial cells (*see* **Notes 1** and **2**). The cells can also be cultured for a few days (to get a ramified morphology) and then the experiments can be performed. For culture in a 96-well plate, a concentration of 20,000 microglial cells/well can be used. During the experiment the serum should be reduced (e.g., 2 %) to maximize the inflammatory response by the microglia.

4 Notes

1. It is common to only use these mixed cultures to harvest one yield of microglia. However, it has been shown by Floden and Combs that it is possible to get out more microglia from subsequent yields [16]. They show that two more yields can be obtained after another 7 + 7 days of culture with fresh culture medium with a yield that is about 20 % of the first yield in terms of cell numbers. The two subsequent yields isolated contain microglia that show similar morphology and immunoreactivity of the microglial markers CD11b, CD68, MHCII, and CD45, and cell survival from yield 1 and 3 was not changed. However, the secretion of two cytokines examined, IL-6 and TNF-α, showed significantly decreased secretion of these cytokines 24 h after an LPS challenge, suggesting that cytokine production or secretory ability is reduced after the first yield. The phagocytic activity, measured by the ability to phagocyte FITC-conjugated bioparticles, was not altered.
2. In this book chapter, I describe a protocol that uses the intrinsic ability of glial cells to proliferate in culture without adding any mitogens (e.g., GM-CSF, M-CSF). With this protocol it is easy to retrieve good yield of microglia that will give robust and reproducible results that can uncover new knowledge about the biology of the microglia.

Acknowledgement

This work has been supported by grants provided by the Swedish Research Council grant no. 2012–2229 by the Gyllenstierna Krapperup, the Royal Physiographic Society, Swedish stroke association, Crafoord, A.E. Berger, Wiberg, Bergvall, G & J Kock foundations.

References

1. Ransohoff RM, Brown MA (2012) Innate immunity in the central nervous system. J Clin Invest 122(4):1164–1171
2. Lambertsen KL et al (2011) Differences in origin of reactive microglia in bone marrow chimeric mouse and rat after transient global ischemia. J Neuropathol Exp Neurol 70(6):481–494
3. Giulian D, Baker TJ (1985) Peptides released by ameboid microglia regulate astroglial proliferation. J Cell Biol 101(6):2411–2415
4. Kalla R et al (2003) Loss of microglial ramification in microglia-astrocyte cocultures: involvement of adenylate cyclase, calcium, phosphatase, and Gi-protein systems. Glia 41(1):50–63
5. Tanaka J et al (1999) Morphological differentiation of microglial cells in culture: involvement of insoluble factors derived from astrocytes. Neurosci Res 34(4):207–215
6. Lambertsen KL et al (2009) Microglia protect neurons against ischemia by synthesis of tumor necrosis factor. J Neurosci 29(5):1319–1330
7. Giulian D et al (1995) Cell surface morphology identifies microglia as a distinct class of mononuclear phagocyte. J Neurosci 15(11): 7712–7726
8. Tanaka J, Maeda N (1996) Microglial ramification requires nondiffusible factors derived from astrocytes. Exp Neurol 137(2):367–375
9. Giulian D, Ingeman JE (1988) Colony-stimulating factors as promoters of ameboid microglia. J Neurosci 8(12):4707–4717
10. Gehrmann J (1996) Microglia: a sensor to threats in the nervous system? Res Virol 147(2–3):79–88
11. Fujita H et al (1996) Effects of GM-CSF and ordinary supplements on the ramification of microglia in culture: a morphometrical study. Glia 18(4):269–281
12. Aloisi F et al (2000) Functional maturation of adult mouse resting microglia into an APC is promoted by granulocyte-macrophage colony-stimulating factor and interaction with Th1 cells. J Immunol 164(4):1705–1712
13. Lee SC et al (1993) Macrophage colony-stimulating factor in human fetal astrocytes and microglia. Differential regulation by cytokines and lipopolysaccharide, and modulation of class II MHC on microglia. J Immunol 150(2):594–604
14. Ponomarev ED et al (2005) Development of a culture system that supports adult microglial cell proliferation and maintenance in the resting state. J Immunol Methods 300(1–2): 32–46
15. Flanary BE, Streit WJ (2004) Progressive telomere shortening occurs in cultured rat microglia, but not astrocytes. Glia 45(1): 75–88
16. Floden AM, Combs CK (2007) Microglia repetitively isolated from in vitro mixed glial cultures retain their initial phenotype. J Neurosci Methods 164(2):218–224

Chapter 5

Isolation of Murine Postnatal Brain Microglia for Phenotypic Characterization Using Magnetic Cell Separation Technology

Ashley S. Harms and Malú G. Tansey

Abstract

To shorten the time between brain harvesting and microglia isolation, and characterization, we utilized the MACS® neural dissociation kit followed by OctoMACS® CD11b magnetic bead isolation technique to positively select for brain microglia expressing the pan-microglial marker CD11b, a key subunit of the membrane attack complex (MAC). This protocol yields a viable and highly pure (>95 %) microglial population of approximately 500,000 cells per pup that is amenable for in vitro characterization within hours or days after being harvested from brain tissue. Primary microglia from C57Bl/6 mice were plated for next-day analyses of morphology and cellular markers by immunocytochemistry or for analysis of gene expression under resting or LPS-stimulated conditions. The ease of isolation enables investigators to perform molecular and cellular analyses without having to wait 1–2 weeks to isolate microglia by conventional methods involving mechanical agitation to dislodge these from astrocyte beds.

Key words Microglia, CD11b, Inflammatory gene expression, Flow cytometry, Immunocytochemistry

1 Introduction

Microglia are the monocyte-derived resident macrophages of the brain in charge of immune surveillance [1]. Activated microglia secrete neurotrophic factors to limit tissue injury, protect vulnerable neuronal populations, and aid in brain repair processes. However, activated microglia can also overproduce prostaglandins, chemokines, cytokines, and reactive oxygen and nitrogen species which can have a deleterious effect on neuronal survival by enhancing oxidative stress and activating cell death pathways (reviewed in ref. 2), raising the interesting possibility that chronic microglia activation may contribute to the etiology and/or progression of neurodegenerative disease (reviewed in ref. 3). In vitro analyses of microglia phenotype and their effector functions are expected to significantly improve the ability to generate testable hypotheses about the in vivo role of microglia in normal and pathophysiological states in

Bertrand Joseph and José Luis Venero (eds.), *Microglia: Methods and Protocols*, Methods in Molecular Biology, vol. 1041, DOI 10.1007/978-1-62703-520-0_5, © Springer Science+Business Media New York 2013

the brain. Yet conventional methods used to obtain homogeneous populations of microglia require maintenance of primary neuron-glia cultures for extended periods of time (from 1 to 2 weeks), increasing the likelihood for loss of morphological and functional microglial phenotype.

2 Materials

2.1 Tissue Harvest and Primary Cell Culture Components

1. Mice at postnatal day 3–5 (P3–P5).
2. DMEM/F12 with 10 % Heat-Inactivated FBS.
3. Lipopolysaccharide (LPS *E. coli* strain O111:B4).

2.2 Magnetic Cell Isolation Components

1. MACS® neural dissociation kit (Miltenyi Biotec).
2. OctoMACS® CD11b magnetic bead separation (Miltenyi Biotec).

2.3 Immuno-cytochemistry Components

1. 4 % paraformaldehyde in 0.01M PBS pH 7.4.
2. Antibodies for CD45 (Serotec) and CD68 (Serotec).
3. Alexa dye-conjugated secondary antibodies for immunocytochemistry (Invitrogen).

2.4 Components for Flow Cytometry

1. F4/80-Texas Red (Caltag) and CD11b-FITC (Miltenyi Biotec).
2. Hoechst 33258 (Invitrogen) for nuclear counterstain.

2.5 Components for Quantitative PCR

1. RNA STAT-60 (Tel-Test, Friendswood, TX).
2. DNase I (Invitrogen).
3. SuperScript II RNase H- Reverse Transcriptase (Invitrogen).
4. SYBR Green PCR Master Mix (Applied Biosystems Inc.).
5. Oligonucleotide forward and reverse PCR primers for TNF, IL-1β, iNOS, MIP-1α, and CD45 (Integrated DNA Technologies).
6. ABI PRISM 7000 Detection System (Applied Biosystems Inc.).

3 Methods

3.1 Positive Selection of Primary Microglia Using CD11b Magnetic Beads for Next-Day Immunocytochemical Analyses

1. Obtain eight postnatal day 3–5 (P3–P5) C57Bl/6 wild-type pups.
2. Isolate primary microglia using the MACS® neural dissociation kit (Miltenyi Biotec).
3. Determine the total cell number by trypan-blue exclusion (~40 million total).

4. To enrich for primary microglia, OctoMACS® CD11b magnetic bead separation (Miltenyi Biotech) can be employed according to the protocol provided by the manufacturer. We recommend shortening the centrifugation time from 10 min to 5 in order to increase cell viability.
5. The number of cells obtained post-magnetic CD11b bead isolation will be ~ 4 million or 10 % of the total number of cells in the single-cell suspension obtained with the MACS neural dissociation kit, which is in the expected range.
6. After [1] a day in vitro in DMEM/F12 with 10 % HI-FBS, the cells can be treated with lipopolysaccharide (LPS *E. coli* strain O111:B4, 10 ng/mL or 1 μg/mL for 24 h) and then fixed in 4 % paraformaldehyde in 0.01M PBS and processed for fluorescence immunocytochemistry as previously described [4] (see **Notes 1**, **2**, and **3**).
7. Acquire images of stained cells with a fluorescence microscope equipped with a digital camera.

3.2 Positive Selection of Postnatal Microglia Using CD11b Magnetic Beads for Gene Expression Analyses Induced by Inflammatory Stimuli

1. Obtain 10 postnatal day 3–5 (P3–P5) C57Bl/6 wild-type pups.
2. Isolate primary microglia using the MACS® neural dissociation kit followed by OctoMACS® CD11b magnetic bead positive selection.
3. Plate cells in 6-well plates at a density of 0.5 million/well.
4. After 1 day in culture in DMEM/F-12 supplemented with 10 % FBS, stimulate cells with lipopolysaccharide (LPS, 1 μg/mL) for 4 h at 37° in a humidified CO_2 incubator.
5. Harvest cells in RNA STAT-60 (Tel-Test, Friendswood, TX) for isolation of total RNA and briefly treat with DNase I (Invitrogen) and then reverse-transcribe into cDNA using SuperScript II RNase H- Reverse Transcriptase (Invitrogen) (*see* **Note 4**).
6. Perform real-time quantitative polymerase chain reaction (qPCR) as previously described [5, 6] using an ABI PRISM 7000 Detection System (Applied Biosystems Inc.) (*see* **Note 5**).

3.3 Positive Selection of Postnatal Microglia Using CD11b Magnetic Beads for Quantitative Flow-Cytometric Analyses of Cell-Surface Markers

1. Obtain eight postnatal day 3–5 pups from wild type or TNF null (or another knockout mouse).
2. Isolate primary microglia using MACS® neural dissociation kit followed by OctoMACS® CD11b magnetic bead separation (Miltenyi Biotech).
3. Label cells with a fluorescently conjugated antibodies specific for the pan-marker CD11b and the activation marker F4/80.

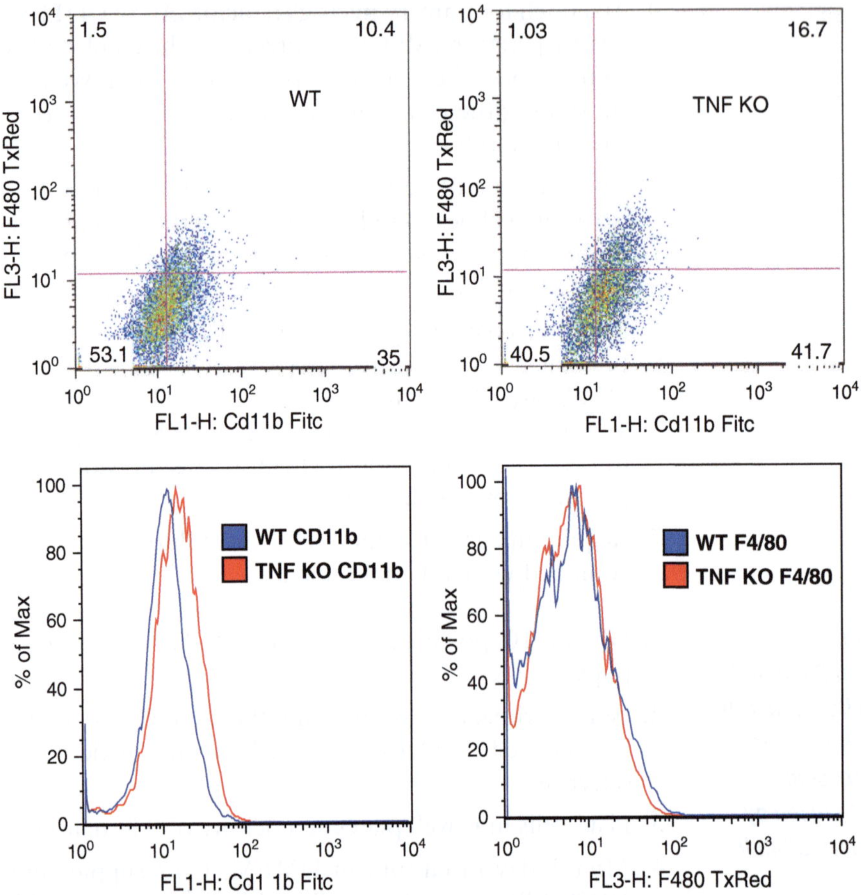

Fig. 1 Primary microglia positively selected with CD11b magnetic bead separation display the expected microglial cell-surface markers as measured by flow cytometry. Primary microglia were isolated from postnatal day 3 (P3) TNF-deficient (TNF KO) and wild-type (WT) pups by MACS® neural dissociation and CD11b magnetic bead separation. Cells were double-labeled with an antibody against the activation marker F4/80 conjugated to the fluorophore Texas Red (Caltag) and an antibody against the microglial marker CD11b conjugated to the fluorophore FITC (Miltenyi Biotec), fixed in 1 % paraformaldehyde and subjected to flow cytometry. FACS analyses of the live cell population revealed the activation status of the microglia isolated from the two different genotypes

4. Fix briefly in 1 % PFA in PBS and sort using flow cytometry as previously described [6].
5. Figure 1 demonstrates a representative scatter plot from fluorescence-activated cell sorting (FACS) analysis from the cell population. The data reveals the expected CD11b expression in the population and relatively low F4/80 expression indicative of minimal activation; no difference between genotypes was noted.

4 Notes

1. The antibody dilutions should be as follows: CD45 (Serotec) 1:250, CD68 (Serotec) 1:250, and Hoechst 33258 (Invitrogen) 1:20,000.
2. The appropriate Alexa-conjugated secondary antibodies (Invitrogen) should be used at a dilution of 1:1,000 in the dark.
3. Figure 2 illustrates images from a representative experiment demonstrating that the number of cells found to be immunoreactive for CD45 was approximately 99 % and for CD68 was 98 %, indicating that the purity of the isolated microglia population is high.

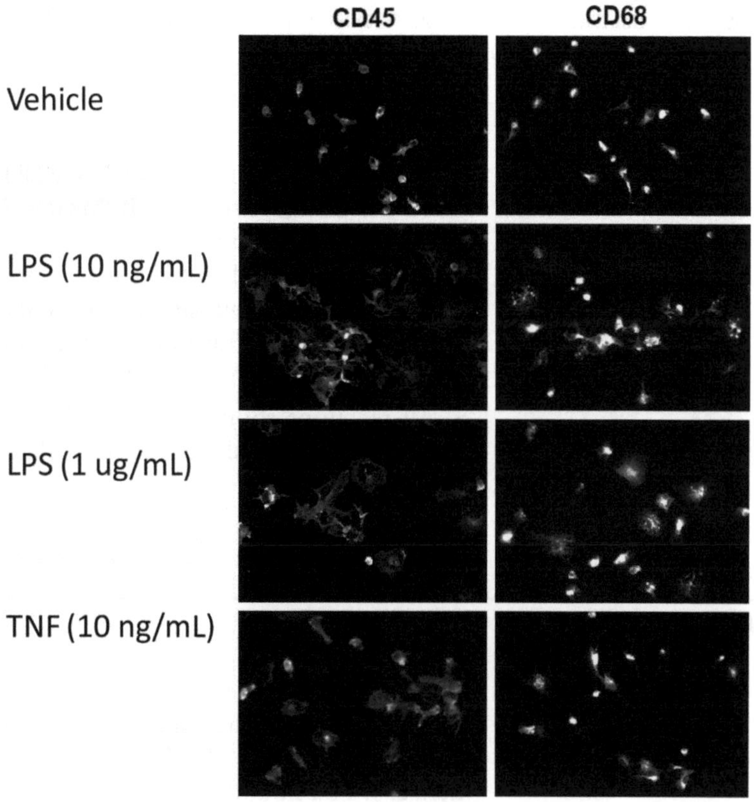

Fig. 2 Primary microglia positively selected with CD11b magnetic beads reveal normal morphological changes upon LPS stimulation as measured by immunocytochemistry. Primary microglial cells were isolated from postnatal day 3–5 (P3–P5) C57Bl/6 pups using the MACS® neural dissociation kit followed by CD11b magnetic bead separation. Cells were plated and treated for 24 h with LPS or TNF, fixed, and stained with anti-CD45 (Serotec) and anti-CD68 (Serotec). Immunocytochemical analysis revealed morphological changes and increased expression of microglial markers CD45 and CD68 after stimulation with LPS or TNF

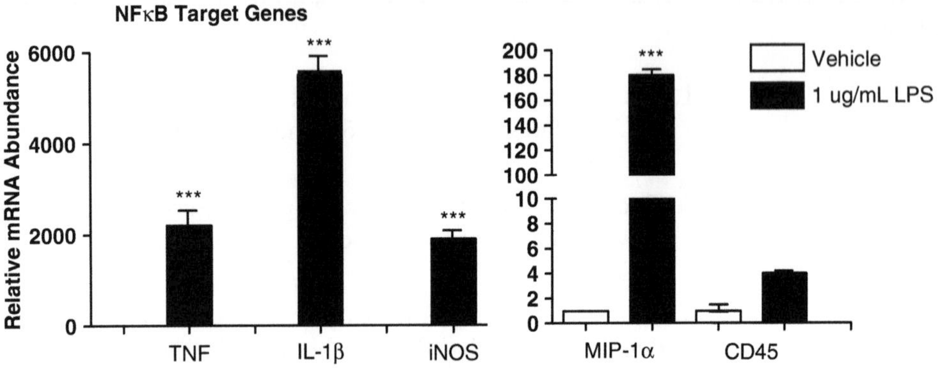

Fig. 3 Primary microglia positively selected with CD11b magnetic bead separation are responsive to inflammatory stimuli as measured by real-time PCR. Primary microglia were isolated from ten postnatal day 3–5 (P3–P5) C57Bl/6 pups by MACS® neural dissociation followed by CD11b magnetic separation and treated for 4 h with 1 μg/mL LPS. Total RNA was harvested using phenol/chloroform extraction and reverse-transcribed into cDNA. Quantitative PCR analysis revealed expression of mRNA for iNOS, MIP1α, IL-1β, TNF, and CD45 at rest and after LPS stimulation

4. All reactions can be done in 384-well format with 50 ng cDNA, 10 μL SYBR Green PCR Master Mix, and 150 nM each forward and reverse primer. Oligonucleotide primers for TNF, IL-1β, iNOS, MIP-1α, and CD45 can be designed using primer design software freely available on the World Wide Web. Figure 3 illustrates changes in mRNA levels for these markers in response to LPS stimulation.
5. All reactions should be performed in triplicate and levels of various mRNAs should be normalized to the geometric mean of several housekeeping genes such as TATA binding protein, cyclophilin, and GAPDH.

Acknowledgement

The content of this article was adapted from one that appeared in the [Date] issue of the Miltenyi Biotec newsletter. This work was supported by NIH/NINDS grant 5R011NS049433 and The Michael J. Fox Foundation for Parkinson's Research.

References

1. Ransohoff RM, Perry VH (2009) Microglial physiology: unique stimuli, specialized responses. Annu Rev Immunol 27:119–145
2. McGeer PL, McGeer EG (2004) Inflammation and neurodegeneration in Parkinson's disease. Parkinsonism Relat Disord 10(Suppl 1):S3–S7
3. Tansey MG, McCoy MK, Frank-Cannon TC (2007) Neuroinflammatory mechanisms in Parkinson's disease: potential environmental triggers, pathways, and targets for early therapeutic intervention. Exp Neurol 208: 1–25

4. McCoy MK, Martinez TN, Ruhn KA et al (2006) Blocking soluble tumor necrosis factor signaling with dominant-negative tumor necrosis factor inhibitor attenuates loss of dopaminergic neurons in models of Parkinson's disease. J Neurosci 26:9365–9375
5. Harms AS, Lee JK, Nguyen TA et al (2012) Regulation of microglia effector functions by tumor necrosis factor signaling. Glia 60(2): 189–202
6. Kurrasch DM, Huang J, Wilkie TM et al (2004) Quantitative real-time polymerase chain reaction measurement of regulators of G-protein signaling mRNA levels in mouse tissues. Methods Enzymol 389:3–15

Chapter 6

Isolation and Culture of Adult Human Microglia Within Mixed Glial Cultures for Functional Experimentation and High-Content Analysis

Amy M. Smith, Hannah M. Gibbons, Claire Lill, Richard L.M. Faull, and Mike Dragunow

Abstract

Microglia are thought to be involved in diseases of the adult human brain as well as normal aging processes. While neonatal and rodent microglia are often used in studies investigating microglial function, there are important differences between rodent microglia and their adult human counterparts. Human brain tissue provides a unique and valuable tool for microglial cell and molecular biology. Routine protocols can now enable use of this culture method in many laboratories. Detailed protocols and advice for culture of human brain microglia are provided here. We demonstrate the protocol for culturing human adult microglia within a mixed glial culture and use a phagocytosis assay as an example of the functional studies possible with these cells as well as a high-content analysis method of quantification.

Key words Human brain tissue, Primary human culture, Mixed glial cultures, Phagocytosis, High-content analysis

1 Introduction

Microglia are specialized immune cells that are prevalent throughout the brain. They maintain homeostasis and respond sensitively to changes in their microenvironment [1]. Microglia influence many brain processes, from neurogenesis to learning. There are numerous reports of microglial activity in human brain disorders, yet we still lack knowledge about the basic molecular mechanisms that govern different activation states and phenotypes of microglia.

Although most microglial research has been carried out using rodent cells and models, it is becoming increasingly clear that there are important differences between rodent microglia and their human counterpart [2, 3]. While neonatal and rodent microglia are often used in studies investigating microglial function, more

Bertrand Joseph and José Luis Venero (eds.), *Microglia: Methods and Protocols*, Methods in Molecular Biology, vol. 1041,
DOI 10.1007/978-1-62703-520-0_6,

research is required to discover which microglial processes and hallmarks are conserved in the adult human brain and where they are divergent from rodent microglia. There is also evidence that immune responsiveness changes with age [4, 5]. Furthermore, microglia may respond differently to peripheral immune cells due to their brain-specific environment.

Thus, isolation and culture of human adult microglia is a unique and valuable tool to complement other approaches to microglial research. Primary adult human microglia are a relevant tool for studying human adult brain disorders due to their tissue and species origin. Cultured primary human microglia can be used with relative ease for a range of studies including functional assays, drug screening and development, and gene profiling. The use of human adult brain cells as a tool for research can play a major role in conjunction with in vivo mechanistic proof-of-principle studies and play an essential role in translation of microglial-based therapies to the clinic.

We culture microglia within a mixed glial culture also containing astrocytes and fibroblast-like cells (the benefits of which are discussed in Subheading 4) [6]. The high yield of microglia obtained from this protocol is ideal for high-throughput analysis when used in conjunction with 96-well microplates and high-content imaging methods [7]. A variety of characterization and functional studies are possible with these primary human cells. We provide a protocol for examining the quintessential microglial function of phagocytosis. The ability of microglia to phagocytose extracellular debris is important throughout life from development to neurodegeneration. This reliable assay can also be used in conjunction with other treatments. We have developed a high-content image analysis protocol to rapidly and reliably quantify microglial phagocytosis. High-content analysis is a useful tool for objective, quick, and accurate examination of human microglial biology [8].

For a review of the different culturing protocols for human microglia in the literature, *see* ref. 9. The protocol outlined here is the one routinely used in our laboratory.

2 Materials

Human Adult Brain Tissue

For our experiments, autopsy adult human brain tissue is obtained from the Neurological Foundation of New Zealand Human Brain Bank. Biopsy adult human brain tissue is obtained from patients undergoing surgery for intractable temporal lobe epilepsy. This research is approved by the Northern Regional Ethics Committee and the University of Auckland Human Participants Ethics Committee, and informed consent was obtained from all tissue donors.

Ethics approval must be obtained from the relevant committees prior to tissue collection and informed consent must be obtained from all donors.

2.1 Isolation and Culture of Human Adult Mixed Glia Including Microglia

Reagents

1. Hank's balanced salt solution (HBSS; Ca^{2+} and Mg^{2+} free, Gibco BRL).
2. Earle's balanced salt solution (EBSS; Ca^{2+} and Mg^{2+} free, Gibco BRL).
3. Hibernate A (Gibco BRL) (*see* **Note 1**).
4. DMEM/F12 media (Gibco BRL) with 10 % Fetal Bovine Serum (FBS) (Gibco BRL), 1 % Penicillin-Streptomycin-Glutamine (Gibco BRL) (final concentrations: penicillin (100 U/ml), streptomycin (100 μg/ml), and l-glutamine (0.29 mg/ml)).
5. DNase (Invitrogen), 1,000 U/ml stock made up in HBSS (can be stored at −20 °C).
6. Papain (Worthington Biochemical Corporation), 20 U/ml stock made fresh in EBSS.

Equipment

1. Ice bucket.
2. 50 ml tube rack.
3. 50 ml tubes (BD Biosciences).
4. 2 × 10 cm dishes (Nunc, Roskilde, Denmark).
5. 2× autoclave-sterilized scalpel handles (Swann-Morton, Sheffield, England).
6. 2× autoclave-sterilized forceps (ProSciTech, Queensland, Australia).
7. 2× scalpel blades (Swann-Morton, Sheffield, England).
8. Waste beaker containing TriGene (Medichem International, UK).
9. 3× T75 flasks (Nunc).
10. 2 × 100 μm filters (BD Biosciences, New Jersey, USA).
11. 10 ml strip pipettes.
12. Pipette gun.

Plating cells for experiments

1. Phosphate-buffered saline (PBS).
2. 0.25 % trypsin/1 mM EDTA (Gibco BRL).
3. Cell scrapers (1 per flask) (BD Biosciences).
4. Hemocytometer.
5. 96-well tissue culture plate (Nunc).

2.2 Phagocytosis Assay for Primary Human Adult Microglia in a Mixed Glial Culture

Reagents

1. Amyloid-β_{1-42} amino acid peptide (Aβ_{1-42}; Bachem, Bubendorf, Switzerland).
2. Thioflavin S (Sigma).
3. Hoechst 33258 (Sigma).
4. Analytical grade ethanol.
5. Phosphate-buffered saline (PBS).
6. 4 % Paraformaldehyde solution (PFA).
7. PBS-Triton 0.2 % (v/v) (PBS-T).
8. TNE buffer (10 mM Tris, 200 mM NaCl, 1 mM EDTA; pH 7.4).

Equipment

1. Aspirator/vacuum pump (a multichannel pipette can be used to remove media/PBS from the wells if an aspirator/vacuum pump is unavailable).
2. Plate rocker.
3. Multichannel pipette.

2.3 High-Content Analysis of Primary Adult Human Microglial Phagocytosis

Equipment

1. Discovery-1 Automated Fluorescence Microscope (Molecular Devices, Sunnyvale, CA, USA).
2. MetaMorph Image Analysis System (6.2.6 software, Molecular Devices).

3 Methods

3.1 Isolation and Culture of Human Adult Mixed Glia Including Microglia

1. Prepare suitable biological safety hood (for safe and sterile handling of samples, *see* **Note 2**), equipment, and reagents prior to collecting tissue. In a sterile tissue culture hood, place 1 × 50 ml tube rack, 2 × 10 cm dishes, 2× autoclave-sterilized scalpel handles, 2× autoclave-sterilized forceps, 2× scalpel blades, waste container with TriGene, 3× T75 flasks, and 2 × 100 μm filters.
2. In 37 °C water bath, pre-warm Hibernate A and DMEM/F12 media with 10 % FBS and 1 % PSG (complete media).
3. For tissue collection, add 20 ml Hibernate A to a 50 ml tube and weigh contents; store on ice.
4. Thaw DNase and allow papain to come to room temperature.
5. Prepare papain by adding 5 ml EBSS to vial and invert to mix (20 units of papain per ml).
6. Collect ~2 g white and gray matter from the middle temporal gyrus (*see* **Note 3**) in ice-cold Hibernate A.

7. Weigh tube containing tissue and calculate weight of tissue by subtracting previously weighed mass of tube and Hibernate A.
8. Wash tissue twice in Hibernate A to remove contaminating blood. Carefully tip waste into waste container with TriGene (*see* **Note 4**).
9. Place tissue in a sterile Petri dish and remove meninges and visible blood vessels aseptically using autoclave-sterilized forceps.
10. Place tissue into a new 50 ml falcon tube.
11. Wash tissue twice to remove contaminating blood and carefully tip waste into waste container with TriGene.
12. Place tissue in a sterile Petri dish and dice tissue into ~1 mm^3 pieces using sterile scalpel blades. Dispose of scalpel blades using a safety blade remover device.
13. Place tissue in a tube containing 50 ml Hibernate A and allow tissue to sink to the bottom of the tube.
14. Prepare enzyme dissociation mix (10 ml/g tissue) with prewarmed 37 °C Hibernate A containing 2.5 U/ml papain and 10 U/ml DNase.
15. Use a 10 ml strip pipette to remove Hibernate A from tube containing tissue and replace with 10 ml warm enzyme dissociation mix per g tissue.
16. Incubate tissue with enzymes for 15 min at 37 °C with agitation (e.g., MACSmix Tube Rotator (Miltenyi Biotec, Bergisch Gladbach, Germany) on continuous full speed in incubator, or place tubes in water bath and regularly invert).
17. Remove from the incubator and triturate tissue ~5× through a 10 ml strip pipette to aid digestion.
18. Return to incubator for a further 15 min (as above).
19. Triturate well through a 10 ml strip pipette (~5× until resistance decreases).
20. Add an equal volume of complete media (DMEM/F12 with 10 % FBS, 1 % PSG) and gently triturate the tissue ~5× more using a 10 ml strip pipette.
21. Using a 10 ml strip pipette, pass the cell suspension through a 100 μm nylon cell strainer into a new 50 ml tube.
22. Centrifuge for 10 min at 160 × *g*.
23. Carefully pipette off the supernatant and resuspend the pellet in 15 ml media.
24. Plate the cell suspension into 75 cm^2 tissue culture flasks. 2 g should be split between 3× T75 flasks for biopsy or short (up to 3 h) postmortem delay tissue (*see* **Note 5**).

25. Add 5 ml cell suspension to each of 3× 75 cm^2 tissue culture flasks. Add 20 ml warm, complete media to each flask for a final volume of 25 ml per flask.
26. Incubate cells at 37 °C in a humidified atmosphere with 95 % air/5 % CO_2.
27. First thing the next morning (~16 h later), refresh the media.
28. For each flask, remove culture media containing debris and unattached cells and transfer to a 50 ml tube. Wash the adhered cells carefully with 10 ml fresh media to remove debris and transfer to the same 50 ml tube.
29. Add 10 ml complete media to the flask of attached cells and incubate at 37 °C. Centrifuge 50 ml tube containing debris and unattached cells for 10 min at 160 × *g*.
30. Carefully pipette off the supernatant and resuspend the pellet in 15 ml medium. Replate onto the adhered cells (for a total of 25 ml per flask) and incubate for a further 24 h.
31. The following day, remove culture media containing debris and unattached cells and discard. Wash the adherent cells carefully 2× with 5 ml complete media to remove debris and add 20 ml fresh complete media to each flask.
32. Culture cells for ~1–2 weeks, with weekly media changes, depending on density and use of cells (*see* **Notes 6** and **7**).

To detach and plate cells for experiments

1. Remove media from flask.
2. Rinse cells with warm PBS.
3. Detach cells with 0.25 % trypsin/1 mM EDTA (~2 ml per flask). Incubate at 37 °C for 5 min.
4. Tap the side of flask to dislodge cells and observe cells under a microscope. If cells are still adherent, return to incubator for 5 min.
5. Tap the side of flask to dislodge cells. Add 5 ml complete media to flask and transfer cells and media to a 50 ml tube. Repeat.
6. Observe flask surface under a microscope. If some cells are still adherent, a cell scraper can be used to detach remaining cells (*see* **Note 8**). Collect scraped cells by rinsing with 5 ml media and transfer to the same 50 ml tube.
7. Centrifuge for 10 min at 160 × *g*.
8. Carefully pipette off the supernatant, resuspend the pellet in medium, and count cells using a hemocytometer.
9. Plate cells by resuspending in medium at 50,000 cells/ml and plate into 96-well plates (100 μl/well) for experiments.
10. Allow cells to settle and recover from trypsinization for at least 48 h before beginning experiments (*see* **Notes 9** and **10**).

3.2 Phagocytosis Assay for Primary Human Adult Microglia in a Mixed Glial Culture

1. Dissolve $A\beta_{1-42}$ (1 mg) in distilled H_2O (444 μl) to a concentration of 500 μM ($A\beta_{1-42}$ solution can be stored at −20 °C).
2. Vortex $A\beta_{1-42}$ solution to break up large crystals.
3. Prepare working stock of $A\beta_{1-42}$ by adding to an equal volume of PBS, i.e., dilute 1:1 in PBS immediately prior to addition to cells.
4. In a suitable biological safety cabinet, add 5 μM $A\beta_{1-42}$ to the cells (*see* **Note 11**), i.e., from a stock concentration of 500 μM, diluted 1:1 with PBS, add 2 μl per well (containing 100 μl media).
5. Incubate for 24 h in normal incubating conditions of 37 °C in 5 % CO_2 (*see* **Note 12**).
6. Wash cells twice with warm PBS to remove extracellular $A\beta_{1-42}$ using an aspirator and multichannel pipette.
7. Fix cells with 4 % PFA for 15 min at room temperature.
8. Wash with PBS-T for 10 min.

Thioflavin S is used to visualize phagocytosed $A\beta_{1-42}$

1. Prepare a 0.01 % solution of Thioflavin S in 50 % ethanol (e.g., 1 mg Thioflavin S in 10 ml 50 % ethanol solution) (*see* **Note 13**).
2. Add 50 μl per well of 96-well plate.
3. Incubate with gentle rocking for 10 min at room temperature in the dark.
4. Wash cells in 50 % ethanol, followed by distilled H_2O, for 10 min each.
5. Add 60 μl PBS-T per well.
6. Visualize by fluorescence microscopy using a fluorescein isothiocyanate (FITC) filter (*see* **Note 14**).

Hoechst 33258 is used to counterstain cell nuclei (see ***Note 15****)*

1. Wash cells for 5 min with TNE buffer containing 10 mM Tris, 200 mM NaCl, 1 mM EDTA; pH 7.4.
2. Prepare a 20 μM Hoechst 33258 solution in TNE buffer.
3. Add 50 μl per well of 96-well plate.
4. Incubate with gentle rocking for 30 min at room temperature in the dark.
5. Wash cells 2 × 5 min with TNE buffer.
6. Add 60 μl PBS-T per well.
7. Visualize under UV light (Fig. 1b) (*see* **Note 16**).

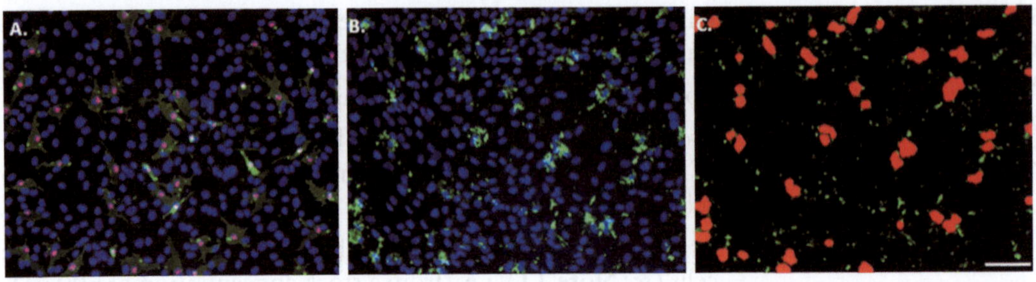

Fig. 1 Adult human microglia in mixed glial cultures can phagocytose $A\beta_{1-42}$ and be quantified by high-content analysis. (**a**) Primary adult human microglia express the transcription factor PU.1 (*pink*) and cell-surface marker CD45 (*green*). All glial nuclei are labelled with Hoechst (*blue*). (**b**) Adult human microglia phagocytose $A\beta_{1-42}$ (*green*, stained with Thioflavin S) in vitro. (**c**) Microglial phagocytosis can be automatically quantified using the "Count Nuclei" application in MetaMorph image analysis software. The corresponding FITC image of $A\beta_{1-42}$ from (**b**) has been thresholded to count the number of phagocytic microglia. Images were acquired using a Discovery-1 automated fluorescence microscope. Scale bar = 100 μm

3.3 High-Content Analysis of Primary Adult Human Microglial Phagocytosis

Images of labelled cells in a 96-well plate format can be acquired by a high-throughput system such as a Discovery-1 microscope

1. Using a 10× objective, take images of 9 sites per well (3 × 3) for high-content analysis.
2. Two (or three) wavelengths can be imaged together for the same site. For Thioflavin S-stained $A\beta_{1-42}$, use a FITC filter set (470Ex/535Em) and an exposure time of approximately 200 ms (*see* **Note 17**).
3. For corresponding fluorescent images of Hoechst-stained cell nuclei, use a UV filter set (403Ex/465Em) and an exposure time of approximately 500 ms.

The images can then be processed with image analysis software such as MetaMorph

4. "Count Nuclei" is a simple application that identifies and isolates cell nuclei, and similarly shaped objects, through image segmentation (*see* **Notes 18–20**).
5. To measure total cell number from Hoechst-stained nuclei:
6. Open the "Count Nuclei" application.
7. Adjust the intensity threshold to segment the nuclei from background.
8. Adjust the cell size thresholds (approximate minimum width and approximate maximum width) to segment individual whole nuclei.
9. Apply "Count Nuclei" analysis to all images to count the number of nuclei per image.

10. Results are logged automatically to Microsoft Excel spreadsheets for easy analysis.

To count the number of microglia that have phagocytosed Thioflavin S-stained $A\beta_{1-42}$:

11. Open the "Count Nuclei" application.
12. Adjust the intensity threshold to segment the Thioflavin S-positive cells from background.
13. Adjust the cell size thresholds (approximate minimum width and approximate maximum width) to segment individual whole cells and exclude extracellular Thioflavin S-stained $A\beta_{1-42}$ blobs.
14. Apply "Count Nuclei" analysis to all images to count the number of fluorescent cells per image (Fig. 1c).
15. Results are logged automatically to Microsoft Excel spreadsheets for easy analysis.

4 Notes

1. Hank's balanced salt solution (HBSS; Ca^{2+} and Mg^{2+} free, Gibco BRL) can be used in place of Hibernate A for tissue collection, washes, and enzymatic dissociation.
2. The highest health and safety practices need to be adhered to when using human cells and tissue. Assume that all tissue is potentially contaminated and work accordingly. All manipulation of cells/tissues should be undertaken in a suitable biological safety cabinet/hood with surgical gloving at all times. All staff should also be immunized against Hepatitis B as a precautionary measure.
3. In our laboratory, we routinely culture from middle temporal gyrus tissue, but other brain regions can also be used for isolation of human adult microglia. Both white and gray matter give rise to viable microglia.
4. Carefully discard all waste that has been in contact with human tissue by treating with TriGene.
5. If there is reason to believe that fewer cells may be obtained from a particular culture (e.g., long postmortem delay), cell suspension can be plated into fewer flasks, i.e., 1 or 2 × 75 cm^2 flasks for 2 g tissue.
6. Changes in the proportions of cell types in mixed glial cultures occur over time due to proliferation of fibroblast-like cells and very limited proliferation of astrocytes and microglia (*see* ref. 6). This should be taken into account for timing of microglial experiments.

7. Some other protocols for microglial cultures use the mitogens macrophage colony-stimulating factor (M-CSF) and granulocyte-macrophage colony-stimulating factor (GM-CSF). We do not routinely use these mitogens in our cultures as they may prime/activate the microglia towards a particular phenotype (unpublished observations).
8. Adult human microglia are very adherent cells (more so than astrocytes or fibroblast-like cells) and do not always detach from the flask with trypsinization. Using a cell scraper does not appear to lead to increased cell death and can increase the microglia yield from the culture.
9. A heterogeneous cell population in mixed glial cultures mimics the in vivo environment better than a single cell population. While being able to characterize and quantify microglia-specific functions (Fig. 1; see phagocytosis assay below), it is also possible to assess how microglia interact with other glial cell types.
10. As each primary mixed glial culture comes from a different donor, each culture of cells will be different (e.g., number of cells isolated, cell composition of culture). Variability in the isolation protocol should be controlled as much as possible, but the heterogeneity of primary human cultures reflects the heterogeneity of individual humans.
11. The final concentration of 5 μM $A\beta_{1-42}$ added to cells is in excess of what they require for phagocytosis and can be adjusted if necessary.
12. The $A\beta_{1-42}$ incubation time may need to be adjusted to suit the cells' phagocytic ability and the test conditions.
13. Thioflavin S is best made fresh before each staining.
14. Thioflavin S may also be visualized through other filters, but a FITC filter gives the strongest signal.
15. Under these culture conditions, phagocytosis of $A\beta_{1-42}$ is rarely seen by cells other than microglia. However, immunocytochemistry can be used to identify microglia using antibodies to microglia-specific antigens, e.g., PU.1 and CD45 [10].
16. Internalization of $A\beta_{1-42}$ can be confirmed by confocal microscopy [11].
17. Exposure times need to be adjusted to acquire the best images for each experiment but must be kept constant between conditions.
18. Count Nuclei application can also be used to quantify other aspects of microglial biology such as immuno-positive markers.
19. A range of assays are available to analyze other end points such as morphology, co-localization, and intensity of staining.
20. For a review of high-content analysis applications to neuroscience and microglial biology, refer to [8].

Acknowledgments

We are very grateful to the tissue donors and their families for their generous and precious gift of brain tissue for our research into brain disorders. We also thank specialist epilepsy nurse Lynair Roberts, neurologist Dr. Peter Bergin, neurosurgeon Dr. Edward Mee, and pathologist Dr. Robyn Oldfield at Auckland City Hospital for providing biopsy tissue. We thank the staff of the Neurological Foundation of New Zealand Human Brain Bank and Centre for Brain Research Biobank for technical assistance. This protocol has been optimized with the help of funding from the National Research Centre for Growth and Development, Coker Charitable Trust, Hugh Green Foundation, and the Health Research Council of New Zealand (Program Grant).

References

1. Hanisch U-K, Kettenmann H (2007) Microglia: active sensor and versatile effector cells in the normal and pathologic brain. Nat Neurosci 10(11):1387–1394
2. Davis MM (2012) Immunology Taught by Humans. Sci Transl Med 4(117)
3. Dragunow M (2008) The adult human brain in preclinical drug development. Nat Rev 7: 659–666
4. Lynch AM, Murphy KJ, Deighan BF et al (2010) The impact of glial activation in the aging brain. Aging Dis 1(3):262–278
5. Streit WJ (2006) Microglial senescence: does the brain's immune system have an expiration date? Trends Neurosci 29(9):506–510
6. Gibbons HM, Hughes SM, Van Roon-Mom W et al (2007) Cellular composition of human glial cultures from adult biopsy brain tissue. J Neurosci Methods 166(1):89–98
7. Narayan P, Gibbons H, Mee E et al (2007) High throughput quantification of cells with complex morphology in mixed cultures. J Neurosci Methods 164:339–349
8. Dragunow M (2008) High-content analysis in neuroscience. Nat Rev Neurosci 9(10): 779–788
9. Gibbons HM, Dragunow M (2010) Adult human brain cell culture for neuroscience research. Int J Biochem Cell Biol 42(6): 844–856
10. Gibbons HM, Smith AM, Teoh HH et al (2011) Valproic acid induces microglial dysfunction, not apoptosis, in human glial cultures. Neurobiol Dis 41(1):96–103
11. Smith AM, Gibbons HM, Dragunow M (2010) Valproic acid enhances microglial phagocytosis of amyloid-b1-42. Neuroscience 169(1):505–515

Part III

Depletion and Transduction of Microglia

Chapter 7

Depletion of Microglia from Primary Cellular Cultures

Lorena Pont-Lezica, Sabrina Colasse, and Alain Bessis

Abstract

Primary cultures are an important in vitro tool to study cellular processes and interactions. These cultures are complex systems, composed of many cell types, including neurons, astrocytes, oligodendrocytes, microglia, NG2 cells, and endothelial cells. For some studies it is necessary to be able to study a pure culture of one cell type, or eliminate a particular cell type, to better understand its function. There exist cell culture protocols for making pure astrocyte or microglia cultures. Here we present two protocols to produce cultures depleted for microglia: in the first case, from a mixed astrocyte–microglia culture and, in the second, for eliminating microglia from neuronal cultures.

Key words Microglia depletion, LME, Conjugated Saporin, Pure primary astrocyte culture, Primary neuronal culture

1 Introduction

Brain function is more than just signaling between neurons. Neuroglia constitute more than half of the cells of the vertebrate adult brain [1] and contribute to many functions of the central nervous system (CNS) including neurotransmission and synaptogenesis, homeostasis of the blood–brain barrier, and immune surveillance (refs within [2]). To get a complete picture of how the brain functions, it is thus important to understand the contribution of the different cell types. Simplified primary cultures of neurons, astrocytes, and microglia from neonate animals have allowed efficient dissection of several complex cell–cell interactions [3–6]. One of the major caveats of primary cultures, however, is that the main cell population can be contaminated by a small number of cells of other cell types.

Astrocyte-enriched cultures are usually prepared from brains dissected from neonate mice. Microglia invade the brain at early stages of development, and so, because the brains of neonates contain microglia, astrocyte cultures are consistently contaminated by these myeloid cells. In initial protocols, microglia were routinely

Bertrand Joseph and José Luis Venero (eds.), *Microglia: Methods and Protocols*, Methods in Molecular Biology, vol. 1041, DOI 10.1007/978-1-62703-520-0_7, © Springer Science+Business Media New York 2013

removed by shaking [7]. Treatment with cytosine β-D-arabinofuranoside hydrochloride (AraC) was used to further deplete microglia, leading to >95 % pure astrocyte cultures [8, 9].

Such astrocyte-enriched primary cultures have been extensively used for several years. However, it recently appeared that despite the fact that microglia make up only 2–5 % of the cell population [2, 10], their presence strongly biased the interpretation of results obtained with these cultures. For instance, NOS2 and TLR4 proteins were detected in astrocyte cultures and their expression was initially wrongly attributed to astrocytes ([11, 12] *see* also ref. 13) until it was shown that they were expressed solely by contaminating microglia [2, 4, 14]. Similarly, work using conditioned medium from astrocyte-enriched cultures showed that TNF-α was important for synaptic scaling [3, 15], but the cellular source of TNF-α is still controversial [16]. These examples show that it is difficult to tease apart the contributions of each cell type when the cultures are not pure, and thus one must use caution when interpreting results in studies using non-purified cell cultures.

Various techniques have been used to eliminate microglia, including repeated mechanical depletion (shaking [8]), genetic depletion [8, 17], pharmacological depletion with L-leucine methyl ester (LME) [9] or clodronate [18], and targeted immunotoxin depletion [19, 20].

Combining mechanical and pharmacological depletion has proven to be an efficient method to get rid of all microglia [4, 9]. LME is a lysomotropic compound that traverses cell and organelle membranes to accumulate in the lysosome. Once within this compartment, LME is hydrolyzed by serine or cysteine proteases to produce free L-leucine [21], and the accumulation of this L-amino acid brings about osmotic swelling and rupture of lysosomes [21, 22]. The LME protocol appears to be highly specific for microglia [9, 10].

The key advantage of using LME to deplete primary astrocyte cultures of microglia is the low cost of the reagent and its specificity for microglia. A possible disadvantage is that, although no obvious deleterious effects have been reported, it is unclear what effect LME might have on the function of other cell types.

The impact of contaminating microglia on neuronal cultures has not been evaluated. However, some of our unpublished data show that, similarly to astrocyte cultures, caution should also be exercised. Since the effect of LME on neurons has not been studied, we adapted another, more specific, strategy to deplete microglia from primary cultures of neurons. Saporin is an unusually stable, ribosome-inactivating protein (RIP) found in the seeds of *Saponaria officinalis.* RIPs enzymatically inactivate ribosomes using their N-glycosidase activity, which removes a single adenine base from the rRNA of the large ribosome subunit, thus effectively shutting down protein synthesis and eventually leading to cell death. Specific targeting and internalization of Saporin have been obtained by conjugating Saporin to antibodies or other molecules that recognize proteins on the surface of the targeted cell [23]. For the elimination of microglia, Saporin can be commercially found conjugated to isolectin B4 or CD11b antibodies. The key advantage of conjugated Saporin is its efficacy (ED50 in the low picomolar range), the possibility of very specifically targeting a particular cell type, and its speed of action. Despite the fact that the targeted cytotoxin can be used at such low concentrations, the main disadvantage of this system is its high cost.

2 Materials

All work must be done in a hood, under sterile cell culture conditions.

1. Stock solution of conjugated Saporin/conj Saporin (Advanced Targeting Systems): dilute to 0.1 μg/μl and store aliquots of stock solution at −80 °C for long-term storage, or at −20 °C for no more than 2 months. If done under sterile conditions, no need to sterilize solutions by filtration.
2. Stock solution cytosine β-D-arabinofuranoside hydrochloride/AraC: make a concentrated stock solution at 100 mM with sterile water and store as 250 μl aliquots at −20 °C. The concentrated stock solution is then diluted in sterile water to make a working solution at 10 mM and stored as 250 μl aliquots at −20 °C.
3. Stock solution of L-leucine methyl ester/LME: stock solution at 500 mM in DMEMc, stored as 2 ml aliquots at −20 °C for no more than a month.
4. Astrocyte culture medium/DMEMc: DMEM supplemented with 10 % Fetal Bovine Serum, 1 % glutamine, 1 % sodium pyruvate, and 0.1 % antibiotics (Invitrogen). Sterilize solutions by filtration.
5. Neuron culture medium/NBc: Neurobasal medium supplemented with B27, 1 % L-GLUTAMINE (stock conc 200 mM), and 0.05 % antibiotics. Sterilize solutions by filtration.

3 Methods

3.1 Protocol 1: Treatment with AraC and LME

Confluent primary astrocytes cultured in 10 cm Petri dishes are first treated with the antimitotic agent AraC for 5 days to eliminate replicating microglia. At the end of the AraC treatment, cultures may still contain microglia embedded in the astrocyte layer. To remove these, we carry out an additional treatment with the lysosomotropic agent LME (*see* **Note 1**). Primary astrocyte cultures become confluent about 7–10 days after culturing.

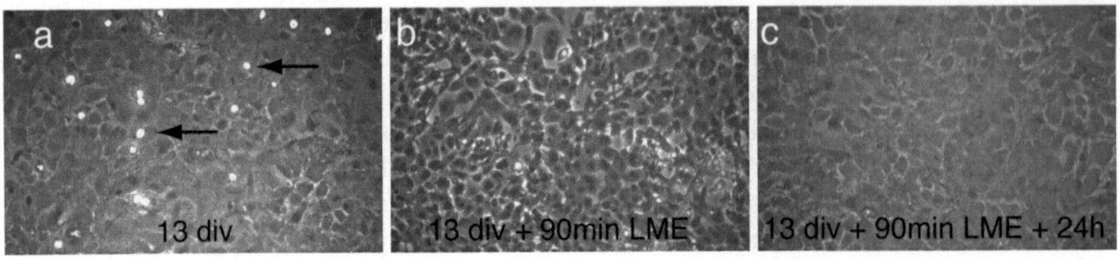

Fig. 1 Effect of AraC and LME treatment on a confluent culture of astrocytes. (**a**) After 13 days in vitro (DIV) including 5 days of AraC treatment, the astrocytes are confluent and microglia are visible as round, refractive cells (*arrows*). (**b**) After 90 min treatment with LME, microglia are not visible anymore; astrocytes are more refractive and their morphology has changed. (**c**) After 24 h, the monolayer of astrocytes has recovered its initial paved-like appearance and is free from microglia

1. AraC treatment.
   (a) Prepare fresh DMEMc culture medium and add AraC from 10 mM working stock to a final concentration of 5 μM. Sterilize solutions by filtration.
   (b) Warm to 37 °C, then replace old medium with AraC-containing DMEMc medium.
   (c) Change medium every day for 5 days with fresh AraC-containing DMEMc medium.
2. LME treatment.
   (a) Dilute LME to 75 mM in DMEMc culture medium. Sterilize solutions by filtration.
   (b) Adjust pH to 7.5 by adding 1N NaOH.
   (c) Warm to 37 °C, then replace astrocyte medium with LME-containing DMEMc medium.
   (d) Put cultures back into the incubator and allow LME to act for 1h30.
   (e) Replace LME-containing DMEMc medium with standard DMEMc culture medium and wait 24 h before using the now pure astrocyte culture (*see* **Note 2** and Fig. 1).

3.2 Protocol 2: Treatment with Conjugated Saporin

Conjugated Saporin can be used to deplete microglia from primary neuronal cultures at any stage of the culture (*see* **Note 3**). Although the ED50 of Saporin is in the hundreds of nanomolar range, the conjugated versions of the cytotoxin are four orders of magnitude more sensitive. The exact ED50 is tested by Advanced Targeting Systems for each lot (*see* **Notes 4** and **5**).

1. Conjugated Saporin treatment.
   Dilute stock solution of conjugated Saporin in NBc medium to a final concentration of 0.35 μg/ml. We use 7 μl of a 0.1 μg/μl stock solution for 2 ml of NBc medium in a 35 mm Petri dish containing five 12 mm coverslips (*see* **Notes 6** and **7**).

We allow the conj Saporin to act for 72 h before using the neuronal cultures. No additional medium changes are necessary unless prescribed for the neurons.

4 Notes

1. A similar protocol was recently published [24]. The key difference is that in our protocol, AraC-containing medium is changed daily over 5 days and that LME is added for no less and no more than 90 min. This seems to result in a much more effective depletion of microglia (compare RT-QPCR results [4] and Western blots in [24]).
2. The morphology of astrocytes will change during LME treatment but will return to normal afterwards (Fig. 1).
3. Isolectin B4 from *Griffonia simplicifolia* conjugated to Saporin targets cells expressing the α-D-galactopyranoside group on their cell surface. This includes activated macrophages [25], microglia [26], as well as endothelial cells [27] and dorsal root ganglion neurons [28]. Thus, if working with spinal cord cultures, it would be more appropriate to use another form of conjugated Saporin, such as CD11b-Saporin.
4. Do not use reducing agents (DTT, β-mercaptoethanol, ascorbic acid) in conjunction with Saporin as these will inactivate the toxin.
5. Conjugated Saporin is an extremely potent cytotoxin and should be handled with appropriate caution (gloves and safety glasses). Saporin can be inactivated by autoclaving or exposure to 0.2M NaOH. All labware that has come into contact with the cytotoxin should be cleaned in this way. For further details, see information from the Advanced Targeting Systems information sheet.
6. If adding conj Saporin to neurons after seeding, dilute into fresh NBc medium. If adding to an existing neuronal culture, add conj Saporin directly into culture. Careful! Neurons are very sensitive to changes in their culture medium: never change all of the medium at once; rather change half or a third of the medium and then add the conj Saporin to the culture.
7. We tested various concentrations (0.1, 0.2, and 0.35 μg/ml) of IB4-Saporin and found that 0.35 μg/ml is sufficient to deplete all microglia from a mixed neuronal culture within 24–48 h, for up to 6 days. We have not examined the presence of microglia at later time points. Furthermore, although this concentration has worked effectively in our hands for isolectin B4-Saporin and CD11b-Saporin, effective concentrations for other Saporin conjugates need to be determined.

Acknowledgments

We thank Sarrah Ben Achour for preliminary experiments and thank Yasmine Belarif-Cantaut for the helpful discussion with the Saporin protocol. The work was supported by a grant from the seventh Framework Program Moodinflame (222963).

References

1. Pelvig DP, Pakkenberg H, Stark AK et al (2008) Neocortical glial cell numbers in human brains. Neurobiol Aging 29(11): 1754–1762
2. Saura J (2007) Microglial cells in astroglial cultures: a cautionary note. J Neuroinflammation 4:26
3. Stellwagen D, Malenka RC (2006) Synaptic scaling mediated by glial TNF-alpha. Nature 440(7087):1054–1059
4. Pascual O, Ben Achour S, Rostaing P et al (2012) Microglia activation triggers astrocyte-mediated modulation of excitatory neurotransmission. Proc Natl Acad Sci USA 109(4): E197–205
5. Saijo K, Winner B, Carson CT et al (2009) A Nurr1/CoREST pathway in microglia and astrocytes protects dopaminergic neurons from inflammation-induced death. Cell 137(1):47–59
6. Bezzi P, Domercq M, Brambilla L et al (2001) CXCR4-activated astrocyte glutamate release via TNFalpha: amplification by microglia triggers neurotoxicity. Nat Neurosci 4(7):702–710
7. McCarthy KD, de Vellis J (1980) Preparation of separate astroglial and oligodendroglial cell cultures from rat cerebral tissue. J Cell Biol 85(3):890–902
8. Holm TH, Draeby D, Owens T (2012) Microglia are required for astroglial Toll-like receptor 4 response and for optimal TLR2 and TLR3 response. Glia 60(4):630–638
9. Hamby ME, Uliasz TF, Hewett SJ et al (2006) Characterization of an improved procedure for the removal of microglia from confluent monolayers of primary astrocytes. J Neurosci Methods 150(1):128–137
10. Giulian D, Baker TJ (1986) Characterization of ameboid microglia isolated from developing mammalian brain. J Neurosci 6(8): 2163–2178
11. Brahmachari S, Fung YK, Pahan K (2006) Induction of glial fibrillary acidic protein expression in astrocytes by nitric oxide. J Neurosci 26(18):4930–4939
12. Alfonso-Loeches S, Pascual-Lucas M, Blanco AM et al (2010) Pivotal role of TLR4 receptors in alcohol-induced neuroinflammation and brain damage. J Neurosci 30(24): 8285–8295
13. Hewett JA, Hewett SJ, Winkler S et al (1999) Inducible nitric oxide synthase expression in cultures enriched for mature oligodendrocytes is due to microglia. J Neurosci Res 56(2): 189–198
14. Sola C, Casal C, Tusell JM et al (2002) Astrocytes enhance lipopolysaccharide-induced nitric oxide production by microglial cells. Eur J Neurosci 16(7):1275–1283
15. Beattie EC, Stellwagen D, Morishita W et al (2002) Control of synaptic strength by glial TNFalpha. Science 295(5563):2282–2285
16. Barres BA (2008) The mystery and magic of glia: a perspective on their roles in health and disease. Neuron 60(3):430–440
17. Burnett SH, Kershen EJ, Zhang J et al (2004) Conditional macrophage ablation in transgenic mice expressing a Fas-based suicide gene. J Leukoc Biol 75(4):612–623
18. Marin-Teva JL, Dusart I, Colin C, Gervais A et al (2004) Microglia promote the death of developing Purkinje cells. Neuron 41(4):535–547
19. Montero Dominguez M, Gonzalez B, Zimmer J (2009) Neuroprotective effects of the anti-inflammatory compound triflusal on ischemia-like neurodegeneration in mouse hippocampal slice cultures occur independent of microglia. Exp Neurol 218(1):11–23
20. Montero M, Gonzalez B, Zimmer J (2009) Immunotoxic depletion of microglia in mouse hippocampal slice cultures enhances ischemia-like neurodegeneration. Brain Res 1291: 140–152
21. Reeves JP (1979) Accumulation of amino acids by lysosomes incubated with amino acid methyl esters. J Biol Chem 254(18): 8914–8921
22. Thiele DL, Kurosaka M, Lipsky PE (1983) Phenotype of the accessory cell necessary for mitogen-stimulated T and B cell responses in

human peripheral blood: delineation by its sensitivity to the lysosomotropic agent, L-LEUCINE methyl ester. J Immunol 131(5): 2282–2290
23. Wiley RG, Oeltmann TN, Lappi DA (1991) Immunolesioning: selective destruction of neurons using immunotoxin to rat NGF receptor. Brain Res 562(1):149–153
24. Uliasz TF, Hamby ME, Jackman NA et al (2012) Generation of primary astrocyte cultures devoid of contaminating microglia. Methods Mol Biol (Clifton, NJ) 814:61–79
25. Maddox DE, Shibata S, Goldstein IJ (1982) Stimulated macrophages express a new glycoprotein receptor reactive with Griffonia simplicifolia I-B4 isolectin. Proc Natl Acad Sci USA 79(1):166–170
26. Streit WJ, Kreutzberg GW (1987) Lectin binding by resting and reactive microglia. J Neurocytol 16(2):249–260
27. Laitinen L (1987) Griffonia simplicifolia lectins bind specifically to endothelial cells and some epithelial cells in mouse tissues. Histochem J 19(4):225–234
28. Fang X, Djouhri L, McMullan S et al (2006) Intense isolectin-B4 binding in rat dorsal root ganglion neurons distinguishes C-fiber nociceptors with broad action potentials and high Nav1.9 expression. J Neurosci 26(27): 7281–7292

Chapter 8

Lentiviral Transduction of Cultured Microglia

Takahiro Masuda, Makoto Tsuda, Hidetoshi Tozaki-Saitoh, and Kazuhide Inoue

Abstract

Microglial cells are the resident immune-related glial cells of the central nervous system (CNS) that are crucial for maintaining homeostasis and sensing pathological alterations in the nervous system. To improve our understanding of the biological function of microglia, gene-transfer techniques have been improved and become widely used over the past several years. Here, we describe lentiviral-mediated transduction as a valuable tool for transduction of cultured microglial cells.

Key words Microglia, Lentiviral vector, Gene transduction

1 Introduction

Microglia are the CNS immune cells that survey around environment by elongating and contracting their processes. During CNS pathologies such as injury, they show rapid responses including change in morphology and cell-surface antigen expression and are crucial for CNS pathology [1–4]. During recent years, gene-transfer techniques have undergone rapid progress; they are now widely used as a useful procedure for the understanding of inherent biological cellular function [5]. However, monomyeloic lineage cells, including microglia, share similar characteristics in that they are difficult to transduce [6]. That is why a limited number of studies utilizing transgene approaches with primary cultured microglia have achieved a satisfactory response. Meanwhile, several types of gene-transfer tools based on viral vectors became more technically advanced and have been spotlighted as a new effective means of transducing such cells [6]. Among them, *Lentivirus*, a genus of retroviruses that can transfer genetic information into the DNA of the host cell, has the unique ability among retroviruses of being able to effectively replicate in both dividing and nondividing cells [6, 7].

Bertrand Joseph and José Luis Venero (eds.), *Microglia: Methods and Protocols*, Methods in Molecular Biology, vol. 1041, DOI 10.1007/978-1-62703-520-0_8, © Springer Science+Business Media New York 2013

Here, we demonstrate the lentiviral-mediated transduction of cultured microglia with high transduction efficiency using some unique methods.

2 Materials

Prepare and store the following at room temperature (unless indicated otherwise). We do not add sodium azide to the reagents.

2.1 For Preparation of Lentiviral Particles

1. Poly-L-lysine hydrobromide MW 30,000–73,000 (Wako): 0.1 μg/mL solution in PBS.
2. 100-mm cell culture dish: for coating, incubate poly-L-lysine solution in a culture dish for at least 3 h. Then, remove the solution and wash twice with PBS.
3. Culture medium for human HEK293T cells: 5 % fetal bovine serum (FBS), penicillin (100 U/mL), streptomycin (100 μg/mL), L-glutamine (2 mM) in Dulbecco's modified Eagle medium (DMEM). Store at 4 °C.
4. Plasmid vectors including targeted vector plasmid, packaging plasmid (pCAG-HIVgp, RIKEN) and envelop plasmid (pCMV-VSV-G-RSV-Rev, RIKEN).
5. 2 mg/mL polyethylenimine (PEI) solution: Store at 4 °C.
6. Forskolin.
7. PEG solution (4×): Make 80 mL of 50 % polyethylene glycol 6,000 (PEG). Add 8 mL of 5 M NaCl and then 4 mL of 1 M HEPES (*see* **Note 1**). Adjust pH with NaOH and dilute to 100 mL with water. Sterilize using an autoclave. Store at 4 °C.
8. Filter: cellulose acetate filter (0.45 μm) (Advantec).

2.2 For Treatment of Lentiviral Particles

1. Primary cultured microglia: prepare according to our previous paper [8].
2. 24-well plate.
3. Culture medium for mixed glial culture: 10 % FBS (Gibco), penicillin (100 U/mL), streptomycin (100 μg/mL), L-glutamine (2 mM) in DMEM (Gibco, Cat. 11965). Store at 4 °C.
4. 10 mg/ml polybrene (hexadimethrine bromide) solution: Store at 4 °C.

3 Methods

Carry out all procedures at room temperature unless otherwise specified.

3.1 Lentiviral Vector Construction

Generate a targeted vector plasmid by cloning target cDNA into the lentiviral vector construct (*see* **Note 2**). Put the fluorescent protein construct [e.g., green fluorescent protein (GFP)] into the vector to determine the transduction efficiency.

3.2 Lentiviral Particles Preparation

1. Seed HEK293T cells in poly-L-lysine-coated 100-mm cell culture dishes at a density of 3.0×10^6 cells per dish in culture medium (*see* **Note 3**).
2. Incubate the cells for 24 h at 37 °C in a humidified incubator with an atmosphere of 5 % CO_2 (about 70 % confluent).
3. Mix three plasmids, targeted vector plasmid (2.76 μg), packaging plasmid (1.62 μg, pCAG-HIVgp), and envelope plasmid (1.62 μg, pCMV-VSV-G-RSV-Rev), in a total volume of 600 μL of serum-free DMEM. Add 14 μL of polyethylenimine (PEI, 2 mg/mL) and mix well (vortex). Incubate the mixture for 15 min at room temperature.
4. Remove the culture medium from HEK293T cells, and add 10 ml of prewarmed fresh medium, and then add the mixture to the dish.
5. Incubate for 16 h at 37 °C in a humidified incubator with an atmosphere of 5 % CO_2.
6. Remove the mixture-containing medium and add 10 ml of prewarmed fresh medium containing 10 μM forskolin.
7. Incubate further for 48 h at 37 °C.
8. Filter the medium containing lentiviral particles through a filter (0.45 μm) and collect in a 50-mL conical tube.
9. Add a one-third volume of PEG solution, mix well, and incubate at 4 °C overnight.
10. Centrifuge the mixed solution at $2{,}600 \times g$ for 30 min at 4 °C and discard the supernatant.
11. Resuspend the pellet in an appropriate solution by pipetting (*see* **Note 4**). Store at –80 °C.

3.3 Preparing Conditioned Medium of the Mixed Glial Culture

Collect the supernatant of the mixed glial culture and filter through a filter (0.45 μm). Store at –20 °C (*see* **Note 5**).

3.4 Treatment of Lentiviral Particles with Primary Cultured Microglia

1. Seed primary microglial cells in 24-well plates at a density of 1.2×10^5 cells per well (*see* **Note 6**).
2. After cells have adhered to the bottom, add an appropriate volume of viral particles and 8 μg/ml polybrene onto cultured microglia (*see* **Note 7**).

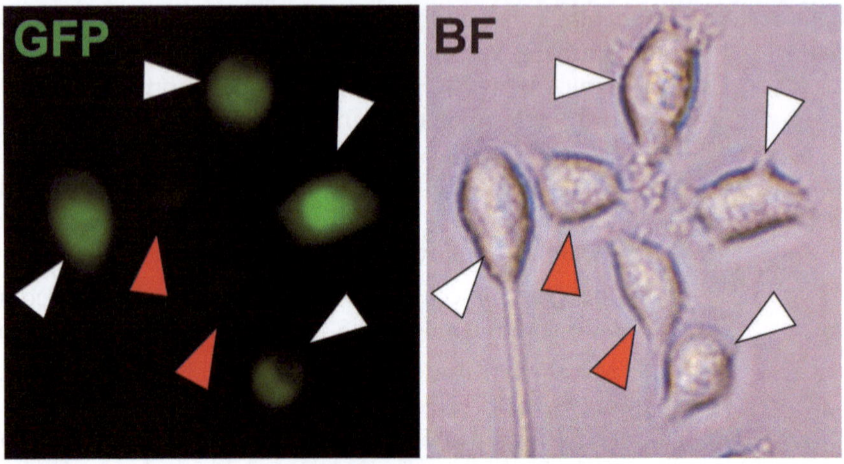

Fig. 1 GFP-positive microglial cells after lentiviral transduction. GFP or bright field (BF) images of cultured microglia transduced with a lentiviral vector encoding GFP. *White* or *red arrowheads* indicate GFP-positive or negative microglial cells, respectively. To determine the transduction efficiency, count GFP+ cells among all microglia cells 72 h after transduction

3. After a 12-h treatment with the lentiviral particles, change culture medium to 1 mL of prewarmed conditioned medium prepared from the mixed glial culture (*see* Subheading 3.3).
4. After a further 60 h, validate the transduction efficiency by counting the microglial cells expressing fluorescent protein, such as GFP (Fig. 1).

4 Notes

1. Mix PEG, NaCl, and HEPES in this order with stirring continually, or it may dissolve poorly (for Subheading 2.1, **item** 7).
2. We usually use the lentiviral CS2 vector (RIKEN) (for Subheading 3.1).
3. For coating 100-mm culture dishes with poly-L-lysine, add poly-L-lysine bromide solution to the dish and incubate overnight (or for at least 3 h) (for Subheading 3.2, **step 1**).
4. We usually use approximately 200 μl of PBS, HBSS, and fresh culture medium for resuspension. Try to avoid introducing air bubbles to keep transduction efficiency high (for Subheading 3.2, **step 11**).
5. We find that it is better to prepare the conditioned medium from the mixed glial culture as freshly as possible (for Subheading 3.3).

6. For using the microglial cell line BV2, seed in 24-well plates at a density of 1×10^4 cells per well and use an appropriate culture medium that is the same in composition as that used before and after transduction (*see* ref. 8) (for Subheading 3.4).
7. During treatment with viral particles, to achieve a high transduction efficiency of the lentiviral vector, reduce the culture medium of microglial cells at least to 500 μl per well (for Subheading 3.4, **step 2**).

Acknowledgment

Dr. Hiroyuki Miyoshi (RIKEN BioResource Center) kindly provided the lentiviral vector and its packaging plasmids.

References

1. Hanisch UK, Kettenmann H (2007) Microglia: active sensor and versatile effector cells in the normal and pathologic brain. Nat Neurosci 10:1387–1394
2. Perry VH, Nicoll JA, Holmes C (2010) Microglia in neurodegenerative disease. Nat Rev Neurol 6:193–201
3. Glass CK, Saijo K, Winner B et al (2010) Mechanisms underlying inflammation in neurodegeneration. Cell 140:918–934
4. Tsuda M, Inoue K, Salter MW (2005) Neuropathic pain and spinal microglia: a big problem from molecules in "small" glia. Trends Neurosci 28:101–107
5. Kamimura K, Suda T, Zhang G et al (2011) Advances in gene delivery systems. Pharm Med 25:293–306
6. Burke B, Sumner S, Maitland N et al (2002) Macrophages in gene therapy: cellular delivery vehicles and in vivo targets. J Leukoc Biol 72:417–428
7. Follenzi A, Naldini L (2002) Generation of HIV-1 derived lentiviral vectors. Methods Enzymol 346:454–465
8. Masuda T, Tsuda M, Yoshinaga R et al (2012) IRF8 is a critical transcription factor for transforming microglia into a reactive phenotype. Cell Rep 1:334–340

Part IV

Analysis of Microglial Cytokine Production

Chapter 9

Microglial Activation: Measurement of Cytokines by Flow Cytometry

Deepak Kumar Kaushik and Anirban Basu

Abstract

Cytokine measurement is a prerequisite to understand the inflammatory state of the body. Quantitative analysis of cytokines by Western blotting and ELISA is a daunting task as these are time-consuming and error-prone protocols. With the advent of flow cytometry, the estimation of cytokines using the classical antigen–antibody reaction has become a popular choice with researchers/clinicians. Here, we describe a protocol for multiple cytokine analysis using flow cytometry.

Key words Microglia, Cytokines, Cytokine bead array, Flow cytometry, Inflammation

1 Introduction

Microglia are the resident macrophages of the central nervous system (CNS) which represent the innate immune arm of the CNS [1]. Activation of microglia is often detected by the levels of cytokines and chemokines secreted in response to any pathogenic stimuli such as lipopolysaccharide (LPS) [2]. Cytokines and chemokines are low-molecular weight proteins which are involved in cell-to-cell communication and execute key regulatory functions in different target cell types [3]. The quantification of cytokines can be achieved using enzyme-linked immunosorbent assay (ELISA) and/or immunoblot techniques. However, both these techniques are time consuming and labor intensive [4]. Flow cytometry is routinely used for quantitative analysis of cell populations for either cell cycle analysis, staining of surface, or intracellular proteins as well as sorting of different populations (achieved by a cell sorter). Cytokine bead array (CBA), a flow cytometry-based application, offers efficient, reliable, and quick quantification of cytokines and chemokines [4, 5]. The CBA principle is based on the antigen–antibody (Ag–Ab) reactions similar to ELISA where the polymer beads are coated with specific antibodies to individual cytokines.

Bertrand Joseph and José Luis Venero (eds.), *Microglia: Methods and Protocols*, Methods in Molecular Biology, vol. 1041,
DOI 10.1007/978-1-62703-520-0_9, © Springer Science+Business Media New York 2013

The detection reagent is also bound to the antibodies which bind to the already-bound cytokines on the beads at different sites. This enables a highly specific binding of the cytokines with the beads. The beads are then run on the cytometer and resolved on different fluorescent channels in order to estimate the levels of different cytokines. CBA offers a flexibility to carry out analysis of several cytokines in a single experiment without much hassle [6, 7]. Various companies offer ready-to-use kits and softwares for the quantification and analysis of these cytokines.

Here, we discuss the methodology of analyzing cytokine analysis using mouse inflammation kit from BD (Becton Dickinson) Biosciences using a BD™ FACSCalibur instrument. The applications and principles are necessarily the same with slight variations observed between different suppliers.

2 Materials

2.1 Brain Homogenate/Cell Culture Supernatant

2.1.1 Brain Homogenate

In vivo analysis of cytokines from mice brain can be detected from whole brain homogenates. However, the source of cytokines can be redundant as astrocytes also secrete cytokines [8, 9]; it is the microglial population that predominates in the CNS in secreting the cytokines and carrying out the innate immune responses during CNS pathologies [10]. The analysis of cytokines from brain homogenates provides the information regarding the inflammatory state of this organ. For the estimation of cytokines, the amount of protein homogenate needs to be standardized; however, a successful sensitive assay can detect as low as 10 ng/L [5]. In practice, we usually take 10 μg protein in 50 μl sample volume for the analysis of cytokines (*see* **Note 1**).

Protocol for isolation of brain homogenate/cell lysate:

1. The brain tissue is homogenized in 1× lysis buffer (1 % Triton-X-100, 10 mM Tris–HCl (pH 8.0), 150 mM NaCl, 0.5 % Nonidet P (NP-40), 1 mM EDTA, 0.2 % EGTA, 0.2 % sodium orthovanadate, and protease inhibitor cocktail).
2. The homogenate is kept on ice with intermittent vortexing for an hour and then centrifuged at 12,000 × *g* at 4 °C for 15 min. Similarly, the protein from cells is also harvested by incubating the cells in 1× lysis buffer and then pelleting them.
3. The supernatant from cell lysate or tissue homogenate is then collected and measured for protein concentration (*see* **Note 2**).

2.1.2 Cell Culture Supernatant

The cytokines are small proteins that act as messengers between different cell types. They are secreted out into the cell culture medium in which the cells are grown. This supernatant can be used for analysis of cytokine secreted by cultured microglia. It is desired that the volume of cell culture medium be as low as possible such that the cytokine concentration is increased per μl (*see* **Note 3**).

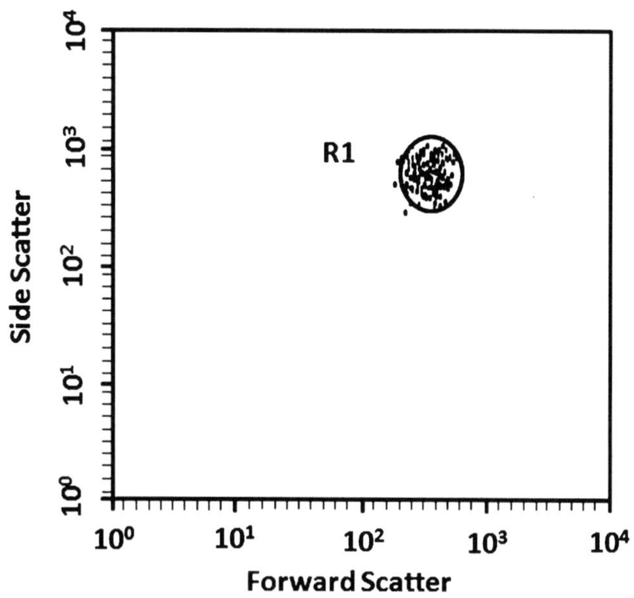

Fig. 1 Gating of bead population on FSC/SSC (forward scatter/side scatter) plot. All the beads having the same shape and size cluster together and can be seen on the dot plot as a single cluster. It is then tightly gated which is represented as R1 (Region 1) in the figure

The volume of supernatant to be taken for each sample may range from 50 to 200 μl depending upon the sensitivity of the assay. The volumes may be standardized before setting up an actual experiment (*see* **Note 4**).

2.2 Cytokine Beads and Detection Reagent

2.2.1 Cytokine Beads

The cytokine beads are precoated with antibodies for detecting different cytokines. Each bead may correspond to a single cytokine. Depending on the provider, these beads may be similar to each other except for the presence of different intensities of fluorophores (enabling them to be detected as separate beads on a different channel) [5]. Some providers have different sizes of beads for different cytokines and can be detected based on their sizes. In the former case, a single bead population is gated on FSC (forward scatter) and SSC (side scatter) plot (Fig. 1), while later may have different bead populations to be gated and analyzed separately. The CBA kit from BD Biosciences have six different beads corresponding to different cytokines and have different intensities of fluorophore that enables the user to identify different bead populations (Fig. 2). However, the number of cytokines that can be detected depends upon the assay system as some systems offer analysis of multiple cytokines and can detect a large number of cytokines simultaneously. For example, the following beads corresponding to the following cytokines are distributed on FL3 channel from brightest to dimmest on a FACSCalibur machine using mouse inflammation CBA kit from BD.

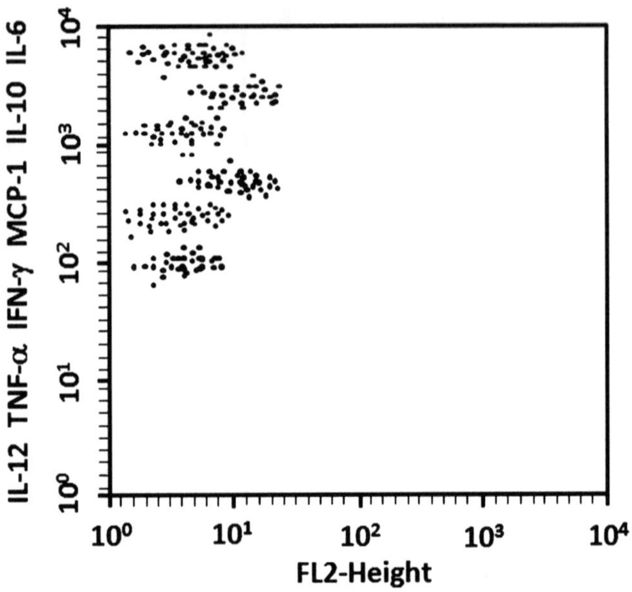

Fig. 2 Resolution of different beads corresponding to different individual cytokines on FL3 channel. There is no change in the levels of these cytokines on FL1 channel as they are not labeled with FITC

Bead population	Cytokines
A1 (Brightest)	IL-6
A2	IL-10
A3	MCP-1
A4	IFN-γ
A5	TNF-α
A6 (Dimmest)	IL-12p70

2.2.2 PE (Phycoerythrin) Detection Reagent

The CBA kit supplied by BD has phycoerythrin-conjugated antibodies that bind to the cytokine beads. The detection is then made on FL2 channel on a BD™ FACSCalibur machine (Fig. 3). Other suppliers may provide other detection reagents.

2.2.3 Recombinant Cytokines for Standard Curve Generation (Inflammation Standards)

The CBA kits from various sources have recombinant cytokines whose concentrations are known. Using serial dilutions of these cytokines, a standard curve is generated which is used as a reference for the estimation of cytokines in a given sample.

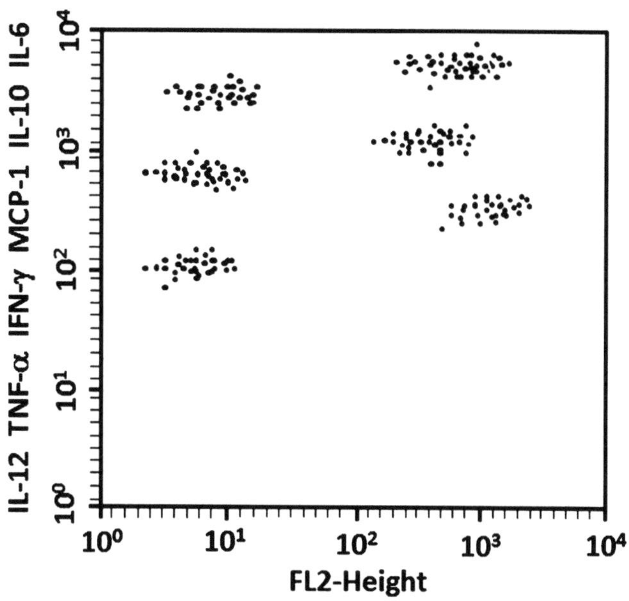

Fig. 3 Levels of cytokines measured by distinct bead population on FL2 channel. Different bead levels corresponding to changes in cytokine levels can be noticed on FL2 channel which detects PE

2.3 Instrumentation and Buffers

2.3.1 FACS Instrument with Cell Analysis Software

A FACS instrument equipped with argon 488 nm laser is required. This enables to detect and distinguish fluorescence emissions at 670 nm on a FACSCalibur machine. Additional lasers may be required if other dyes emitting at higher wavelength are used. For example, phycoerythrin (PE) emits at 575 nm; therefore, using a machine which is equipped with either 3-color or 4-color laser is a prerequisite for running this application (this application may be required if the beads are clustered as different populations and coated with different dyes).

2.3.2 Instrument Setup Beads

These beads are provided for setting up the instruments. Using FACSComp™ software from BD Biosciences, this can easily be performed in a stepwise fashion according to the manufacturer's protocol. Instrument setup generates a file which is stored in the "instrument setup" folder, and these settings are used while CBA is carried out.

2.3.3 Sheath Fluid for Sample Running

Sheath fluid is a proprietary product provided by the manufacturers. Many workers also use filtered 1× PBS as sheath fluid which may also contain antifungal/antibacterial agents such as sodium azide. Some labs also use 0.9 % saline solution with no antimicrobial agents for this application. Always refer to the recommended buffers from the manufacturer in order to prolong the life of the machine (*see* **Note 5**).

2.3.4 Wash Buffer

Wash buffer is usually supplied with the kit. Wash buffer can also be made in the laboratory: To make 500 ml of wash buffer, take 480 ml of 1× PBS, pH 7.4, and add 20 ml of fetal bovine serum (FBS) to it. To this add 0.09–0.1 % sodium azide, mix well, and store at 4°C till further use (*see* **Note 5**).

2.3.5 Buffer for Cleaning the Fluidics

FACS Clean™ Solution provided by BD Biosciences is a routinely used for cleaning the instrument after the sample has been analyzed. Ten percent bleach solution (sodium hypochlorite) may also be used after the analysis for cleaning of the fluidics of flow cytometer. This step is crucial for the proper cleaning of the fluidics in order to avoid accumulation of debris from previous runs. This also ensures uniform flow rates during a machine's run. The solution is run for 5 min at high flow rates.

3 Method

3.1 Instrument Setup Using Instrument Setting Beads

For CBA application, instrument setup beads are usually supplied by the provider along with the CBA kit. Using interactive software provided by the manufacturer, these setup beads are run on the machine during which the various parameters are set automatically. During the instrument setup, the various parameters such as threshold and compensations are adjusted. This is then saved separately as an "instrument setup" file in "applications" folder. This path may be different for different manufacturers and machines.

3.2 Instrument Calibration

Instrument calibration is routinely carried out using the beads conjugated to different dyes in order to set the threshold and compensations as well as check the instrument sensitivity for proper instrument functioning. BD Biosciences offer BD™ Calibrite beads for a regular instrument calibration. These beads are usually polymers. The Calibrite beads provided with BD CBA kit are made of polymethylmethacrylate microspheres of approximately 6 μm diameter. The kit provides the following beads for calibration of the instrument:

1. A two-color calibrite kit: contains unlabeled beads, fluorescein isothiocyanate (FITC)-labeled beads, and phycoerythrin (PE)-labeled beads.
2. A three-color calibrite kit: contains unlabeled, FITC-labeled, PE-labeled, and peridinin chlorophyll (PerCP)-labeled beads.
3. A four-color calibrite kit: contains unlabeled, FITC-labeled, PE-labeled, PerCP-labeled, and allophycocyanin (APC)-labeled beads.

The following PMTs carry out following signal detection on FACSCalibur (*see* **Note 6**):

FL1 (fluorescence 1): FITC (yellow-green)
FL2 (fluorescence 2): PE (red-orange)
FL3 (fluorescence 3): PerCP (red)
FL4 (fluorescence 4): APC (red)

3.3 Standard Curve

3.3.1 Preparation of Standards

Recombinant cytokines are provided in the kit or can be procured separately from various companies as recombinant proteins. In order to carry out the standards, these recombinant cytokines are diluted in assay diluent in recommended volumes. These cytokines are then serially diluted. Five thousand pg of each cytokine is provided in the kit. Using CBA kit from BD Biosciences, the standards are double diluted in a stepwise fashion to make the second standard of 2,500 pg and so on such that the last dilution has 20 pg of cytokines. Assay diluent (a solution provided by the manufacturer) is also run on the cytometer which has 0 pg of standards. These are then run on the instrument and standard curve for the following concentrations are generated:

Concentration (pg)	Dilution
0	Negative control (assay diluent)
20	(1:250)
40	(1:128)
80	(1:64)
156	(1:32)
312.5	(1:16)
625	(1:8)
1,250	(1:4)
2,500	(1:2)
5,000	Top standard

3.3.2 Acquiring the Data

1. Gating the bead population: On a FSC and SSC dot plot, bead population may be represented by a cluster of cells (beads in this case) which is then gated using the gating tool. Care must be taken to avoid scattered population on the plot as it may represent degraded beads or nonspecific bead population (*see* **Note** 7).
2. Setting the parameters and events: Go to the cytometer application, select the latest instrument setup file, and apply it to the current settings. Set minimum event rate to 3,000 of the total bead population to be analyzed. A further increase in the event rate may offer better results. Using the "setup" mode, the voltages are adjusted to put the values of the highest dilution of

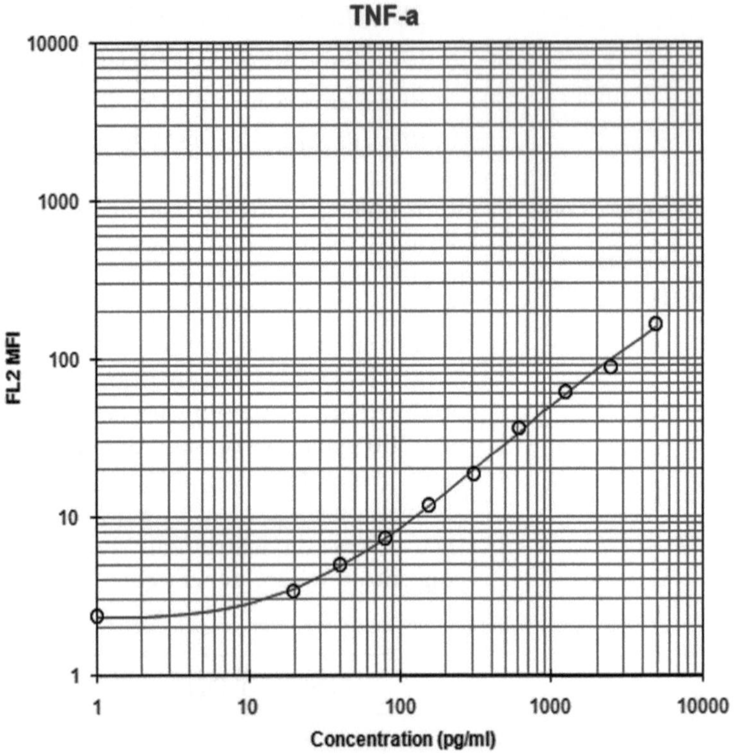

Curve-Fitting Model: 4-Parameter Logistic

y = ((A - D)/(1 + (x/C)^B)) + D

A = 0.37 B = 2.94 C = 3.40 D = 3.63 R^2 = 0.9984

Fig. 4 A four-parametric standard curve generated for TNF-α, one of the cytokines in mouse inflammation kit. *X*-axis labels the concentration in pg/ml while *Y*-axis represents the mean fluorescent intensities (MFI) on FL2 channel

standard such that it occupies a left quadrant on the dot plot between FL3 (separates different beads) and FL2 (detects the levels of individual cytokines) channel. Once the voltages are adjusted, the standards are run without the "setup" mode and the data is acquired.

3. Generating the standard curve: Standard curves may either be created manually from the readings obtained for individual cytokines and then generating a line equation on a simple excel sheet. This equation can be then used to analyze the levels of various cytokines individually. Automatic analysis files or softwares are also available with the instrument. A template file for CBA analysis such as FCAP™ software may offer analysis of individual cytokines with reference to the standard files. An example of a typical 4-parametric logistic graph generated by the software can be seen in Fig. 4. (The graph represents the standard curve for TNF-α cytokine.)

4. Running the samples: The samples are run similar to the standard samples. The samples are first gated and voltages are set such that the values from untreated (the sample which is expected to have a low cytokine level) samples are on the left quadrant of the FL3-FL2 plot (on a FACSCalibur machine; remember phycoerythrin is detected on FL2 channel) (*see* **Note 8**).

4 Detailed Protocol for Running of Samples

1. Preparation of samples: The beads corresponding to different cytokines are mixed in 1:1 ratio. The bead vials must be vortexed before use (*see* **Note 9**). Depending upon the detection capability, the volume of the bead mixture is defined. BD CBA kit recommends 50 μl of bead mix for each sample. For a six-cytokine bead array, the volume of each bead required is

$$50\ \mu l \div 6 = 8.333\ \mu l.$$

   This number is then multiplied by the number of samples to be analyzed (*see* **Note 10**). For example, in order to analyze four samples, the volume of each bead required is

$$8.333\ \mu l \times 4 = 33.333\ \mu l.$$

   Once the bead mixture is prepared, 50 μl of sample is added to each vial (*see* **Note 11**) along with 50 μl of phycoerythrin (PE) detection reagent (*see* **Note 12**) to a 50 μl bead mixture. This step may be carried out in 1.5 ml plastic centrifuge tubes.
2. Incubation of samples: The mixture is then incubated for recommended time durations. Often, these time ranges from 2 to 3 h depending upon the manufacturer or the kit being used. For analysis of mouse cytokines using the BD CBA bead array, the samples must be incubated for 2 h at room temperature under dark conditions (*see* **Note 13**).
3. Washing the samples: Upon incubation, the samples are then washed using the wash buffer. Approximately, 1 ml of wash buffer is added to each sample mixture which is then centrifuged at 200 × *g* for 5 min (*see* **Note 14**).
4. The wash buffer is now carefully aspirated without disturbing the bead pellets and then additional 300 μl of wash buffer is added to each vial to resuspend the pelleted bead population.
5. The samples are ready to be analyzed on a flow cytometer.

4.1 Instrument Start-up and Data Acquisition

1. Instrument is started and flow check is performed. Calibration of the machine (explained in the Subheading 3.2) may also be carried out at this juncture if not carried out recently (*see* **Note 15**).
2. CellQuest Pro™ software is launched and BD CBA instrument setup template is opened (generated as explained in

Subheading 3.1). The software may vary from manufacturer to manufacturer but the basic principles remain the same.

3. Samples are then run at low to medium flow rates in the "setup" mode first to adjust the settings such that the bead population is observed as a single cluster on FSC/SSC dot plot. Once the gating is carried out of this population, the data is then acquired without the "setup" mode as described in earlier sections. These settings are kept constant during the acquisition of the rest of the samples.

   Parameter description: Using this application of the CellQuest Pro, the directory (folder) is assigned to a run along with file names for each sample.

   Acquisition and storage: Using this application, events pertaining to either total count or the gated population can be set.

   Counters: The counters provide information regarding the total cells being analyzed and flow rates as events per second.

4. Analysis of results: The results are analyzed with reference to the standard curves generated (Fig. 4). Using the line equation from the standard curve (shown in Fig. 4), the amount of cytokines can be measured in pg/ml. Analysis of result is also carried out automatically by automatic softwares such as FCAP CBA flex from BD Biosciences.

5 Notes

1. The amount of protein depends upon the levels of cytokines secreted. Usually 10 μg of total protein is sufficient to analyze several cytokines that are measured in pg/ml.
2. In case the samples are obtained from infected animals/cells, they must be handled with care and any direct contact must be avoided. Discarding such samples must be done according to the institutional chemical discard guidelines. A 10 % bleach solution may come handy and should be used to treat such samples before discarding them.
3. The samples must not be too concentrated as this may result in high background. In a case where sample is too concentrated, number of washes must be increased.
4. The sample volume may depend on the amount of cytokine secreted in the supernatant and also depends upon the cell type in culture conditions. For example, while BV-2 mouse microglia cell line often exhibits increased levels of MCP-1, IL-6, and TNF-α, there is a relatively decreased expression of IL-10, IL-12p50, and IFN-γ. Therefore, the amount of supernatant depends upon the cell phenotype and the cytokines secreted that are to be analyzed.

5. The beads, the wash buffer, and assay diluents contain sodium azide. It is therefore advised not to come directly in contact with these reagents and wear gloves and lab coats during performance of these assays.
6. The fluorescence channels and their detection wavelength may vary among different models of instruments. It is therefore imperative that information about different PMTs of an instrument is known to the user.
7. Beads representing a single population must be tightly gated. This avoids false analysis of nonspecific bead population which is often found scattered around a single cluster of bead population.
8. The detection limit of cytokine analysis is 5,000 pg/ml and therefore samples having high concentration of individual cytokines may need to be diluted such that the cytokines are within the range of linearity.
9. The bead vials must be thoroughly vortexed before aspirating out the beads because the beads are often adhered to the plastic surfaces and get aggregated.
10. It is advised to take additional beads. For example, if the sample number is n, then beads must be calculated for $n+1$ samples such that there is no shortage of bead mix due to pipetting errors.
11. The samples to be analyzed must be fresh and kept on ice such as to avoid cytokine degradation. Avoid freeze–thaw cycles of the samples.
12. PE must be added in dark conditions and the incubation with PE must be carried out under dark conditions at room temperature. The light conditions and incubation time limits must be carefully read from the manufacturer's guidelines. In case this information is not accessible to the user, the user may standardize these time points.
13. Incubation of samples with bead mix and detection reagent usually results in increased debris or nonspecific binding of beads if kept for longer durations. It is therefore necessary to keep in mind that the incubation must not exceed the recommended time.
14. Beads must be not be centrifuged beyond recommended gravitational forces (measured in relative centrifugal field) as this may result in rupture of the beads. Once the beads are settled after the centrifugation, the wash buffer is gently discarded without disturbing the beads.
15. It is recommended that calibration of the machine is carried out regularly just before the samples are analyzed.

References

1. del Rio-Hortega P (1932) Cytology and cellular pathology of the nervous system. In: Penfield W (ed) Microglia. P. B. Hoebaer, New York, NY, pp 483–534
2. Kaushik DK, Gupta M, Das S et al (2010) Kruppel-like factor 4, a novel transcription factor regulates microglial activation and subsequent neuroinflammation. J Neuroinflammation 7:68
3. Borish LC, Steinke JW (2003) 2. Cytokines and chemokines. J Allergy Clin Immunol 111:S460–S475
4. Chen R, Lowe L, Wilson JD et al (1999) Simultaneous Quantification of Six Human Cytokines in a Single Sample Using Microparticle-based Flow Cytometric Technology. Clin Chem 45:1693–1694
5. Oliver KG, Kettman JR, Fulton RJ (1998) Multiplexed analysis of human cytokines by use of the FlowMetrix system. Clin Chem 44:2057–2060
6. McHugh TM (1994) Flow microsphere immunoassay for the quantitative and simultaneous detection of multiple soluble analytes. Methods Cell Biol 42(Pt B): 575–595
7. Carson RT, Vignali DA (1999) Simultaneous quantitation of 15 cytokines using a multiplexed flow cytometric assay. J Immunol Methods 227:41–52
8. Carpentier PA, Begolka WS, Olson JK et al (2005) Differential activation of astrocytes by innate and adaptive immune stimuli. Glia 49:360–374
9. Ransohoff RM, Brown MA (2012) Innate immunity in the central nervous system. J Clin Invest 122:1164–1171
10. Kaushik DK, Gupta M, Basu A (2011) Microglial response to viral challenges: every silver lining comes with a cloud. Front Biosci 17:2187–2205

Chapter 10

In Situ Hybridization of Cytokine mRNA Using Alkaline Phosphatase-Labelled Oligodeoxynucleotide Probes

Bettina Clausen, Christina Fenger, and Bente Finsen

Abstract

In situ hybridization is a powerful tool for visualizing cellular gene expression in morphologically preserved brain tissue giving precise information on the regional expression of specific mRNA sequences in cells of diverse phenotype. Here, we describe a sensitive, simple, and robust method using alkaline phosphatase (AP)-labelled oligodeoxynucleotide probes to detect cytokine mRNA in the acutely injured or inflamed mouse CNS.

Key words In situ hybridization, mRNA, Enzyme-conjugated oligodeoxynucleotide probes, Tumor necrosis factor, Interleukin-1, Cytokines, CNS

1 Introduction

Hybridization of specific cytokine mRNA in situ is an important supplement to immunohistochemistry for unequivocal identification of the site of cellular transcription/translation, as the majority of cytokines are cleaved or released into the extracellular space. We have for the detection of cytokine transcripts in the CNS optimized our in situ hybridization procedure for the use of alkaline phosphatase (AP)-labelled, short 26–30 mer oligodeoxynucleotide, probes [1–3]. The major advantages of this non-isotopic technique are that it provides similar high resolution [1–4] as digoxigenin-labelled probes [5–7] and reaches the same high sensitivity as radiolabelled probes [8, 9]. Thus, the threshold levels of detection are in the region of 10–20 mRNA copies per cell [10–12]. Hybridization occurs when the AP-labelled oligodeoxynucleotide probe recognizes and binds by the means of hydrogen bonding to the target mRNA sequence in the tissue section [13], followed by probe-signal development using an AP substrate solution. Critical for successful in situ hybridization is to avoid premature degradation of target mRNA by nucleases. The stability of a given probe-mRNA hybrid depends on its melting temperature (T_m), which for a given

Bertrand Joseph and José Luis Venero (eds.), *Microglia: Methods and Protocols*, Methods in Molecular Biology, vol. 1041,
DOI 10.1007/978-1-62703-520-0_10, © Springer Science+Business Media New York 2013

Table 1
In situ hybridization probes for TNF and IL-1β mRNAs

Probe ID	Probe sequence	No. of bases	Target region	% GC
TNF-I	5′CT TCT CAT CCC TTT GGG GAC CGA TCA CC	28	305–332 [17]	57
TNF-II	5′CG TAG TCG BBB CAG CCT TGT CCC TTG AA	28	570–597 [17]	60
TNF-III	5′CT TGA CGG CAG AGA GGA GGT TGA CTT TC	28	644–671 [17]	53
IL-1β	5′CT GTC CAT TGA GGT GGA GAG CTT TCA GCT C	30	501–530 [18]	57
IL-1β	5′GC TTG TGA GGT GCT GAT GTA CCA GTT GGG G	30	779–808 [18]	53
GADPH	5′CC TGC TTC ACC ACC TTC TTG ATG TCA	26	846–871 [19]	50

hybrid is a function of its cytosine (C) and guanine (G) content. High GC content equals high T_m of the probe-mRNA hybrid, since these nucleobases bind more strongly by the means of three hydrogen bonds, as compared to adenine (A) and thymidine (T). The strength of the hydrogen bonds is influenced by the stringency factors [13, 14], which are extrinsic factors affecting the melting temperature (T_m) of the newly formed hybrid. These factors are pH, salt concentration, temperature, and organic solvents such as formamide. Increased salt concentration stabilizes the hybrid while increased pH, temperature, and high formamide content destabilize the hybrid [13]. Here, we provide a detailed protocol for specific and sensitive detection of microglial- and macrophage-produced TNF and IL-1β mRNA in the injured or diseased mouse CNS using short AP-labelled oligodeoxynucleotide probes (Table 1) [2, 3, 14].

2 Materials

2.1 Probe Design

1. Identify the specific mRNA sequence in the species of interest (here Mus musculus) using the nucleotide search tool at the NCBI website.
2. Identify the exon–exon junctions in the selected cytokine mRNA (*see* **Note 1**) and use the FASTA format of a cDNA sequence spanning one or more exon–exon junction(s) as input sequence for the probe design software (for instance, Oligo-design software v. 6.0, Molecular Biology Insight, CO, USA).
3. Design antisense oligodeoxynucleotide (~26–30 bases) probe(s) that span an exon–exon junction site.
4. The CG content must be approx. 50 % and the DNA T_m must be approx. 37 °C in the used hybridization conditions (1× SSC, 50 % formamide, 165 mM NaCl) (*see* **Note 2**).

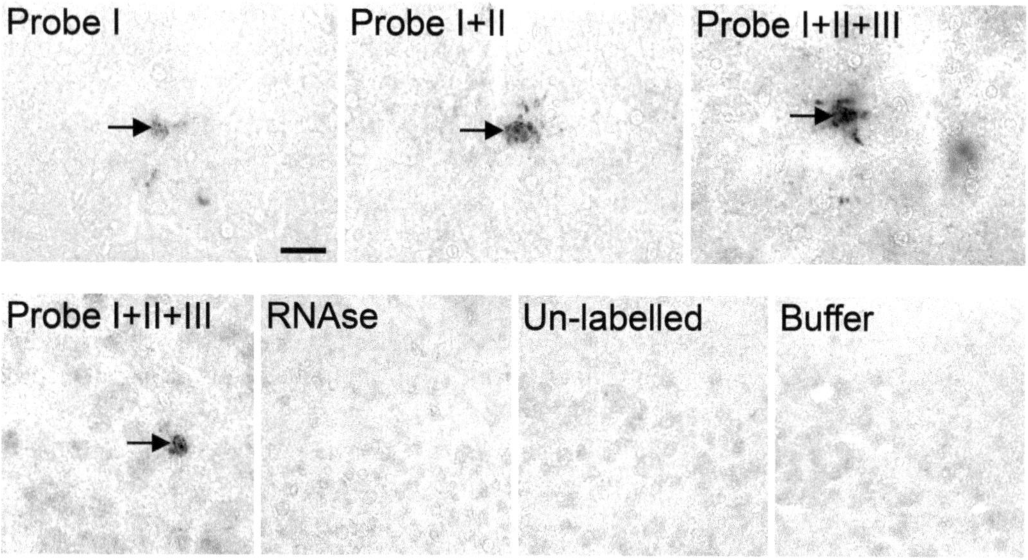

Fig. 1 In situ hybridization specificity and sensitivity illustrated on stroke-lesioned mouse brain showing TNF mRNA expression in microglial-like cells. (*Upper row*) In situ hybridizations performed with Probe I, Probe II, or Probe III show identical cellular and regional localization of the in situ signal (shown for Probe I, only). Furthermore, hybridization with Probe I+II or Probe I+II+III yields stronger in situ signal than hybridization with a single probe only, here Probe I. Photomicrographs were obtained from the border zone of the infarct 12 h after induction of the stroke lesion, at which time single-cell expression is very high. (*Lower row*) Hybridization with Probe I+II+III on RNAse-treated sections yields no signal. Similarly, hybridization with 100-fold excess of each of the unlabelled probes, and incubation with hybridization buffer, yields no signal. Photomicrographs were obtained of the cingulate cortex 4 h after a stroke lesion in a mouse (for details, *see* ref. 3). *Arrows point* to TNF mRNA$^+$ cells. Scale bar: 20 μm

5. Reject suggested probes with tendency to make stable dimers or hairpins ($\Delta G \leq -5$ kcal/mol).
6. Confirm that the probe(s) selectively recognize(s) the target mRNA by using standard nucleotide BLAST (blastn) tool at the NCBI website.

2.2 Solutions and Reagents Used for In Situ Hybridization

1. AP-labelled antisense oligodeoxynucleotide probes written in the direction from 5′ to 3′ (Table 1) are purchased from DNA Technology Aps, Denmark. The probe is synthesized with a modified 5′ amino terminal which is coupled to AP thereby generating the AP-probe conjugate [14]. Using two or more probes targeting different, non-overlapping sequences of the target mRNA increases signal strength in an additive way (Fig. 1, upper row).
2. Diethylpyrocarbonate (DEPC)-treated water: 1 % DEPC solution in dH_2O (*see* **Note 3**). Stir overnight and autoclave the DEPC-water to remove excess DEPC.

3. Hybridization buffer: 500 ml 50 % formamide, 200 ml 20× SSC, 25 ml 40× Denhardt, 200 ml 50 % dextran sulfate, 100 μl single-stranded salmon sperm DNA (*see* **Note 4**).
   (a) *Formamide*: 100 ml formamide and 5 g Amberlite for 1 h in the dark at room temperature (RT), filter and store at −20 °C.
   (b) *20× Saline sodium citrate (SSC):* 150 mM NaCl and 15 mM sodium citrate dissolved in 500 ml DEPC-water. Store at 4 °C.
   (c) *40× Denhardt:* 0.04 g Ficoll, 0.04 g polyvinylpyrrolidone, and 0.04 g bovine serum albumin dissolved in 5 ml DEPC-water (*see* **Note 5**).
   (d) *50 % dextran sulfate:* 5 g dextran sulfate dissolved in 10 ml DEPC-water. Stir at RT.
   (e) *Single-stranded salmon sperm DNA:* 50 μg single-stranded salmon sperm DNA dissolved in 50 ml Tris–EDTA (TE) buffer.
4. *TE buffer:* 0.5 ml 2 M Tris buffer (pH 7.4) adding up to 100 ml DEPC-water, 20 μl 0.5 M EDTA. Adjust pH to 8.
5. *Tris buffer 2 M:* 24.2 g Tris(hydroxymethyl)aminomethane TRIS (Sigma) dissolved in 100 ml DEPC-water. Add concentrated HCl to pH 7.4.
6. *EDTA:* 4.65 g EDTA dissolved in 100 ml DEPC-water. Adjust pH to 8.
7. 1× SSC: 50 ml of 20× SSC and 950 ml dH_2O.
8. Tris–HCl buffer (rinsing buffer): 0.01 M Trizma hydrochloride and 0.15 M NaCl dissolved in 1 L dH_2O. Adjust pH to 9.5.
9. Tris–HCl–$MgCl_2$ (developer buffer): 0.01 M Trizma hydrochloride, 0.01 M NaCl, and 0.05 magnesium chloride ($MgCl_2{\cdot}6H_2O$) dissolved in 1 L dH_2O. Adjust pH to 9.5.
10. AP substrates: Nitroblue tetrazolium (NBT) and 5-bromo-4-chloro-3-indolyl phosphate (BCIP).
    (a) *NBT stock solution:* 70 mg NBT dissolved in 1 ml 70 % *N*,*N*-dimethylformamide.
    (b) *BCIP stock solution:* 50 mg BCIP dissolved in 1 ml 100 % *N*,*N*-dimethylformamide (*see* **Note 6**).
11. AP developer: 100 ml Tris–HCl–$MgCl_2$ (developer buffer) containing 450 μL NBT stock solution and 350 μL BCIP stock solution.
12. RNAse A solution (control reaction): 200 μg/ml RNAse A diluted 1:3 in RNAse buffer to give a final concentration of 50 μg/ml (*see* **Note 7**).
13. *RNAse buffer:* 0.1 M Tris–HCl, 0.5 M NaCl, 1 mM EDTA diluted in 800 ml DEPC-water.

3 Methods

3.1 Tissue Processing

The PBS perfused or unperfused brain is quickly removed and frozen by the use of CO_2 snow (*see* **Note 8**). The brain is cut into series of 30 μm (*see* **Note 9**) thick cryostat sections and mounted on RNAse-free SuperfrostPlus glass slides (Hounisen) (*see* **Note 10**) and stored at −80 °C until further used (*see* **Note 11**).

3.2 Hybridization, Posthybridization Rinsing, and Development

Day 1

1. Place glass slides containing the fresh frozen tissue sections on a paper tray (RNAse-free) and dry the sections at 55 °C for 10 min.
2. Immediately after, place the sections in 96 % ethanol for a minimum of 30 min (*see* **Note 12**).
3. Air dry the section at RT for approx. 20 min.
4. Add probe(s) to the hybridization buffer and mix gently, but well (*see* **Note 13**). Keep probes on ice while pipetting. The optimal probe concentration must be determined every time the procedure is set up for a new target mRNA (*see* **Note 14**). The volume of hybridization buffer per slide depends on the number of sections per slide (*see* **Note 15**). Carefully add the hybridization buffer containing the probe onto the slide, but next to the sections (*see* **Note 16**). Cover the section with a 60 mm long mounting glass, while gently pushing the drop of liquid over the sections, thereby avoiding air bubbles (*see* **Note 17**).
5. Place the cover-slipped sections in a humid plastic hybridization chamber (Life Technology) containing 10 ml distilled H_2O at the bottom, which reduces evaporation of hybridization buffer.
6. Seal the plastic chamber using aluminum foil/parafilm and place it in a heating chamber at 37 °C overnight.
7. Prepare 1× SSC and place in a 55 °C heating chamber overnight. Cover to prevent evaporation.

Day 2

1. Remove the mounting glass and rinse the hybridized section.
2. Rinse sections 3 × 30 min in 1× SSC, pH 8, at 55 °C.
3. Rinse for 2 × 10 min in Tris–HCl, pH 9.5 (*see* **Note 18**).
4. Mix Tris–HCl–$MgCl_2$ buffer, pH 9.5, with the AP substrates (AP developer). Place sections in upright rack and add the AP developer. Wrap the rack in aluminum foil and allow signal development to take place in darkness at RT.

Days 4–5

1. Arrest signal development after 48–72 h and rinse well in running tap water (*see* **Note 19**).
2. Cover slip sections using Aqua-Mount™. When successful hybridization has occurred, the target mRNA will appear as blue formazane deposits within the mRNA-expressing cells.

3.3 Control Reactions

Specificity controls are performed on sections parallel to those of interest. A control panel for in situ hybridizations for TNF mRNA in stroke-lesioned mouse brain is shown in Fig. 1.

1. *Probe specificity.* Confirmed by the use of two or more probes, alone and in combination. This is illustrated for detection of TNF in Fig. 1 (upper row) and Table 1. The probability that Probe I, II, or III yields identical regional and cellular localization of the in situ hybridization signal is negligible, if one of the probes should happen to detect a different mRNA than TNF mRNA. Besides adding to the sensitivity, yielding a stronger signal in sections hybridized with the probe mixtures (Probe I+II or Probe I+II+III) compared to a single probe also provides evidence of probe specificity (Fig. 1, upper row).
2. *RNAse pretreatment.* The binding specificity of the probe for mRNA is tested on RNAse pretreated sections (*see* **Note 7**). Absence of signal suggests that the probe binds to RNA within the tissue section (Fig. 1, bottom row). For RNAse pretreatment, 1–2 control sections are incubated with RNAse A solution for 1–2 h(s) at 37 °C. Rinse 2 × 10 min in 2× SSC and 10 min in DEPC-water at RT and dehydrate sections in 96 % ethanol before reentering the main protocol again, at **step 3** (Day 1).
3. *Competition with unlabelled probe.* Hybridization with AP-labelled probes and a 100-fold excess of unlabelled probes results in absence of hybridization signal (Fig. 1, bottom row) (*see* **Note 20**).
4. *In situ hybridization for GAPDH mRNA.* Finally, as a test for tissue quality, hybridization with a probe specific for GAPDH mRNA is routinely included, since GAPDH mRNA is highly expressed throughout the brain (*see* **Note 21**).
5. *Detection of the cytokine encoded by the mRNA of interest.* Detection of the cytokine itself by immunohistochemistry in the cells with the same morphology and location reinforces the specificity of the in situ hybridization reaction as illustrated in Lambertsen et al. [3]. However, absence of protein does not mean that the in situ hybridization signal for the mRNA is unspecific (*see* **Note 22**).

4 Notes

1. To make sure that the oligodeoxynucleotide probes selectively recognize a specific mRNA member of protein family, design probes against regions of high variability within the protein family. Furthermore, make sure that the selected exon–exon junction(s) is/are present in all the mRNA isoforms or the single mRNA isoform of interest, depending on the purpose.
2. The T_m is the temperature at which half the DNA is present in a single-stranded form. However, the RNA:DNA stability is higher than the DNA:DNA stability. Therefore, we use a $T_{hyb} \approx T_m$ (DNA:DNA).
3. Addition of DEPC inactivates RNAses present in the water. In situ hybridization should be carried out under RNAse-free conditions until the hybridization step is completed. To secure RNAse-free condition, glass articles should be heated at 260 °C for 6–12 h, and solutions for the hybridization buffer need to be made based on DEPC-water.
4. Formamide reduces the thermal stability of the hydrogen bonds allowing hybridization to be carried out at a lower temperature. The monovalent cations of SSC decrease the electrostatic repulsion between the two strands of the DNA–RNA duplex. EDTA removes free divalent cations from the hybridization solution that strongly stabilize the DNA–RNA duplex. Dextran sulfate accelerates the hybridization by generating a local increase of the probe concentration. Denhardt's solution and salmon sperm DNA decrease the chance of nonspecific binding of the oligodeoxynucleotide probe. Hybridization buffer can be stored at −20 °C up to 12 months.
5. Add reagents in the indicated order.
6. AP developer should be mixed just before use and be light yellow. NBT and BCIP stock solutions should be shielded from light.
7. Perform the RNAse control reaction away from the rest of the experiment using separate glass articles, pipettes, and equipment.
8. Liquid nitrogen or dry ice can be used as alternatives to CO_2 snow. The usage of paraformaldehyde-perfused, sucrose-immersed tissue introduces a requirement for enzymatic treatment of the tissue sections prior to hybridization.
9. Sections can be cut thicker or thinner; however, they must fit into the small hybridization chamber, which is formed when covering slipping the RNAse-free SuperfrostPlus glass slide. This small hybridization chamber facilitates a uniform hybridization.

10. It is possible to perform in situ hybridization on standard gelatin-coated glass slides.
11. It is possible to perform in situ hybridization on tissue stored for >5 years, at −80 °C and −20 °C.
12. Tissue can be immersed longer for convenience. However, immersion in ethanol for longer than 24 h leads to precipitation of the mRNA and thereby to reduced hybridization signal.
13. The hybridization buffer is stored at −20 °C and is quite viscous. Therefore, leave at RT for 30 min before adding the probe. Good mixing (do not vortex) facilitates even hybridization.
14. Start with testing the probe at a concentration of 2, 4, 6, 8, and 10 pmol/ml hybridization buffer. Then continue with the concentration giving the best signal as compared to background.
15. For 2 mouse brain sections per glass slide, dilute probes into 100 μl hybridization buffer and apply approx. 95 μl per slide. For 4 or more brains per glass slide, apply up to 120 μl per slide.
16. Important that the drop does not touch the tissue to avoid hybridization artifacts caused by quick evaporation of the hybridization buffer due to its content of formamide.
17. Avoid toughing the cover slip on the tissue slide. If air bubbles, carefully press them out.
18. If necessary, the sections can rest here.
19. The optimal time of signal development depends on the amount of mRNA in the tissue. Low mRNA expression equals longer time of signal development. Sections can (quickly) be inspected macroscopically (GAPDH mRNA) or microscopically (GAPDH and cytokine mRNA) during development.
20. Competition control ensures that the in situ hybridization signal is not due to nonspecific binding of the AP-probe conjugate to the tissue section [14]. Excess unlabelled probe prevents (by competition for binding sites and excess amount) binding of the AP probe to the target mRNA and thereby leaves the section devoid of signal.
21. For illustration of GADPH mRNA expression in murine CNS, see Clausen et al. [15].
22. Cytokines are subject to posttranscriptional processing and are not necessarily translated into immunohistochemically detectable levels of protein. This is illustrated in Fenger et al. [16].

Acknowledgment

Lene Jørgensen and Sussanne Petersen are acknowledged for their excellent technical assistance. The experimental material shown in Fig. 1 was kindly provided by Dr. Kate Lambertsen. The work was supported by grants from the Lundbeck Foundation, the Novo Nordisk Foundation, and the Danish Multiple Sclerosis Society.

References

1. Gregersen R, Lambertsen KL, Finsen B (2000) Microglia and macrophages are major sources of tumor necrosis factor in permanent middle cerebral occlusion in mice. J Cereb Blood Flow Metab 20:53–65
2. Clausen BH, Lambertsen KL, Meldgaard M et al (2005) A quantitative in situ hybridization and polymerase chain reaction study of microglial-macrophage expression of interleukin-1beta mRNA following permanent middle cerebral artery occlusion in mice. Neuroscience 132:879–892
3. Lambertsen KL, Clausen BH, Babcock AA et al (2009) Microglia protect neurons against ischemia by synthesis of tumor necrosis factor. J Neurosci 29:1319–1330
4. Christensen JE, Simonsen S, Fenger C et al (2009) Fulminant lymphocytic choriomeningitis virus-induced inflammation of the CNS involves a cytokine-chemokine-cytokine-chemokine cascade. J Immunol 182:1079–1087
5. Buttini M, Appen K, Sauter A et al (1996) Expression of tumor necrosis factor alpha after focal cerebral ischemia in the rat. Neuroscience 71:1–16
6. Meltzer JC, Sanders V, Grimm PC et al (1998) Production of digoxigenin-labelled RNA probes and the detection of cytokine mRNA in rat spleen and brain by in situ hybridization. Brain Res Prot 2:339–351
7. Woodroofe MN, Cuzner ML (1993) Cytokine mRNA expression in inflammatory multiple sclerosis lesion: detection by non-radioactive in situ hybridization. Cytokine 5:583–588
8. Kiefer R, Funa K, Schweitzer T et al (1996) Transforming growth factor-beta 1 in experimental autoimmune neuritis. Cellular localization and time course. Am J Pathol 148:211–223
9. Turrin NP, Rivest S (2006) Tumor necrosis factor a but not interleukin 1β mediates neuroprotection in response to acute nitric oxide excitotoxicity. J Neurosci 26:143–151
10. Höfler H, Childers H, Montminy MR et al (1986) In situ hybridization methods for the detection of somatostatin mRNA in tissue sections using antisense RNA probes. Histochem J 18:597–604
11. VandenBroecke C, Tovey MG (1991) Expression of the genes of class I interferons and interleukin-6 in individual cells. J Interferon Res 11:91–103
12. Lu J, Tsourkas A (2009) Imaging individual microRNAs in single mammalian cells in situ. Nucleic Acids Res 37:e100
13. Tecott LH, Eberwine JH, Barchas JD et al (1987) Methodological considerations in the utilization of in situ hybridization. In: Eberwine JH, Barchas JD, Valentino KL (eds) In situ hybridization: amplification to neurobiology. Oxford Press, New York, NY, pp 3–24
14. Finsen B, Gregersen R, Lehrmann E et al (2004) In situ hybridization. In: Evans SM, Janson AM, Nyengaard JR (eds) Quantitative methods in neuroscience—*a neuroanatomical approach*. Oxford University Press, New York, NY, pp 115–145
15. Clausen BH, Lambertsen KL, Finsen B (2006) Glyceraldehyde-3-phosphate dehydrogenase versus toluidine blue as a marker for infarct volume estimation following permanent middle cerebral artery occlusion in mice. Exp Brain Res 175:60–67
16. Fenger C, Drøjdahl N, Wirenfeldt M et al (2006) Tumor necrosis factor or its TNFp55 and -p75 receptors are not required for axonal lesion-induced microgliosis in mouse fascia dentata. Glia 54:591–605
17. Pennica D, Hayflick JS, Bringman TS et al (1985) Cloning and expression in Escherichia coli of the cDNA for murine tumor necrosis factor. Proc Natl Acad Sci USA 82:6060–6064
18. Gray BW, Glaister D, Chen E et al (1986) Two interleukin 1 genes in the mouse: cloning and expression of the cDNA for murine interleukin-1b. J Immunol 137:3644–3648
19. Sabath DE, Broome HE, Prytowsky MB (1990) Glyceraldehyde-3-phosphate dehydrogenase mRNA is a major interleukin-2-induced transcript in a cloned T-helper lymphocyte. Gene 91:185–191

Chapter 11

Use of Meso-Scale Discovery™ to Examine Cytokine Content in Microglia Cell Supernatant

Miguel A. Burguillos

Abstract

Cytokine production by activated microglia is one of the hallmarks of inflammatory response in the CNS. The cytokines released by microglia cells can be very different depending on the proinflammatory stimulus.

Traditionally, to quantify these different cytokines, the "Sandwich"-enzyme-linked immunosorbent assay (Sandwich-ELISA) has been used. In this chapter we will discuss and describe an improved protocol of the Sandwich-ELISA developed by Meso-Scale Discovery based on an electrochemiluminescence detection system, which allows the ultralow detection of multiple cytokines in microglia cell supernatant.

Key words Cytokine, ELISA, Antibody, SULFO-TAG, Inflammation, Electrochemiluminescence

1 Introduction

Inflammatory signals produced within the central nervous system (CNS) and peripheral tissues regulate diverse biological processes. Surveying microglia generate these signals upon several stimuli, for instance, tissue damage, cellular dysfunction, and infection. These inflammatory signals trigger intracellular signaling cascades that eventually lead to immune cell activation, proliferation, cell recruitment, or cellular demise. Of particular importance in the genesis of inflammatory events are the immunomodulators referred to as cytokines. These immunomodulators have been shown to play very important roles in several processes such as acute neurodegeneration [1], immunodeficiency syndromes [2], and even during the development of CNS [3].

Cytokines are heterogeneous, nonstructural proteins with molecular weights ranging from 8 to 40,000 Da. There is no amino acid sequence motif or three-dimensional structure shared among the different cytokines. They are grouped depending on their biological activity [4].

Bertrand Joseph and José Luis Venero (eds.), *Microglia: Methods and Protocols*, Methods in Molecular Biology, vol. 1041,
DOI 10.1007/978-1-62703-520-0_11, © Springer Science+Business Media New York 2013

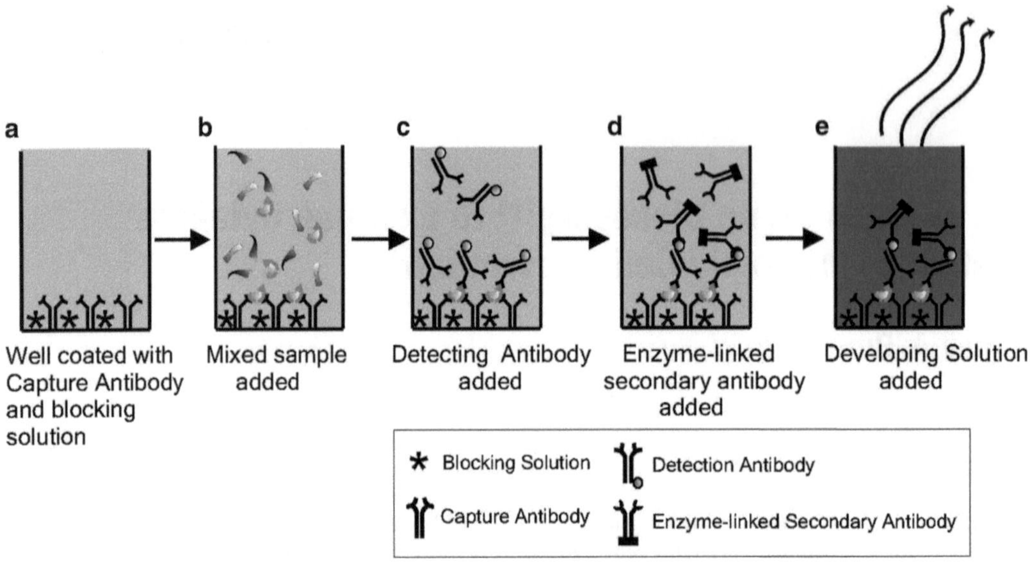

Fig. 1 Principle of the Sandwich-ELISA. (**a**) The surface of the well is applied with a known amount of the capture antibody. (**b**) The antigen-containing sample to the surface. (**c**) A specific antibody against our antigen is added forming a "sandwich"—antibody-sample-antibody. (**d**) We add an enzyme-linked secondary antibody that binds specifically to the nonspecific region antibody (Fc) of the previous antibody used in step c. (**e**) A chemical solution is applied and converted by the enzyme into a color, fluorescent, or electrochemical signal that can be quantified

Cytokines can be divided as:

- *Th1-type cytokines*: include interleukin 1-β, IFN-γ, and TNF-α. They generate the proinflammatory phenotype responsible for killing intracellular parasites and for perpetuating autoimmune responses.
- *Th2-type cytokines*: include interleukins 4, 5, and 13, which are associated with the promotion of IgE and eosinophilic responses in atopy, and also interleukin 10, which contributes to an anti-inflammatory response.
- *Th17-type cytokines*: include interleukin 17, which plays a role in host defense against extracellular pathogens by mediating the recruitment of neutrophils and macrophages to infected tissues.

The ratio between Th1/Th2/Th17 cytokines is what defines the type of activation of our microglia. The quantification of the cytokine content is a very useful tool to characterize our microglia population. Generally, these measurements have been performed using the "Sandwich"-enzyme-linked immunosorbent assay (or "Sandwich"-ELISA). This technique is based on the detection of an analyte (in this case a cytokine) dissolved in a certain solution (Fig. 1). Sandwich-ELISA is able to separate some components of the analytical mixture by adsorbing these components onto a solid phase, which is physically immobilized. The principle of the technique is the following:

1. On top of a stationary solid phase (for instance, a 96-well plate), a molecule with special binding properties (the capture antibody) is added.
2. Then, the blocking solution is added to decrease unspecific binding. The plate is washed several times to take away the rest of the blocking solution (Fig. 1a).
3. The sample is poured into the well and binds to the capture antibody (Fig. 1b). Additional washes take away the rest of proteins that have not bound to the antibody.
4. There is an incubation step with another antibody, the detecting antibody that will bind to our protein, forming a "sandwich"—antibody/sample/antibody (Fig. 1c).
5. A secondary antibody with an enzyme linked in its Fc region recognizes and binds to the detecting antibody (Fig. 1d).
6. Finally, a developing solution is supplied, that is, a substrate for the enzyme conjugated to the secondary antibody. The detection's sensitivity depends on the amplification of the signal during the analytic reactions. In order to obtain accurate readings, fixed proportions of enzymes must be linked to the detection reagents.

Meso-Scale Discovery (MSD) MULTI-ARRAY® using an improved electrochemiluminescence detection system is able to detect ultralow levels of cytokines (up to five logs of linear dynamic range). In addition, the assay can be performed with lower amounts of sample as compared with the traditional ELISA. Finally, it presents the advantage of measuring several cytokines from a single sample with a very simple protocol.

The Meso-Scale Discovery MULTI-ARRAY technology is based on an electrochemiluminescence method to quantify the detection antibody. This detection antibody is linked to SULFO-TAG™ labels that emit light upon electrochemical stimulation initiated at the electrode surface of MULTI-ARRAY®–MULTI-SPOT® microplates. One advantage of this array is that the emission light wavelength is at 620 nm, which avoid the color quenching problems. Only labeled antibodies located near to the electrode are excited and detected.

MSD provides a plate coated with anti-species antibody (capture antibody) that allows the immobilization of the protein/s of interest in the sample. Later on, the detecting antibody is added together with the SULFO-TAG™ label. The user adds an MSD Read Buffer (with the appropriate chemical environment for the electrochemiluminescence reaction plus co-reactants that enhance the signal) and quantifies the emitted light into an MSD SECTOR™ instrument.

You can choose between several predetermined mouse/human/rat cytokine assays (is able to measure from 1 to 10 different cytokines in a 96-well or 384-well format MULTI-ARRAY® and MULTI-SPOT® cytokine assays).

2 Materials

2.1 Materials Included in 1 Kit (and Storage Conditions)

1. Cytokine calibrators (stock concentration: 1 μg/ml). Store at −80 °C.
2. Detection antibody mix (Anti-cytokine antibodies already prelabeled with SULFO-TAG reagent. 50× stock solution). Store it at 2–8 °C.
3. MULTI-ARRAY or MULTY-SPOT plate. Store it at 2–8 °C.
4. Two different blocking solutions or "blockers" with stabilizing agents (depending on the cytokine content).
   (a) Blocker B (if the kits measure IL-12p40; room temperature).
   (b) Blocker D-B (only for kits that include TNF-α; −20 °C).
5. Diluent 1 (RPMI-based medium to dilute the calibrators; 2–8 °C).
6. Diluent 100 (blocking solution with stabilizing agents; 2–8 °C).
7. Read Buffer T (4× read buffer T with surfactant; room temperature).

2.2 Materials and Equipment Required

1. Deionized water for diluting concentrated buffers.
2. 15 and 50 ml tubes for reagent preparation.
3. Microcentrifuge tubes for preparing serial dilutions.
4. Phosphate-buffered saline (pH 7.4) + 0.05 % Tween-20 (wash buffer).
5. Accurate pipette to dispense volume from 10 to 150 μl into a 96-well microtiter plate.
6. Automated plate washer of multichannel pipette.
7. Adhesive plate seals.
8. Rotating plate shaker.
9. SECTOR® Imager apparatus from Meso-Scale company (Check the different models at Meso-Scale website: www.mesoscale.com).

2.3 Buffers Needed

1. *Blocker B solution* (Only for assays that detect IL-12p40) Prepare 20 ml of the solution per plate of 20 mg Blocker B in 20 ml of PBS (0.1 % w/v).

2. *Calibrators and control solutions*
   Dilute the calibrators using Diluent 1 solution in the following way to generate a standard curve with the following range (2.4 pg/ml up to 10,000 pg/ml):
   10,000 pg/ml: Mix 10 μl of the 1 μg/ml stock + 990 μl of Diluent 1.
   2,500 pg/ml: Mix 50 μl of the 10,000 pg/ml calibrator + 150 μl of Diluent 1.
   625 pg/ml: Mix 50 μl of the 2,500 pg/ml calibrator + 150 μl of Diluent 1.
   156 pg/ml: Mix 50 μl of the 625 pg/ml calibrator + 150 μl of Diluent 1.
   39 pg/ml: Mix 50 μl of the 156 pg/ml calibrator + 150 μl of Diluent 1.
   9.8 pg/ml: Mix 50 μl of the 39 pg/ml calibrator + 150 μl of Diluent 1.
   2.4 pg/ml: Mix 50 μl of the 9.8 pg/ml calibrator + 150 μl of Diluent 1.
   0 pg/ml: 150 μl of Diluent 1.

   Diluent 1 is an RPMI medium with 10 % serum. Other mediums can be used, instead of RPMI 10 % serum, if the experiments have been performed with it (i.e., DMEM 2 % serum). In general, the presence of some protein in the diluent solution is desirable to prevent loss of the analyte through adsorption with other surfaces like tubes and pipettes.

3. *Detection antibody solutions*
   (CAUTION: Some of the antibodies are light sensitive. Keep this solution in darkness.)

   The antibodies are premixed at a 50× concentration (50 mg/ml). The final concentration to work with should be 1× (1 mg/ml).

   (a) For one plate you need to mix in a 15 ml tube: 60 μl of the detection antibody mix solution + 2.94 ml of Diluent 100.
   (b) If the assay measures TNF-α, then add 210 μl of Blocker D-B (10 %) + 2.73 ml Diluent 100 + 60 μl of the detection antibody mix solution in a 15 ml tube.

4. *Read Buffer T solution*
   The Read Buffer T must be diluted in deionized water to a final concentration of 2× Read Buffer T. For one plate, mix 10 ml of stock Read Buffer T (4×) + 10 ml of deionized water in a 50 ml tube.

5. *MSD plate*
   The plate can be used as delivered (the plate has been already pre-coated with capture antibodies specific for the cytokines). It has a stabilizing treatment to guarantee the stability of the

antibodies, but it has not been pre-blocked. Some cytokines need a blocking step to improve the measurement (i.e., IL-12 p40).

6. *Wash buffer*
   PBS (pH 7.4)—0.05 % Tween-20

2.4 Sample Preparation

This assay can be used to measure the cytokine content in tissue/cell culture supernatant and tissue lysates (*see* **Note 1**). The following notes must be considered:

1. For most of the samples, there is no need of a dilution step unless the experimental conditions promote an extremely high production of cytokines. If a dilution step is necessary, it should be done by a factor of 2–10.
2. Normally avoid multiple freeze/thaw cycles since the sensitivity for the detection of cytokines decrease greatly after the first round of thawing due to the labile nature of some of them. Also keep the samples at −80 °C, if they are not measured immediately (*see* **Note 2**).
3. In tissue lysates, the lysis buffer should contain low levels of denaturing detergents (i.e., less than 0.1 % SDS) and reducing agents (i.e., less than 1 mM DTT). Also a carrier protein should be added (i.e., 1 % BSA) to avoid loss of the analyte with tubes or pipette tips.

3 Methods

(Carry out all procedures at room temperature unless otherwise specified.) A schematic protocol of this method is presented in Fig. 2.

3.1 Addition of the Blocker Solution (Only for Assays Containing IL-12p40 Cytokine)

Dispense 150 μl of Blocker B solution per well. Seal the plate with an adhesive plate seal and incubate with vigorous rotational shaking (300–1,000 rpm) for 1 h at room temperature. Wash three times with the wash buffer.

3.2 Addition of Sample and Calibrator

Add 25 μl of either your sample or calibrator into a separate well of the MSD plate (it is recommended to perform the analysis in duplicates) (*see* **Notes 3** and **4**). Seal the plate with an adhesive plate seal and incubate with vigorous rotational shaking (300–1,000 rpm) for 1–2 h at room temperature. This step can be done without shaking but in this case the incubation time is longer (4 h or longer) in order to get the same sensitivity.

3.3 Addition of Detection Antibody Solution

Add 25 μl of the 1× detection antibody solution into each well of the MSD plate. Seal the plate with an adhesive plate seal and incubate with vigorous rotational shaking (300–1,000 rpm) for 1–2 h

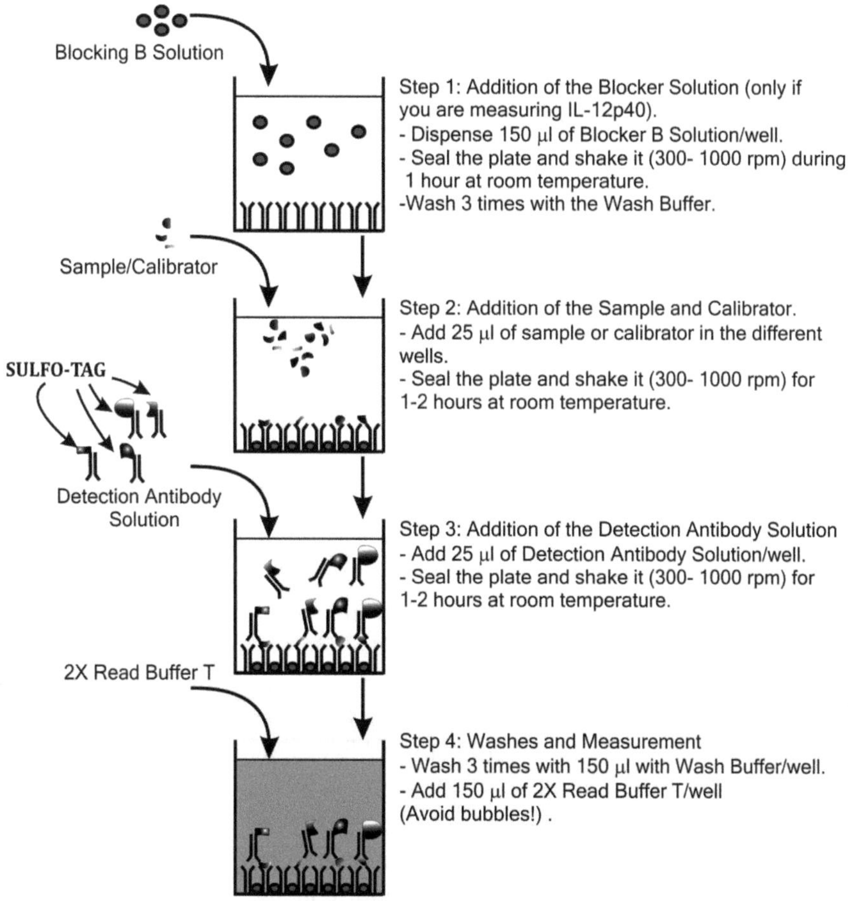

Fig. 2 Schematic protocol for measuring cytokines with the MULTI-ARRAY and MULTI-SPOT cytokine assay

at room temperature. This step can be done without shaking but in this case the incubation time is longer (4 h or longer) in order to get the same sensitivity.

3.4 Wash and Measurement of the Plate

Wash three times with the wash buffer and add 150 µl of the 2× Read Buffer T to each well (Avoid bubble formation!). Plates may be read immediately after adding the reading buffer. Quantify the plate with SECTOR® Imager (*see* **Notes 5–7**).

4 Notes

1. In our experience, there's a much higher production of cytokines in primary microglial cultures (for instance, primary microglia extract from the cortex) than in microglial cell line (e.g., the murine microglial cell line BV2).

2. If you are measuring cytokine content from supernatant obtained from cell cultures, it is good to fast freeze them once (i.e., by putting the eppendorfs with the supernatant immediately in dry ice and keep the samples in −80 °C until they are used).
3. Do not touch the pre-coated electrode surface with the pipette.
4. Use the reverse pipetting technique to deliver accurately the different solutions and obtain reproducible measurements.
5. In each wash, remove the buffer thoroughly and tap the empty plate on top of paper towels in order to get rid of as much buffer as possible.
6. The stability of the signal after adding the reading buffer decreases with time (around 20 % the first hour). Read within the first 2 h after adding the reading buffer to the plate.
7. In order to increase the sensitivity of the assay, some modifications in the protocol can be done:
   (a) Increasing incubation time (up to 4 h with shaking and at least 12 h without shaking).
   (b) Increase the sample volume up to 50 μl. In this case it is very important to get a more efficient shaking or longer incubation time. Also the concentration of the detection antibody must be adjusted to the larger volume of sample.
   (c) Wash sample prior the addition of the detection antibody. This step is useful if you are using larger volumes of sample. In this way you avoid a dilution of the detection antibody. CAUTION: Not all the cytokines will improve with this step, but also some could even lose some sensitivity.

Acknowledgement

This work has been supported by a grant provided by the Swedish Research Council.

References

1. Allan SM, Rothwell NJ (2001) Cytokines and acute neurodegeneration. Nat Rev Neurosci 2:734–744
2. Leonard WJ (2001) Cytokines and immunodeficiency diseases. Nat Rev Immunol 1:200–208
3. Deverman BE, Patterson PH (2009) Cytokines and CNS Development. Neuron 64:61–78
4. Peck A, Mellins ED (2009) Plasticity of T-cell phenotype and function: the T helper type 17. Immunology 129:147–153

Part V

Analysis of Microglia Activation

Chapter 12

Analysis of Microglial Production of Reactive Oxygen and Nitrogen Species

Urte Neniskyte and Guy C. Brown

Abstract

Reactive oxygen and nitrogen species are both regulators and effectors of microglial activation, and assays of these oxidants can be used as a measure of acute and chronic activation of microglial cells. Here we describe quick methods to assess the production of superoxide, hydrogen peroxide, nitric oxide, and peroxynitrite by microglia.

Key words Reactive oxygen species, Reactive nitrogen species, Superoxide, Hydrogen peroxide, Nitric oxide, Peroxynitrite, Phagocytic NADPH oxidase, Inducible nitric oxide synthase, 3-Nitrotyrosine

Abbreviations

iNOS	Inducible NOS
LPS	Lipopolysaccharide
LTA	Lipoteichoic acid
nNOS	Neuronal NOS
NO	Nitric oxide
NOS	Nitric oxide synthase
PHOX	Phagocytic NADPH oxidase
RNS	Reactive nitrogen species
RONS	Reactive oxygen and nitrogen species
ROS	Reactive oxygen species

1 Introduction

Reactive oxygen species (ROS) and reactive nitrogen species (RNS) are molecules derived from oxygen and nitric oxide, respectively, that are reactive, i.e., they react directly with other molecules without requiring catalysis by enzymes. Reactive oxygen and nitrogen species (RONS) include superoxide (O_2^-), hydrogen peroxide

Bertrand Joseph and José Luis Venero (eds.), *Microglia: Methods and Protocols*, Methods in Molecular Biology, vol. 1041, DOI 10.1007/978-1-62703-520-0_12,

(H_2O_2), nitric oxide (NO), and peroxynitrite ($OONO^-$). RONS are both regulators and effectors of inflammation. Inflammatory agents (such as LPS, β-amyloid, ATP, and cytokines) acutely activate the phagocyte NADPH oxidase (PHOX, found mainly in microglia but also in astrocytes and some neurons) to produce superoxide, which then dismutates to hydrogen peroxide. This hydrogen peroxide contributes to the inflammatory activation of the microglia through NF-κB, resulting in the expression of inflammatory proteins such as inducible nitric oxide synthase (iNOS). iNOS when expressed produces high levels of NO, which may react with oxygen to produce NO_2 and thence nitrite and nitrate or react with superoxide to produce peroxynitrite. Thus, measurements of ROS can be used as a measure of acute activation of microglia (seconds to minutes), while measurements of iNOS, NO, or its relatively stable product nitrite can be used as measurements of chronic activation of microglia (hours to days).

Superoxide production within cells can be assessed by nitro-blue tetrazolium reduction, leading to the formation of a dark formazan precipitate [1, 2]. Meanwhile extracellular superoxide may be detected by the reduction of cytochrome c added to the medium [3]. Extracellular superoxide readily dismutates into hydrogen peroxide, whose formation rate can be measured in a continuous Amplex Red fluorometric assay [4].

Nitric oxide production can be assessed either directly by NO electrode [5] or indirectly by measurement of nitrite (a stable breakdown product of nitric oxide) by the Griess reaction [6] or the more sensitive fluorometric 2,3-diaminonaphthalene assay [7]. In addition, nitric oxide synthase activity can be visualized by NADPH diaphorase staining [6].

Rates of peroxynitrite production can be determined by peroxynitrite-mediated oxidation of nonfluorescent dihydrorhodamine-1,2,3 to fluorescent rhodamine-1,2,3 [5]. Sites of peroxynitrite production and reaction can be visualized by 3-nitrotyrosine immunostaining, detecting tyrosine residues in proteins that have reacted with peroxynitrite [6].

2 Materials

2.1 Superoxide Production Measured with NBT

1. Nitro-blue tetrazolium (Sigma).
2. Phosphate buffered saline (PBS): 10 mM phosphate, 150 mM sodium chloride, pH 7.4 (available in tablets and solutions or made from scratch).
3. (*Optional*) 2 M potassium hydroxide (KOH).
4. (*Optional*) Dimethyl sulfoxide (DMSO).

2.2 Superoxide Production Measured with Cytochrome c

1. Buffered salt solutions:
   (a) Krebs-HEPES: 1 mM $CaCl_2$, 11 mM glucose, 25 mM HEPES, 4.8 mM KCl, 1.2 mM KH_2PO_4, 1.2 mM $MgSO_4$, 118 mM NaCl, pH 7.4.
   (b) Hanks buffered salt solution (HBSS): 1.2 mM $CaCl_2$, 0.5 mM $MgCl_2$, 0.4 mM $MgSO_4$, 0.5 mM KCl, 0.4 mM KH_2PO_4, 138 mM NaCl, 0.3 mM Na_2HPO_4, 5 mM glucose, pH 7.4 supplemented with 5 mM glucose (*see* **Note 1**).
2. Cytochrome c (Sigma).
3. Catalase from human erythrocytes (Sigma C3556).
4. Superoxide dismutase (Sigma).

2.3 Cellular H_2O_2 Production Measured by Amplex Red Assay

1. Amplex Red (Invitrogen).
2. Dimethyl sulfoxide (DMSO).
3. Horseradish peroxidase (Sigma).
4. HBSS: 1.2 mM $CaCl_2$, 0.5 mM $MgCl_2$, 0.4 mM $MgSO_4$, 0.5 mM KCl, 0.4 mM KH_2PO_4, 138 mM NaCl, 0.3 mM Na_2HPO_4, 5 mM glucose, pH 7.4 supplemented with 5 mM glucose.

2.4 Direct Measurement of NO Generation Using a NO Electrode

1. Krebs buffer: 126 mM NaCl, 2.5 mM KCl, 25 mM $NaHCO_3$, 1.2 mM NaH_2PO_4, 1.2 mM $MgCl_2$, 2.5 mM $CaCl_2$, 10 mM glucose, pH 7.2 (*see* **Note 2**).
2. L-arginine (Sigma).
3. Clark-type NO electrode (World Precision Instruments, WPI).
4. (*Optional*) Oxygen electrode (Rank Brothers) (*see* **Note 3**).

2.5 Assessment of NO Generation by Griess Reaction

1. (*Optional*) Nitrate reductase (NAD[P]H) from *Aspergillus niger* (Sigma).
2. (*Optional*) NADPH (Sigma).
3. Sulfanilamide: 2 mM in 1.2 M hydrochloric acid (Sigma).
4. *N*-1-(1-naphthyl)ethylenediamine: 3 mM in double-distilled water (Sigma).
5. Nitrite standard solution (Sigma).

2.6 Assessment of NO Generation by Fluorometric 2,3-Diaminonaphthalene Assay

1. 2,3-Diaminonaphthalene: 0.05 mg/ml in 0.62 M hydrochloric acid (Sigma).
2. NaOH: 0.28 M.
3. Nitrite standard solution (Sigma).

2.7 Diaphorase Staining as a Measure of iNOS Expression

1. 4 % Paraformaldehyde (PFA) in PBS: 10 mM phosphate, 150 mM sodium chloride, pH 7.4 (available in tablets and solutions or made from scratch) (*see* **Note 4**).

2. PBS.
3. Triton X-100 solution: 0.3 % in PBS (*see* **Note 5**).
4. NADPH (1 mg/ml) and nitro-blue tetrazolium (0.2 mg/ml) solution in PBS with 0.3 % Triton X-100.

2.8 Peroxynitrite Production Measured by Dihydrorhodamine Oxidation

1. Dihydrorhodamine-1,2,3 (Sigma).
2. Dimethyl sulfoxide (DMSO).
3. Krebs-HEPES buffer: 1 mM $CaCl_2$, 11 mM glucose, 25 mM HEPES, 4.8 mM KCl, 1.2 mM KH_2PO_4, 1.2 mM $MgSO_4$, 118 mM NaCl, pH 7.4 (Sigma).

2.9 Peroxynitrite Production Measured by 3-Nitrotyrosine Immunocytochemistry

1. PBS: 10 mM phosphate, 150 mM sodium chloride, pH 7.4 (available in tablets and solutions or made from scratch).
2. 4 % Paraformaldehyde (PFA) in PBS (*see* **Note 4**).
3. Anti-nitrotyrosine monoclonal antibody (Upstate, Millipore).
4. Secondary antibody conjugated with horseradish peroxidase or fluorescent tag (e.g., secondary antibodies from Jackson ImmunoResearch Laboratories).

3 Methods

3.1 Superoxide Production Measured with NBT

1. Dissolve nitro-blue tetrazolium (NBT) in double-distilled water to obtain 10 mg/ml concentration.
2. At specific time points after stimulation of cells, add 30 μl of NBT solution to 270 μl of medium taken from individual wells of cultures in 24-well plate to obtain final concentration of 1 mg/ml. This will help to avoid local NBT concentration differences.
3. Remove the remaining medium from the wells.
4. Add NBT solution to the respective wells and incubate for 30 min at 37 °C, 5 % CO_2. Superoxide within cells reduces nitro-blue tetrazolium, leading to the formation of a dark formazan precipitate.
5. Wash cells with PBS.
6. The number of formazan-positive cells is counted under a transmission light microscope.
7. (*Optional*) Alternatively, the reduction of nitro-blue tetrazolium may be measured in a plate reader. Allow cells to dry in the air. Dissolve intracellular formazan deposits with 240 μl of potassium hydroxide and 280 μl of dimethyl sulfoxide with gentle shaking for 10 min at room temperature. Transfer the dissolved formazan solution to a 96-well plate and read the absorbance at 630 nm.

3.2 Superoxide Production Measured with Cytochrome c

1. Suspend cells at a density of about 10^6 cells/ml in appropriate medium (*see* **Notes 1** and **6**).
2. Dissolve 60 mg of cytochrome c in 1 ml of double-distilled water to obtain stock solution of 5 mM. Add 10 μl of cytochrome c solution to 1 ml of cell suspension to obtain final cytochrome c concentration of 50 μM.
3. (*Optional*) Add 100 U/ml catalase to prevent reoxidation of reduced cytochrome c by H_2O_2 (but remember that catalase is also a hemoprotein).
4. Stir the cells before and during measurement in a spectrophotometer or plate reader, which ideally should be thermostated at 37 °C (*see* **Note 7**).
5. Superoxide directly reduces oxidized (Fe^{3+}) cytochrome c to reduced (Fe^{2+}) cytochrome c, which absorbs light at 550 nm, while oxidized cytochrome c has little absorbance at this wavelength. Follow the reduction continuously, measuring the absorbance at 550 nm. If you have a dual-wavelength or multi-wavelength spectrophotometer or plate reader, measure absorbance at both 550 and 540 nm (where there is no change in absorbance on reduction of cytochrome c), and subtract the absorbance at 540 nm from the absorbance at 550 nm to correct for any absorbance changes not due to cytochrome c reduction.
6. Repeat measurements with and without 50 U/ml superoxide dismutase to ensure that the signal is due to extracellular superoxide only.

3.3 Cellular H_2O_2 Production Measured by Amplex Red Assay

1. Suspend the cells in HBSS with glucose at a density of 3.5×10^5 cells/ml.
2. Dissolve 5 mg of Amplex Red reagent in 1.94 ml of DMSO to obtain stock solution of 10 mM. Prepare 1,000 U/ml of horseradish peroxidase solution in double-distilled water.
3. Add 10 μl of Amplex Red and 10 μl of horseradish peroxidase to 1 ml of cell suspension to obtain final concentration of 100 μM and 10 U/ml, respectively.
4. Horseradish peroxidase catalyzes H_2O_2 oxidation of Amplex Red to fluorescent resorufin. Measure the rate of hydrogen peroxide production in a stirred cuvette in a spectrofluorophotometer or in a plate reader (use black plates to avoid reading fluorescence from adjacent wells), which ideally should be thermostated at 37 °C. Use excitation wavelength of 560 nm and emission wavelength of 587 nm. H_2O_2 passes rapidly through membranes, so extracellular H_2O_2 reflects both extracellular and intracellular production and breakdown.
5. Repeat the measurements with and without cells as well as with and without catalase to ensure that fluorescence is due to microglial H_2O_2 production only.

3.4 Direct Measurement of NO Generation Using a NO Electrode

1. Prepare activated cell suspension in Krebs buffer at a density between 10^5 and 10^6 cells/ml (*see* **Notes 8–10**).
2. Add L-arginine (200 μM) to see NO generation. Glucose (10 mM, a component of Krebs buffer) is required for the cells to be able to produce NADPH.
3. Put cells in a stirred, thermostated, and sealed incubation chamber (supplied by World Precision Instruments, WPI) with the NO electrode. NO readily passes through cell membranes and therefore NO from cellular iNOS is measurable by this method. Before adding the cells, it is necessary to have a steady baseline reading from the NO electrode. For this, all components of the system (including particularly the electrode) need to come to thermal equilibrium. Note that warming up medium or the electrode can cause small bubbles to form, which can cause electrode noise. After the measurement add a small amount of hemoglobin to remove all NO from the chamber so that the location of the electrode baseline can be reestablished.
4. Calibrate the NO electrode using either NO-saturated water (NO gas bubbled through distilled water in gas-tight vial inside a gas hood) or acidified nitrite (*see* WPI manual).
5. (*Optional*) It may be advisable to combine the measurement with an oxygen electrode (*see* **Note 3**).

3.5 Assessment of NO Generation by Griess Reaction

1. (*Optional*) To avoid the underestimation of NO production, it may be advisable to convert any nitrate formed to nitrite by incubating cell-culture supernatants with nitrate reductase (0.5 U/ml) and NADPH (80 μM) for 30 min at room temperature. However, nitrate is not specific to NO production and may arise from other sources. Therefore, the necessity of this step has to be determined empirically.
2. In a 96-well plate, mix 50 μl of cell-culture supernatant with 50 μl of ice-cold sulfanilamide in hydrochloric acid and incubate for 10 min at room temperature, protected from light.
3. Add 50 μl of N-1-(1-naphthyl)ethylenediamine and incubate the mixture for another 10 min at room temperature, protected from light.
4. Measure the absorbance of the resulting red azo dye in a plate reader at 550 nm.
5. Calculate nitrite concentrations using standards prepared from nitrite standard solution.

3.6 Assessment of NO Generation by Fluorometric 2,3-Diaminonaphthalene Assay

1. Mix 100 μl of cell-culture supernatants with 20 μl of 2,3-diaminonaphthalene solution in hydrochloric acid and incubate for 10 min at room temperature, protected from light.
2. Add 100 μl of NaOH to stop the reaction.

3. Measure fluorescence intensity in a spectrofluorophotometer or a plate reader at an excitation wavelength of 363 nm and an emission wavelength of 426 nm.
4. Calculate nitrite concentrations using standards prepared from nitrite standard solution.

3.7 Diaphorase Staining as Measure of iNOS Expression

1. After stimulation fix cells with 4 % PFA in PBS for 30 min at 4 °C, followed by three washes with PBS, 15 min each.
2. Permeabilize fixed cells with 0.3 % Triton X-100 for 5 min at room temperature.
3. Incubate cells in NADPH and nitro-blue tetrazolium solution with Triton X-100 for 2 h at 37 °C. Nitric oxide synthase acts as a NADPH diaphorase that reduces the chromogen nitro-blue tetrazolium to form a dark formazan precipitate.
4. After incubation wash cells once with PBS and detect cells with NOS activity under a transmission light microscope. nNOS-expressing neurons have a light staining, whereas iNOS-expressing glia can be very dark.

3.8 Peroxynitrite Production Measured by Dihydrorhodamine Oxidation

1. Suspend cells in pre-warmed (37 °C) Krebs-HEPES buffer at a density of 10^6 cells/ml.
2. Prepare 5 mM dihydrorhodamine stock solution in DMSO.
3. Add 10 μM of dihydrorhodamine stock solution to 1 ml of cell suspension to obtain final 5 μM concentration.
4. Measure the rate of continuous fluorescence increase in a stirred cuvette in a spectrofluorophotometer or a plate reader at an excitation wavelength of 500 nm and an emission wavelength of 536 nm. Peroxynitrite mediates the oxidation of nonfluorescent dihydrorhodamine-1,2,3 to fluorescent rhodamine-1,2,3, and the rate of fluorescence increase is equal to the rate of peroxynitrite production (*see* **Note 11**).

3.9 Peroxynitrite Production Measured by 3-Nitrotyrosine Immunocytochemistry

1. Fix stimulated cells with 4 % PFA in PBS for 20 min at room temperature, followed by three washes with PBS, 15 min each.
2. Incubate cells in PBS with 5 % of normal serum of the host species of the secondary antibody to block unspecific epitopes.
3. Incubate fixed cells with 10 μg/ml of anti-nitrotyrosine monoclonal antibody (1:100 dilution in PBS) for 1 h at room temperature. Wash three times with PBS, 15 min each.
4. Detect primary antibody with secondary antibody conjugated with horseradish peroxidase or fluorescent tag (1:200 dilution in PBS, 30 min at room temperature).
5. Visualize 3-nitrotyrosine-positive cells under a transmission light microscope (for horseradish peroxidase immunocytochemistry) or fluorescence microscope (for fluorescent secondary antibodies) (*see* **Note 12**).

4 Notes

1. Any other medium that is not colored (i.e., without phenol red), does not scatter light, does not scavenge superoxide, and does not reduce cytochrome c (i.e., without ascorbate) can be used as well.
2. It is best to use a relatively simple incubation medium such as Krebs buffer rather than a cell culture medium such as DMEM, as components of the latter including riboflavin react with NO in the presence of light.
3. Rank Brothers supply oxygen electrode and chamber. Drill a 2 mm hole through the chamber top to allow the NO electrode in and seal it carefully, because the chamber needs to be gas tight and bubble free.
4. To dissolve 4 g of PFA in 100 ml of PBS, add 1 ml of 1 M NaOH to increase pH and stir the solution in the water bath at 70 °C until it is clear. Filter the solution through the filter paper and adjust the pH to 7.4. Aliquot 4 % PFA to avoid repeated freezing–thawing cycles and store it at −20 °C.
5. Since Triton X-100 is very viscous, it is easier to measure it by weight rather than volume.
6. In a plate reader, extracellular superoxide production by adherent cells may be measured as well.
7. At this density the cells may run out of oxygen after a few minutes—it may be an idea to saturate the medium with 95/100 % oxygen, but bear in mind that superoxide production may depend on the oxygen level.
8. Direct NO measurement by NO electrode is difficult or impossible for cells expressing nNOS or eNOS, but relatively easy for any cells expressing iNOS (such as microglia), because iNOS produces a lot more NO continuously.
9. For example, cells may be activated with 100 ng/ml LPS and 10 ng/ml IFN-γ. Addition of IFN-γ ensures strong iNOS expression, which peaks between 8 and 24 h after activation.
10. It may be advisable to use arginine-free culture medium for the activation or add arginase to the medium, because the NO released by iNOS may damage the cells and limit iNOS expression.
11. Other intra- and extracellular oxidants can oxidize dihydrorhodamine; therefore, it is important to include controls with specific peroxynitrite scavengers, such as urate (100 μM) or FeTPPS (5,10,15,20-tetrakis(4-sulfonatophenyl)porphyrinato iron (III) chloride, 25 μM).

12. To determine the site of peroxynitrite production and action, 3-nitrotyrosine immunostaining can be combined with staining for markers of specific cell types, such as isolectin B_4 (Invitrogen) or antibodies against OX-42 to label microglia.

Acknowledgement

Relevant research in our laboratory has been funded by the Wellcome Trust (Grant RG50995).

References

1. Wyss-Coray T (2006) Inflammation in Alzheimer disease: driving force, bystander or beneficial response? Nat Med 12:1005–1015
2. Ransohoff RM, Perry VH (2009) Microglial physiology: unique stimuli, specialized responses. Annu Rev Immunol 27:119–145
3. Colton CA, Gilbert DL (1987) Production of superoxide anions by a CNS macrophage, the microglia. FEBS Lett 223:284–288
4. Jekabsone A, Mander P, Tickler A et al (2006) Fibrillar beta-amyloid peptide Abeta1-40 activates microglial proliferation via stimulating TNF-alpha release and H_2O_2 derived from NADPH oxidase: a cell culture study. J Neuroinflammation 3:24
5. Bal-Price A, Matthias A, Brown GC (2002) Stimulation of the NADPH oxidase in activated rat microglia removes nitric oxide but induces peroxynitrite production. J Neurochem 80: 73–80
6. Mander P, Brown GC (2005) Activation of microglial NADPH oxidase is synergistic with glial iNOS expression in inducing neuronal death: a dual-key mechanism of inflammatory neurodegeneration. J Neuroinflammation 2:20
7. Jekabsone A, Neher J, Borutaite V et al (2007) Nitric oxide from neuronal nitric oxide synthase sensitises neurons to hypoxia-induced death via competitive inhibition of cytochrome oxidase. J Neurochem 103:346–356

Chapter 13

Quantification of Active Caspase-3 and Active Caspase-8 in Microglia Cells

Edel Kavanagh

Abstract

During microglia activation the levels of active caspase-3, caspase-7 and caspase-8 are increased, which leads to the transcription of proinflammatory cytokines and factors. As such, the induction of caspase activity in microglia can be used as a marker for activation. The use of sensitive and quantitative techniques has made it possible to reproducibly detect these low levels of active caspases. This chapter outlines the materials and methodology for three different ways to detect caspase activation in microglia.

Key words Flow cytometry, Luminometry, Fluorometry, Caspase-3, Caspase-8, Antibodies, Luminogenic substrate, Fluorogenic substrate

1 Introduction

Cysteine-dependent aspartate proteases (caspases) contain a cysteine in their active site, which is required for their activity, and cleave their substrates after an aspartate residue preceded by a recognition sequence [1]. Currently, there are 14 caspases described in the mammalian genome, with a variety of signalling functions including apoptosis, differentiation, and inflammatory response [2]. During the activation of microglia, the levels of active caspase-1, caspase-3, caspase-7, and caspase-8 increase [3]. The activity of each specific caspase, or specific groups of caspases, can be experimentally calculated based on their substrate recognition sequences [4]. Active caspase-1 cleaves its substrates after the peptide sequence YVAD; caspase-3 and caspase-7 both share the same substrate recognition sequence of DEVD, while active caspase-8 cleaves after IETD [4]. This characteristic of caspases is harnessed to experimentally measure activity in the first two methods described in this chapter.

All caspases have a similar structure, consisting of an N-terminal prodomain, a large subunit, and a small subunit. During activation, cleavage occurs between the large and small subunits, followed by cleavage between the prodomain and the large subunit [5]. A tetramer

Bertrand Joseph and José Luis Venero (eds.), *Microglia: Methods and Protocols*, Methods in Molecular Biology, vol. 1041, DOI 10.1007/978-1-62703-520-0_13, © Springer Science+Business Media New York 2013

consisting of two large and two small subunits comprises the active caspase [5]. During activation, the caspase subunits which undergo cleavage and conformational changes expose new epitopes. The final method to measure caspase activity described in this chapter is based on antibodies specifically designed to bind to these newly exposed epitopes after caspase activation.

1.1 Chemiluminescence Detection of Caspase Activity

In this assay, active caspases cleave a luminogenic substrate containing their specific substrate recognition sequence. Following cleavage, a substrate for luciferase (aminoluciferin) is released and reacts with luciferase to produce light. This assay can be used to detect either DEVDase activity (for caspase-3 and caspase-7) or IETDase activity (for caspase-8 and caspase-10) (*see* **Note 1**). The reagents described in this assay are supplied as a kit called Caspase-Glo from Promega.

1.2 Caspase Assay Using a Fluorescent Caspase Substrate

The activity of caspases can also be determined fluorometrically. A peptide substrate is synthesized with an attached fluorescent tag, acetyl-L-aspartyl-L-glutamyl-L-valyl-L-aspartic acid α-(4-methyl-coumaryl-7-amide) (DEVD-MCA). When the peptide substrate is cleaved, the fluorescent 7-amino-4-methylcoumarin (AMC) is released and can be measured over time to calculate the enzymatic activity of the caspase.

1.3 Quantification of Cleaved Caspases by Flow Cytometry

Antibodies, specific for active or cleaved caspases, have been developed against the amino acid residue which is only exposed after cleavage between the large and small caspase subunits. Coupled with the sensitivity, quantitative power, and speed of flow cytometry, antibodies against cleaved caspases can be used to accurately measure small differences in levels of cleaved caspases.

2 Materials

2.1 Chemiluminescence Detection of Caspase Activity

1. Caspase-Glo 3/7 reagent or Caspase-Glo 8 reagent (Promega).
2. White-walled 96-well plate.
3. Plate reader with chemiluminescence detection capability, e.g., Wallace plate reader.
4. Hemocytometer.

2.2 Caspase Assay Using a Fluorescent Caspase Substrate

1. Ac-Asp-Glu-Val-Asp-α-(4-methyl-coumaryl-7-amide) (DEVD-AMC) and acetyl-Ile-Glu-Thr-Asp-4-methylcoumaryl-7-amide (IETD-MCA) available from companies such as Peptide Institute or BD Pharmingen.
2. Black 96-well microtiter plates.

3. Plate reader with the correct excitation and emission filters, e.g., Wallace.
4. Substrate buffer (100 mM *N*-2-hydroxyethyl-piperazine-*N′*-2-ethanesulphonic acid (HEPES) pH 7.0, 10 % sucrose, 0.1 % 3-[(3-cholamidopropyl)-dimethylammonio]-1-propanesulfonate (CHAPS)).
5. 5 mM D dithiothreitol (DTT).
6. 0.0001 % Nonidet P-40 (make a working stock of 0.1 %).
7. Phosphate buffered saline (PBS 10×: 1.37 M NaCl, 27 mM KCl, 100 mM Na_2HPO_4, 20 mM KH_2PO_4).
8. Liquid nitrogen.

2.3 Quantification of Cleaved Caspases by Flow Cytometry

1. Primary antibodies against cleaved caspase-3 (Cell Signalling cat no. 9664 against mouse, human, and rat) and cleaved caspase-8 (Cell Signalling cat no. 8592 mouse specific or cat no. 9496 human specific).
2. Fluorescent secondary antibodies (must be compatible with flow cytometer lasers, such as Alexa488).
3. Ice-cold methanol.
4. Phosphate buffered saline (PBS 10×: 1.37 M NaCl, 27 mM KCl, 100 mM Na_2HPO_4, 20 mM KH_2PO_4).
5. Blocking buffer (0.5 % BSA dissolved in PBS).
6. 4 % Paraformaldehyde (dissolved in PBS).
7. Water bath set to 37 °C.
8. Centrifuge.
9. Vortex.
10. Orbital shaker.
11. Flow cytometer.
12. Flow cytometer tubes.

3 Method

3.1 Chemiluminescence Detection of Caspase Assay

1. Equilibrate the Caspase-Glo buffer and lyophilized Caspase-Glo substrate to room temperature before use.
2. Transfer the contents of the Caspase-Glo buffer bottle into the bottle containing Caspase-Glo substrate. Mix until the substrate is thoroughly dissolved to form the Caspase-Glo reagent.
3. Store the reconstituted Caspase-Glo reagent at 4 °C and use within 4 weeks.
4. Allow the Caspase-Glo reagent to equilibrate to room temperature. Mix well.

Table 1
Generation of an AMC standard curve

nmol AMC	AMC solution (μl)	Substrate buffer (μl)
0	0	100
0.1	5	95
0.2	10	90
0.5	25	75
1.0	50	50
1.5	75	25
2.0	100	0

5. Grow and perform treatments on 100 μl cells in white-walled 96-well plates. Remove 96-well plates containing cells from the incubator. Alternatively, grow and treat in different-sized plates but detach cells from those plates and transfer to the white-walled 96-well plates.
6. Add 100 μl of Caspase-Glo reagent to each well of a white-walled 96-well plate containing 100 μl treated cells in culture medium. Because of the sensitivity of this assay, be careful not to touch pipette tips to the wells containing samples to avoid cross-contamination. Cover the plate with a plate sealer or lid (*see* **Note 2**).
7. Gently mix contents of wells using a plate shaker at 300–500 rpm for 30 s. Incubate at room temperature for 1 h.
8. Measure the luminescence of each sample in a plate-reading luminometer as directed by the luminometer manufacturer (*see* **Notes 3** and **4**).

3.2 Caspase Assay Using a Fluorescent Caspase Substrate

1. Set up a program on the plate reader following the manufacturer's instructions as follows: Assay temperature at 37 °C, the number of plate reads at 25, delay between reads is 1 min, excitation filter is 355–380 nm, and emission filter is 430–460 nm.
2. To generate an AMC standard curve, dissolve 7-amino-4-methylcoumarin (AMC) in substrate buffer to a final concentration of 20 μM. Add increasing concentrations of AMC solution to substrate buffer in wells of a black 96-well plate (Table 1). Measure the fluorescence generated as relative fluorescence units (RFU). Plot RFU (*y*-axis) against substrate concentration (*x*-axis) and draw a linear standard curve. Record the RFU value generated with 1 nmol AMC.
3. 2×10^5 cells are needed for this assay per sample.

4. After the appropriate treatment, scrape cells into culture medium and centrifuge the resulting cell suspension at 1,000 × *g* for 5 min to pellet the cells. Discard the supernatant.
5. Wash cell pellet once in 1 ml PBS and centrifuge at 1,000 × *g* for 5 min. Discard the supernatant.
6. Resuspend the cell pellet in 50 μl PBS. Lyse cells in eppendorfs by snap freezing in a bath of liquid nitrogen. The suspension can be stored at −80 °C at this stage until further analysis.
7. Thaw the cell suspensions on ice and divide equally between two wells of a black microtiter plate.
8. Just before use, add 5 mM DTT, 0.0001 % Nonidet P-40, and 50 μM of DEVD-MCA/IETD-MCA/YVAD-MCA to the substrate buffer. Store in the dark and on ice. Use within 30 min.
9. When the plate reader temperature reaches 37 °C, add the substrate buffer to the samples and start the program immediately. Assay will take approximately 30 min to complete (*see* **Note 5**).
10. Using the assay results, the fluorescent units (RFU, *y*-axis) vs. time (min, *x*-axis) for each sample are graphed as an XY scatter plot. Calculate the rate at which fluorescence (free AMC) increases over time using the linear portion of the curve. This value is the rate of AMC released/min.
11. RFU are converted to nmoles of AMC released per minute using the standard curve generated with free AMC. Divide value from the previous step by the RFU value which corresponds to 1 nmol AMC. The new value obtained is nmol AMC released/min.
12. Protein concentration in each sample is calculated using a standard protein assay such as Bradford assay. Divide the previous value by the amount of protein (in mg) present in each sample. The new value obtained is nmoles AMC released/min/mg, which corresponds to the caspase activity in that sample (*see* **Note 6**).

3.3 Quantification of Cleaved Caspases by Flow Cytometry

Carry out all procedures at room temperature unless otherwise specified.

1. Grow and treat microglia cells (2×10^5).
2. Detach cells from plates/flasks by pipetting and transfer cells with medium to 15 ml tubes. Collect cells by centrifugation at 1,000 × *g* for 5 min at 4 °C and discard supernatant.
3. Resuspend cell pellet in 0.5 ml of 4 % paraformaldehyde and incubate tubes at 37 °C for 10 min.
4. Place tubes on ice for 1–2 min.

5. Using the vortex at a low speed, slowly add 4.5 ml ice-cold methanol to the fixed cells to achieve a final methanol concentration of 90 % (*see* **Note 7**). Incubate the cells on ice for 30 min. Alternatively, the procedure can be stopped here by storing the cells at −20 °C (*see* **Note 8**).
6. Centrifuge cells at 1,500 × *g* for 5 min and discard supernatant.
7. Wash cell pellet in 1 ml blocking solution (freshly prepared) and centrifuge cells at 1,500 × *g* for 5 min. Discard supernatant. Repeat this wash step once (*see* **Note 9**).
8. Resuspend pellet in 100 μl blocking solution (*see* **Note 10**) and incubate at room temperature for 10 min.
9. Dilute the primary antibody in blocking solution to a concentration of 1:1,600 for anti-cleaved caspase-3, 1:800 for cleaved caspase-8 (mouse specific), or 1:100 for cleaved caspase-8 (human specific).
10. Add 100 μl primary antibody solution to cells and incubate with gentle agitation for 1 h.
11. Wash cell pellet in 1 ml blocking solution and centrifuge cells at 1,500 × *g* for 5 min. Discard supernatant. Repeat this wash step once.
12. Dilute the appropriate secondary antibody in blocking buffer at 1:200 concentration (*see* **Notes 11–14**).
13. Add 100 μl antibody solution to each sample and incubate in the dark with gentle agitation for 30 min.
14. Wash cell pellet in 1 ml blocking solution and centrifuge cells at 1,500 × *g* for 5 min. Discard supernatant. Repeat this wash step once.
15. Resuspend pellet in 0.5 ml PBS and transfer to flow cytometry tubes. Analyze immediately by flow cytometry or store in the dark at 4 °C for up to 1 week.
16. Display the acquired data in the format of a histogram overlay. A cell population which is undergoing apoptosis will display two peaks, one corresponding to the healthy population (no expression of cleaved caspase-3) and one corresponding to the dying population (high expression of cleaved caspase-3). On the other hand, activated microglia display only one peak, but the whole population shows a low level of cleaved caspase-3 expression and appears shifted to the right when compared with unstimulated microglia (Fig. 1) (*see* **Note 15**).
17. K/S statistics on the histogram peaks can be performed to measure the difference between two experimental samples of microglia.

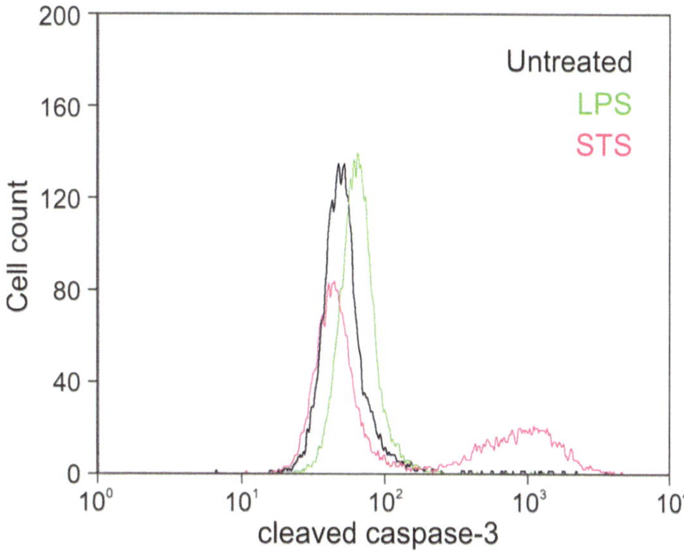

Fig. 1 Histogram showing the distribution of cleaved caspase-3 in microglia cells upon lipopolysaccharide (LPS) or staurosporine (STS) treatment. BV2 mouse microglia cells were treated with 1 μM LPS for 6 h or 0.2 μM STS for 3 h. Cells were stained with cleaved caspase-3 (Cell Signalling cat. no. 9664) and analyzed by flow cytometry (FACS Calibur, Becton Dickinson)

4 Notes

1. In human cells IETDase activity assays also measure the activity of caspase-10, which is not present in mouse cells.
2. The assay works equally well using as little as 50 μl Caspase-Glo reagent mixed with 50 μl cells in culture medium.
3. The luminescence values obtained must be normalized to the number of cells in each sample. This can be done in two ways. One way is to count the number of cells in each sample using a hemocytometer or cell counter. Alternatively, lyse the remainder of the cells from the sample wells, extract the protein, and perform a protein assay. Then divide the luminescence values by the protein concentration.
4. As a control, measure the background luminescence from the cell culture medium (fresh unused medium) and subtract that value from the experimental samples.
5. Since there are no protease inhibitors in the lysis buffer, it is vital that before the assay begins, the samples are kept on ice and placed in the plate reader only when it has reached 37 °C.
6. When calculating the rate of free AMC over time, ensure that only the linear portion of the curve is used. At the start of the assay, there will be a lag phase with little or no cleavage of substrate while the reagents reach the optimal temperature.

At the end of the assay, the substrate or enzyme may be completely used up causing a plateau in activity.

7. For convenience, store a stock of 100 % methanol at −20 °C.
8. Alternative permeabilization step. After fixation, the cells can be centrifuged at 1,500 × *g* and paraformaldehyde discarded. Then the pellet can be resuspended in 1 ml methanol. The advantage of this is the rest of the protocol can be carried out in 1.5 ml tubes instead of 15 ml tubes.
9. There is only one wash step between antibody incubations to avoid loss of cells.
10. Blocking solution can be prepared in advance and stored at 4 °C for up to 1 week. Discard if the solution is cloudy.
11. Ensure the fluorescent secondary antibodies are compatible with the lasers of the flow cytometer. Information can be found on the BD Biosciences website (http://www.bdbiosciences.com/research/multicolor/spectrum_viewer/index.jsp).
12. Incubate one sample with a primary antibody isotype control instead of the cleaved caspase primary antibody to show nonspecific binding of the primary antibody.
13. Incubate one sample with just secondary antibody to show nonspecific binding of the secondary antibody.
14. Leave one sample unstained to show autofluorescence of the cells.
15. An advantage of this method is the ability to detect and quantify the cleavage of each specific caspase, compared with DEVDase or IETDase activity assays, which measures both caspase-3 and caspase-7 or caspase-8 and caspase-10, respectively. Another advantage is the ability to quantify the relative quantities of cleaved caspase in each individual cell. As a result, this technique easily distinguishes apoptotic microglia cells from activated microglia cells.

References

1. Alnemri ES, Livingston DJ, Nicholson DW, Salvesen G, Thornberry NA, Wong WW, Yuan J (1996) Human ICE/CED-3 protease nomenclature. Cell 87(2):171. doi:S0092-8674(00)81334-3 [pii]
2. Lamkanfi M, Declercq W, Kalai M, Saelens X, Vandenabeele P (2002) Alice in caspase land. A phylogenetic analysis of caspases from worm to man. Cell Death Differ 9(4):358–361. doi:10.1038/sj/cdd/4400989
3. Burguillos MA, Deierborg T, Kavanagh E, Persson A, Hajji N, Garcia-Quintanilla A, Cano J, Brundin P, Englund E, Venero JL, Joseph B (2011) Caspase signalling controls microglia activation and neurotoxicity. Nature 472(7343):319–324. doi:nature09788 [pii]10.1038/nature09788
4. Timmer JC, Salvesen GS (2007) Caspase substrates. Cell Death Differ 14(1):66–72. doi:4402059 [pii] 10.1038/sj.cdd.4402059
5. Pop C, Salvesen GS (2009) Human caspases: activation, specificity, and regulation. J Biol Chem 284(33):21777–21781. doi:R800084200 [pii] 10.1074/jbc.R800084200

Chapter 14

Quantification of Microglial Phagocytosis by a Flow Cytometer-Based Assay

Refik Pul, Kandiyil Prajeeth Chittappen, and Martin Stangel

Abstract

Microglia represent the largest population of phagocytes in the CNS and have a principal role in immune defense and inflammatory responses in the CNS. Their phagocytic activity can be studied by a variety of techniques, including a flow cytometry-based approach utilizing polystyrene latex beads. The flow cytometry-based microglial phagocytosis assay, which is presented here, offers the advantage of rapid and reliable analysis of thousands of cells in a quantitative fashion.

Key words Phagocytosis, Microglia, Polystyrene latex beads, Flow cytometry, Assay

1 Introduction

The first extensive experiments on quantification of phagocytosis by using solid particles were published in 1912 by the Dutch physiologist Hartog Jacob Hamburger [1]. His usual procedure for measuring phagocytosis was to incubate suspensions of neutrophils and carbon mixed in a test tube for a given length of time and to quantify with a microscope the percentage of neutrophils that ingested carbon. In 1921 Fenn measured phagocytosis by counting the number of quartz particles phagocytosed per leucocyte [2]. According to Fenn phagocytosis was the result of surface forces that exist in a system containing phagocytes and particles. Whether phagocytosis will occur in a given system was held to be governed by the balance of these forces [2]. Even though the theory of phagocytosis at that time was purely mechanistic, such simple quantification methods are still widely in use, for example, by noting the ingestion of cells, such as bacteria, red blood cells, or other damaged (e.g., apoptotic) cells, as well as noncellular foreign particulate matter, including crystals, immiscible vacuolated liquids, and polystyrene latex beads ([3], Fig. 1).

Bertrand Joseph and José Luis Venero (eds.), *Microglia: Methods and Protocols*, Methods in Molecular Biology, vol. 1041, DOI 10.1007/978-1-62703-520-0_14,

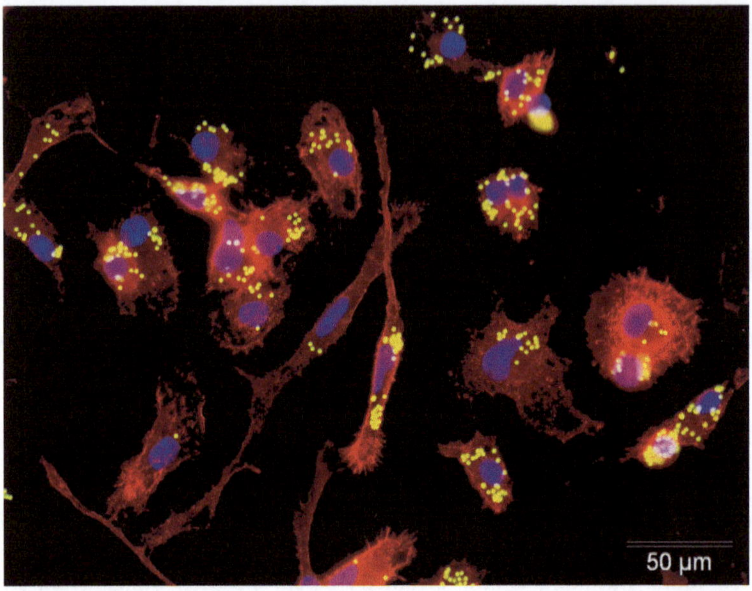

Fig. 1 Immunocytochemistry illustrates phagocytosis of fluorescent latex beads (*yellow–green*) by microglia (*red*, labelled with anti-OX42). Nearly all microglia cells contain beads after 1 h of phagocytosis

Polystyrene latex beads, initially used to produce experimentally regional ischemia by intravascular injection, were used for the first time by Sbarra and Karnovsky in 1959 to quantify phagocytosis [4, 5]. They observed that, in contrast to starch particles, polystyrene microspheres do not require a serum or phagocytosis-promoting factor which the authors ascribed to the lipophilic nature of these particles [5]. Since iodination (iodine-131) of polystyrene beads appeared to be not practicable, Wilkins reported in 1964 that these beads can be tagged with a green fluorescent dye ("dimethyl-6:12-coeroxenol acetate") [6]. His method made use of the fact that during the preparation of polystyrene latex by emulsion polymerization, seed particles from an initial polymerization are used to grow the particles to larger sizes in successive steps. If an oil-soluble dye dissolved in fresh monomer is added to the seed suspension during one of these steps, it will be incorporated into the polymer particles [6]. This paved the way for the use of fluorescent polystyrene beads in flow cytometric assays which enabled the facile determination of the percent of phagocytic cells and bead uptake [7–9]. In this regard, the mean fluorescence intensity directly correlates with the binding and uptake of beads because of uniformity in bead size and fluorescence distribution throughout the outer radii of these beads [8]. Bead uptake was also found to be dependent on bead size, concentration, and time. As a general trend, cells incubated with either the 0.2 or 0.5 µm microspheres resulted in a greater uptake of beads as compared to the larger 2.0 μm beads [10]. Maximal values were approached with increasing bead/cell ratios; rates were approximately linear up to 60 min. Effective inhibitors (4 °C, colchicine, and energy source deprivation) prevented phagocytosis and uptake was not present when assayed at zero time [9]. Interestingly, this method was applied to microglia as recently as 1997 by von Zahn et al., who showed a slight reduction of the phagocytic activity by TGF-β1 and probably by IL-4 [11]. Since this method provides reliable, fast, and objective measurements, flow cytometry-based microglial phagocytosis assay has been used in numerous studies [12–17].

2 Materials

1. Dulbecco's minimum essential medium supplemented with 10 % fetal bovine serum (FBS) and 1 % penicillin plus streptomycin (DMEM+).
2. Phosphate-buffered saline (PBS).
3. PBS containing 2 mM EDTA.
4. Polypropylene, 50-ml sterile, conical centrifuge tubes (Becton Dickinson Labware).
5. Polystyrene, 5.5-ml tubes (Becton Dickinson Labware).
6. 6-well plate.
7. Fluorescent-labelled Fluoresbrite microspheres, 1 μm diameter (Polysciences, Inc., Warrington, PA).
8. Trypan blue solution.
9. Propidium iodide solution 1 mg/ml.
10. Flow cytometer.
11. Incubator.

3 Methods

3.1 Plating and Preparation of Microglia

1. Microglia isolation is described elsewhere in this volume. The procedure, how microglia are isolated, is not relevant to the phagocytosis assay.
2. Centrifuge isolated microglia at 475 ×*g* for 6 min and discard the supernatant.
3. Resuspend the pellet in fresh DMEM+ and determine the viable cell number by trypan blue staining.
4. Adjust the cell density to 2.5×10^5 cells per ml. Add 2 ml of cell suspension (i.e., 5×10^5 cells) to each well of a 6-well plate and incubate for 2 h at 37 °C in a 5 % CO_2 humidified incubator. This procedure allows the viable microglia to adhere to the cell

culture plastic while the dead cells remain floating in the supernatant.

5. Aspirate the supernatants with floating cell debris and add 2 ml fresh DMEM+ to each well. Return the plates to the incubator for another 12–16 h.

3.2 Phagocytosis of Latex Beads

In order to distinguish the fluorescence resulting from active phagocytosis or from passive attachment of beads to the cells surface, the assay is performed at 4 °C and 37 °C in parallel.

1. Considering the cell/beads ratio to be used in the assay, prepare appropriate dilutions of latex beads in DMEM+ (*see* **Note 1**).
2. Remove the 6-well plates seeded with microglia from the incubator and aspirate the medium (*see* **Note 2**).
3. Add 1 ml of appropriate bead dilution to the cells in 6-well plates and incubate at 37 °C in a 5 % CO_2 humidified incubator for 1 h. As a control incubate an additional plate with beads at 4 °C for the same period of time (*see* **Note 3**).
4. After this incubation period, place the plates immediately on ice. Further steps have to be strictly followed at 4 °C (*see* **Note 4**).
5. To remove non-internalized beads, wash the wells by adding 2 ml of PBS and then aspirate it completely. Repeat this washing step 5 times (*see* **Note 5**).
6. At the end of the last washing step, add 2 ml of ice-cold PBS containing 2 mM ethylenediaminetetraacetic acid (EDTA) to each well and incubate on ice for 10 min.
7. Of note, microglia strongly adhere to plastic (*see* **Note 6**). To release the cells, vigorously pipette the PBS–EDTA including each edge of the well. Collect the cell suspension into 5 ml polystyrene tubes (*see* **Note 7**).
8. Centrifuge these tubes at 475 × *g* for 6 min at 4 °C.
9. Discard the supernatant and resuspend the cell pellet in 300 μl of ice-cold PBS containing 2 mM EDTA. Store the samples at 4 °C until further use.

3.3 Flow Cytometric Analysis

1. Prior to sample acquisition, adjust the flow cytometer photomultiplier (PMT) voltage settings by using the microglia sample that does not contain any beads.
2. Just before acquisition, add propidium iodide (PI) to each sample and vortex thoroughly. Propidium iodide will later enable discrimination of live and dead cells in data analysis.
3. Acquire and store ~20,000 live events for each sample (*see* **Note 8**).

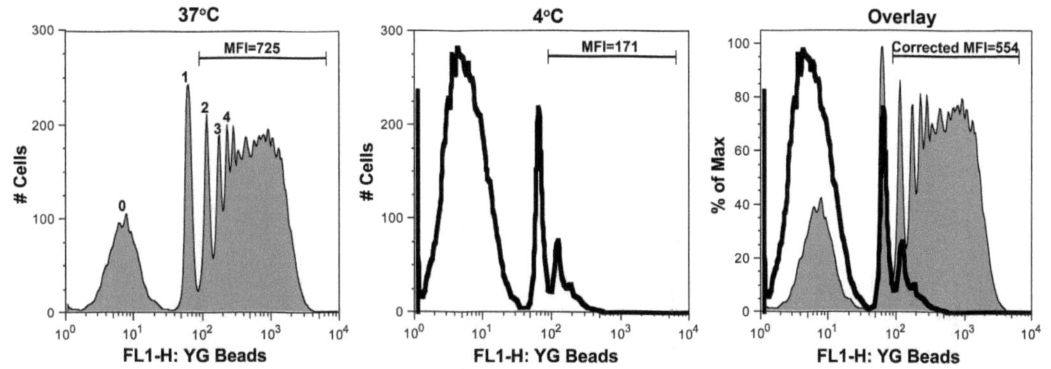

Fig. 2 Representative flow cytometry histograms display different mean fluorescence intensities (MFI) at 4 and 37 °C. Each peak represents the microglia population which has ingested 1, 2, 3, and 4 beads, respectively. Due to overlapping, further discrimination of peaks is not possible. However, since the first peak is also detectable at 4 °C, it may just result from beads attached to the cell surface. Thus, the MFI at 37 °C minus the MFI at 4 °C is the amount of incorporated fluorescent latex particles phagocytosed by a given number of microglia

4. For analysis, histogram profiles are generated for each sample with log scale of corresponding fluorescence intensity taken on the *x*-axis and the cell counts taken on the *y*-axis (Fig. 2).
5. The final data can be represented either as the percentage of cells that have phagocytosed the fluorescent beads or as the mean fluorescence intensity of the gated events. The fluorescence observed in the 4 °C control samples is subtracted from the 37 °C samples to obtain the final result (Fig. 2).

4 Notes

1. Prior to the assay, it is important to determine the optimal ratio of cell/beads for optimal results. Addition of excess beads may lead to unwanted "background noise," whereas insufficient number of beads may lower the significance of the findings. We recommend for microglia a cell/beads ratio of 1:100.
2. Of course, other plates, i.e., 12- or 24-well plates, can be used as well. However, then lower density seeding is required. Cell numbers may then be insufficient to record an appropriate number of life events.
3. Regardless of the uptake mechanism, it is considered plausible that polystyrene beads spend some time on the exterior of the cell membrane prior to translocation to the intracellular environment. To distinguish between uptake and adsorption, other methods can be used as well:
   (a) "Quenching" of adsorbed fluorescent markers with agents like ammonium acetate or trypan blue was proposed [18, 19].

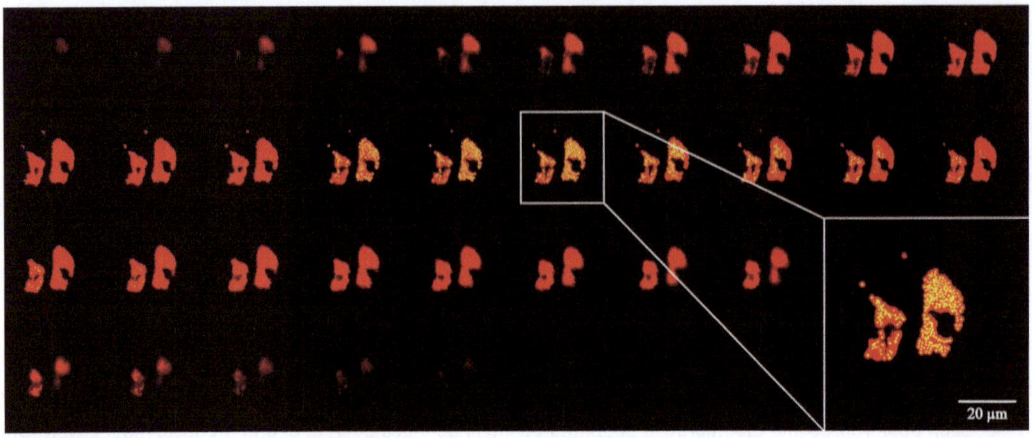

Fig. 3 Representative z-stack series of confocal microscopic images of two microglia cells (*red*, labelled with anti-OX42) definitely demonstrate that the particles were completely internalized and not merely attached to the outer membrane

(b) Phagocytosis but not adsorption of the beads to the cells can be completely abolished by cytochalasin D (5 µg/ml), an inhibitor of actin polymerization [13]. The use of cytochalasin D is a very elegant approach to correct for the fluorescence/number of the adsorbed beads.

We recommend the use of confocal microscopy for a qualitative distinction between beads taken up by cells or merely attached to the plasma membrane ([13], Fig. 3). It should be noted that fluorescent latex beads may be phagocytosed in a different way and at a different rate than other particle types. Unfortunately, little work has been carried out on exploiting the mechanism of the intracellular uptake of polystyrene latex beads. To our knowledge, there is only one report proposing that CD11c may be involved in the uptake of non-opsonized polystyrene beads [20].

4. For better results it is important to maintain the cells at 4 °C for further steps. All solutions used and the centrifuge have to be precooled and maintained at 4 °C.
5. The number of washes is sufficient to remove unattached beads since further washing did not reduce the adsorption-corrected fluorescence. However, thorough washing is important to eliminate all the non-phagocytosed beads that are either floating in the medium or attached to the cell surface. Residual beads can influence the final result as they can be detected on the flow cytometer and lead to false signals.
6. Vigourous pipetting can result in death of few cells. The dead cells can be excluded from the analysis by using propidium iodide or any other viability tracking dyes (e.g., 7-AAD). We strongly recommend to perform the phagocytosis assay

with adherent microglia because the bead uptake by cells in suspension is very low.

7. There are also other methods to detach cells, for example, trypsin–EDTA. Usage of trypsin–EDTA requires incubation of cells at 37 °C, which indeed cannot be used for the above described assay but may be applied to a cytochalasin D-based phagocytosis assay (*see* **Note 3**). However, trypsin may digest some surface antigens and thereby impair analysis of surface receptors. We do not recommend the usage of cell scraper as the cell viability of the cells is very low.
8. For a proper analysis, at least 10.000 events should be recorded.

References

1. Hamburger HJ (1912) Physikalisch-chemische untersuchungen über phagozyten. Ihre bedeutung von allgemein biologischem und pathologischem gesichtspunkt. In: Bergmann JF (ed). Wiesbaden
2. Fenn WO (1921) The phagocytosis of solid particles : I. Quartz. J Gen Physiol 3: 439–464
3. Harvath L, Terle DA (1999) Assay for phagocytosis. Methods Mol Biol 115:281–290
4. Crismon JM, Fuhrman FA (1947) Production of regional ischemia by intravascular injection of glass and plastic microspheres in graded sizes. Fed Proc 6:93
5. Sbarra AJ, Karnovsky ML (1959) The biochemical basis of phagocytosis. I. Metabolic changes during the ingestion of particles by polymorphonuclear leukocytes. J Biol Chem 234:1355–1362
6. Wilkins DJ (1964) Fluorescent Labelling of Polystyrene Latex for Tracing in Biological Systems. Nature 202:798–799
7. Dunn PA, Tyrer HW (1981) Quantitation of neutrophil phagocytosis, using fluorescent latex beads. Correlation of microscopy and flow cytometry. J Lab Clin Med 98:374–381
8. Steinkamp JA, Wilson JS, Saunders GC et al (1982) Phagocytosis: flow cytometric quantitation with fluorescent microspheres. Science 215:64–66
9. Schroeder F, Kinden DA (1983) Measurement of phagocytosis using fluorescent latex beads. J Biochem Biophys Methods 8:15–27
10. Sanchez-Martin RM, Muzerelle M, Chitkul N et al (2005) Bead-based cellular analysis, sorting and multiplexing. Chembiochem 6: 1341–1345
11. von Zahn J, Moller T, Kettenmann H et al (1997) Microglial phagocytosis is modulated by pro- and anti-inflammatory cytokines. Neuroreport 8:3851–3856
12. Pul R, Moharregh-Khiabani D, Skuljec J et al (2011) Glatiramer acetate modulates TNF-alpha and IL-10 secretion in microglia and promotes their phagocytic activity. J Neuroimmune Pharmacol 6:381–388
13. Stangel M, Joly E, Scolding NJ et al (2000) Normal polyclonal immunoglobulins ('IVIg') inhibit microglial phagocytosis in vitro. J Neuroimmunol 106:137–144
14. Tambuyzer BR, Casteleyn C, Van Cruchten S et al (2012) Interferon-gamma modulates the functional profile of in-vitro-cultured porcine microglia. Neuroreport 23:519–524
15. Dijkstra S, Geisert EE Jr, Dijkstra CD et al (2001) CD81 and microglial activation in vitro: proliferation, phagocytosis and nitric oxide production. J Neuroimmunol 114:151–159
16. Nagai A, Mishima S, Ishida Y et al (2005) Immortalized human microglial cell line: phenotypic expression. J Neurosci Res 81:342–348
17. Nagano T, Kimura SH, Takemura M (2010) Prostaglandin E2 reduces amyloid beta-induced phagocytosis in cultured rat microglia. Brain Res 1323:11–17
18. Lorenzi R, Brickell PM, Katz DR et al (2000) Wiskott-Aldrich syndrome protein is necessary for efficient IgG-mediated phagocytosis. Blood 95:2943–2946
19. Aslam R, Kim M, Speck ER et al (2007) Platelet and red blood cell phagocytosis kinetics are differentially controlled by phosphatase activity within mononuclear cells. Transfusion 47:2161–2168
20. Keizer GD, Te Velde AA, Schwarting R et al (1987) Role of p150,95 in adhesion, migration, chemotaxis and phagocytosis of human monocytes. Eur J Immunol 17:1317–1322

Chapter 15

Quantification of Microglial Proliferation and Apoptosis by Flow Cytometry

Alicia A. Babcock, Martin Wirenfeldt, and Bente Finsen

Abstract

Microglia are innate immune cells that survey the central nervous system (CNS) and respond almost immediately to any disturbance in CNS homeostasis. They are derived from primitive yolk sac myeloid progenitors and in the mouse colonize the CNS during fetal development. As a population, microglia have the potential to expand rapidly in response to inflammatory stimuli, injury, or any other pathological changes, due to a high capacity for proliferation. In addition, apoptotic mechanisms can be evoked to retract the microglial population, as reactivity declines. In the normal CNS, a low rate of proliferation and apoptosis maintain a low rate of microglial turnover. Here, we describe quantitative analysis of proliferation and apoptosis of microglial cells isolated from individual adult mice by flow cytometry, which allows distinction from perivascular or infiltrating macrophages, based on differential expression of CD45. These methods can be applied to analyze microglial turnover in various models of neuroinflammation.

Key words Flow cytometry, Proliferation, Apoptosis, Microglia, Macrophages, Brain

1 Introduction

Microglia are the resident innate immune cells of the CNS. Despite sometimes referred to as "resting" under normal conditions, microglia are highly active cells that continuously survey defined regions with their processes [1–4]. When damage is sensed, microglial processes converge, and a reactive response program that includes proliferative expansion is initiated [5–7]. As a population, microglia have the potential to numerically expand very rapidly after inflammatory stimuli or other pathological changes, due to a high capacity for proliferation. For example, 30–40 % of microglia proliferate 3 days after acute neural injury, a response that begins 24 h after injury [8–12]. At the same time, apoptotic mechanisms occur at low levels in normal CNS and can be induced to retract an expanded microglial population, as reactivity to damage subsides [8, 13]. Without pathological triggers, apoptosis

Bertrand Joseph and José Luis Venero (eds.), *Microglia: Methods and Protocols*, Methods in Molecular Biology, vol. 1041,
DOI 10.1007/978-1-62703-520-0_15,

and proliferation are held in check. A low rate of microglial turnover occurs in the normal CNS [14].

Cellular proliferation is often analyzed by measuring incorporation of bromodeoxyuridine (BrdU), a synthetic thymidine analogue that becomes incorporated into the newly synthesized DNA of proliferating cells. Incorporated BrdU is then detected with specific anti-BrdU antibodies, thus indicating cells ($BrdU^+$) that were replicating their DNA at the time the chemical was administered [15, 16].

One of the earliest events in apoptosis is the externalization of phosphatidylserine residues from the inner leaflet of the plasma membrane to the outer leaflet [17]. This occurs prior to the loss of membrane integrity that occurs in cell death during late apoptosis or necrosis. Externalized phosphatidylserine can be detected on the cell surface with fluorochrome-conjugated Annexin V, a calcium-dependent phospholipid-binding protein with a high affinity for phosphatidylserine [18]. Annexin V is typically used in combination with a live/dead discriminator dye, such as the DNA-binding dye 7-Aminoactinomycin D (7-AAD), which permeates the membranes of dead cells, but is excluded by viable cells with intact membranes. This distinguishes early apoptotic events (7-AAD^- Annexin V^+) from dead cells (7-AAD^+).

Though microglial proliferation and apoptosis are frequently assessed using immunohistochemical methods, both processes may be monitored using flow cytometry [8, 10, 19]. Flow cytometry is an invaluable technique for multiparametric analysis of single cells. Using a panel of fluorescently conjugated reagents simultaneously allows the detection and quantification of activation markers or assessment of functional states by cells identified as microglia by labeling with specific antibodies [8, 10, 19–22]. As myeloid-lineage cells, microglia express CD11b. However, perivascular macrophages that express CD11b are present, even in perfused CNS. $CD11b^+$ microglia can be distinguished from $CD11b^+$ macrophages by differential expression of CD45 [23]. Studies in bone marrow chimeric animals show that the vast majority of $CD45^{dim}$ cells are resident microglial cells and that the vast majority of $CD45^{high}$ cells are derived from the periphery [8, 19–22].

Here, we describe the analysis of proliferation and apoptosis of microglial cells from individual adult mice by flow cytometry. Unlike most protocols that use a Percoll gradient for cell isolation and may require whole CNS or even pooling tissue from groups of mice [19, 23–28], the method described in this chapter allows investigation of isolated regions of the CNS in individual mice. Since only a small amount of tissue is required, a single brain can often be used for several different types of analyses. Our methods can be applied to analyze microglial turnover in various models of neuroinflammation.

2 Materials

2.1 BrdU Injections

1. ≥2 adult mice (*see* **Note 1**).
2. Scale.
3. BrdU (10 mg/ml solution): Dissolve BrdU in room temperature PBS in a laboratory chemical hood at a concentration of 10 mg/ml and filter sterilize (*see* **Note 2**).
4. 1 cc syringe and 26 gauge needles.

2.2 Perfusion/Tissue Collection

1. Pentobarbital.
2. 70 % ethanol.
3. Perfusion block and absorbent material.
4. 23 gauge needles.
5. 20 cc syringe and 26 gauge needles.
6. Ice.
7. Phosphate buffered saline (PBS) for perfusion, preferably ice-cold.
8. Surgical instruments for perfusion: blunt-ended large scissors, forceps, hemostat, and sharp-pointed scissors.
9. Surgical instruments for removal of brain: small straight-edged scissors, forceps (preferably with teeth), and scoop.
10. Surgical instruments for dissection of brain: blunt-ended forceps and scalpel with #10 blade.
11. Petri dishes.
12. Hanks balanced salt solution (HBSS) or RPMI-1640 medium (*see* **Note 3**).
13. Tubes for collecting samples: Eppendorf tubes or 24-well plate for brain samples and 15 ml centrifuge tube for spleen. Samples should be collected into a small amount of buffer (*see* **Note 3**).

2.3 Preparation of Cell Suspensions

1. HBSS or RPMI (*see* **Note 3**), dispensed by squirt bottle or plastic bulb pipettes.
2. 35 mm Petri dishes.
3. 70 μm cell strainers.
4. Glass Pasteur pipettes and latex bulbs.
5. Plungers from 1 cc syringes.
6. 12 × 75 mm sample acquisition tubes for a flow cytometer.
7. Rack for 12 × 75 mm tubes, such as a test tube peg rack.
8. 15 ml centrifuge tubes.
9. Refrigerated centrifuge with racks for 12 × 75 mm tubes and 15 ml centrifuge tubes.
10. 0.83 % solution of ammonium chloride.

2.4 Staining and Flow Cytometry

1. Blocking buffer: HBSS containing 1 μg/ml anti-FcgIII/II receptor antibody (clone 2.4G2), 50 μg/ml Syrian hamster Ig, and 2 % FCS (*see* **Note 4**).
2. Staining buffer: HBSS containing 2 % FCS (*see* **Note 4**).
3. 10 % sodium azide solution: Dissolve 10 g of sodium azide in 100 ml of distilled H_2O (*see* **Note 4**). The solution should be prepared with care inside a chemical laboratory hood.
4. Refrigerated centrifuge with racks for 12×75 mm tubes and 15 ml centrifuge tubes.
5. Vortexer.
6. Aluminum foil.
7. PE-conjugated CD45 (clone 30-F11) and rat IgG_{2b} isotype control.
8. FITC-conjugated CD11b (clone M1/70) and rat IgG_{2b} isotype control.
9. PerCP-Cy™ 5.5-conjugated CD11b (clone M1/70) and rat IgG_{2b} isotype control.
10. 10× Annexin V binding buffer. Dilute to 1× with distilled water prior to use.
11. 7-Aminoactinomycin D (7-AAD).
12. APC-conjugated Annexin V.
13. APC BrdU Flow Kit (BD Bioscience): contains BD Perm/Wash™ Buffer (10×, diluted to 1× in deionized water prior to use), BD Cytofix/Cytoperm™ Buffer, BD Cytoperm™ Permeabilization Buffer Plus, DNase I (1 mg/ml stock solution, *see* **Note 5**), and APC-conjugated anti-BrdU antibody.
14. Dulbecco's PBS (D-PBS).
15. Ice.
16. FACS Calibur equipped with 488 nm and 635 nm lasers, capable of detecting FITC, PE, 7-AAD, PerCP-Cy™ 5.5, and APC, or a similar instrument.
17. CellQuest Pro or other software for analysis of flow cytometry data.

3 Methods

3.1 Administration of BrdU

1. Weigh the mice and calculate the volume of BrdU solution to administer (i.e., 90 mg/kg; 0.18 ml of a 10 mg/ml solution to a 20 g mouse).
2. Use a 1 cc syringe and 26 gauge needle to administer BrdU i.p. at 2 h, 10 h, and 22 h prior to perfusion (*see* **Note 6**).
3. Include one control mouse that has not been injected with BrdU in the experiment.

3.2 Tissue Isolation

1. Anesthetize a mouse with an i.p. overdose of pentobarbital.
2. Pin the anesthetized mouse to a perfusion board on its back, with limbs spread, using 23 gauge needles through the feet.
3. Wet the mouse with 70 % ethanol.
4. Carefully open the abdominal cavity using blunt-ended scissors and forceps.
5. Cut the diaphragm, cut the ribs upwards, and clamp the sternum with a hemostat, placing its handles near the mouse's head to expose the heart.
6. Place the tip of a 26 gauge needle connected to a 20 cc syringe filled with ice-cold PBS into the left ventricle.
7. Cut the right atrium using sharp-pointed scissors so that blood flows from the heart.
8. Slowly inject the PBS into the left ventricle, making sure not to damage the heart (*see* **Note 7**). Blood flowing from the right atrium will become clear.
9. Decapitate the mouse using a pair of large scissors.
10. Wet the head with 70 % ethanol.
11. Cut through the skin from the back of the neck to the scalp and fold the skin down under the nose of the mouse for grip, exposing the skull.
12. Insert the tip of a small, straight pair of scissors into the cisterna magna and make small cuts through the bone, along the midline, keeping the tip of the scissors pointed upwards to not damage the brain.
13. Using a pair of forceps, carefully pull the skull bone away to each side to reveal the brain.
14. Gently scoop out the brain into a small Petri dish containing PBS (*see* **Note 8**).
15. Using forceps to grab the brain by the cerebellum, place the rinsed brain onto the lid of the Petri dish.
16. Use a scalpel and a small pair of blunt-edged forceps to micro-dissect the brain region of interest (*see* **Notes 9–11**).
17. Collect the brain region of interest by placing each individual sample into labeled Eppendorf tubes or 24-well culture plates containing a small volume of buffer (*see* **Note 3**).
18. Collect the spleen for $CD45^{high}/CD45^{dim}$ controls (*see* **Note 12**). Place the mouse on its right side and locate the spleen in the left side of the abdominal cavity. Make a small cut to detach it, remove any associated fat tissue, and place the spleen into HBSS.

3.3 Preparation of Cell Suspensions from Brain

1. Place a 70 μm cell strainer into a 35 mm Petri dish, and place the plunger from a 1 cc syringe into the lid of the Petri dish.

2. Using a short glass pipette fitted with a latex bulb, pick up a brain sample, and place it on top of the cell strainer (*see* **Note 13**).
3. Place the pipette temporarily into a 12×75 mm tube labeled to match the tissue sample.
4. Fill a new, empty 12×75 mm tube nearly full with HBSS/RPMI (*see* **Note 3**). Never place the pipette into this tube, or you will have to change it every time.
5. Decant approximately half the HBSS onto the cell strainer in the Petri dish.
6. Lift the cell strainer slightly, and using the 1 cc plunger, gently spread the tissue over the cell strainer. Submerge in HBSS. Repeat with the plunger.
7. Use the glass pipette and bulb to gently move the tissue down through the cell strainer (*see* **Note 14**). All clumps should be fully homogenized.
8. Collect the suspension into the previously labeled 12×75 mm tube.
9. Wash the cell strainer with the remaining HBSS and transfer to the previously labeled 12×75 mm tube.
10. Prepare other samples.
11. Spin all samples at 400×*g* for 10 min at 4 °C (*see* **Note 15**).

3.4 Preparation of Cell Suspensions from Spleen

1. Prepare a single cell suspension using a 70 μm cell strainer and plunger, as for brain cells.
2. Collect cell suspension into a 15 ml centrifuge tube.
3. Spin at 4 °C for 7 min at 400×*g*. Decant and vortex.
4. Resuspend well in 2 ml 0.83 % ammonium chloride. Keep at room temperature for 10 min, rotating the tube every few minutes.
5. Fill the centrifuge tube to 15 ml with HBSS.
6. Spin at 4 °C for 7 min at 400×*g*. Decant and vortex (*see* **Note 16**).
7. Repeat HBSS washes three additional times.

3.5 Blocking Non-specific Staining and Setting Up Samples/Controls

1. Resuspend spleen cells in 1 ml staining buffer. Vortex.
2. Resuspend brain cells in 50 μl of blocking buffer per stain (*see* **Note 9**). Vortex.
3. Incubate at room temperature for 20–30 min. During this time, split brain cells into equivalent fractions (*see* **Notes 9, 17**) by measuring the volume of the suspension with a pipette. Separate tubes are required for analysis of proliferation and apoptosis.
4. Set up all controls, including compensation controls (Table 1) and staining controls (Table 2) (*see* **Note 18**).

Table 1
Controls for adjusting photomultiplier tubes and compensation

		Voltage				Compensation					
Analysis	**Antibody/reagent**	**FL1**	**FL2**	**FL3**	**FL4**	**FL2-%FL1**	**FL1-%FL2**	**FL3-%FL2**	**FL2-%FL3**	**FL4-%FL3**	**FL3-%FL4**
Annexin V	Unstained cells	x	x	x	x						
	CD11b FITC	x				x					
	CD45 PE		x				x	x			
	7-AAD			x					x	x	
	Annexin V APC				x						x
BrdU	Unstained cells	(x)	x	x	x						
	(FITC or GFP)	(x)				(x)					
	CD45 PE		x				x	x			
	CD11b PerCP-Cy™5.5			x					x	x	
	BrdU APC				x						x

5. Wash the cells with 1–2 ml of staining buffer.
6. Spin at 4 °C at 400 × *g* for 10 min (*see* **Note 19**).
7. Decant supernatant, then vortex well.

3.6 Annexin V Staining

1. Stain for surface antigens (CD45 PE and CD11b FITC—or IgG_{2b} controls) by adding 50 μl of a mixture of antibodies diluted in staining buffer (*see* **Note 20**).
2. Vortex, and incubate for 20–30 min at room temperature, covered in foil.
3. Wash with 1–2 ml cold 1× binding buffer.
4. Spin at 4 °C at 400 × *g* for 10 min.
5. Decant supernatant and vortex.
6. Add 100 μl of 1× binding buffer and 10 μl of a 1:1 mix Annexin V APC/7-AAD (*see* **Note 21**).
7. Incubate 15 min at room temperature, under foil.
8. Add 200 μl of 1× binding buffer, and store on ice under foil.
9. Acquire using a flow cytometer within 1 h.

3.7 BrdU Staining

1. Stain for surface antigens (CD45 PE and CD11b PerCP-Cy™5.5—or IgG_{2b} controls) by adding 50 μl of a mixture of antibodies diluted in staining buffer (*see* **Notes 20** and **22**).
2. Vortex and incubate for 20–30 min at room temperature, covered in foil.
3. Wash with 1–2 ml staining buffer.
4. Spin at 4 °C at 400 × *g* for 10 min.

Table 2
Staining controls for determining positive versus negative fluorescence levels

		Fluorescence channels			
Analysis	**Controls**	**FL1**	**FL2**	**FL3**	**FL4**
Annexin V	Positive control	CD11b FITC	CD45 PE	7-AAD	Annexin V APC
	CD45^{high}/CD45^{dim} control[a]	CD11b FITC	CD45 PE	7-AAD	Annexin V APC
	FL1 negative control	Rat IgG$_{2b}$ FITC	CD45 PE	7-AAD	Annexin V APC
	FL2 negative control	CD11b FITC	Rat IgG$_{2b}$ PE	7-AAD	Annexin V APC
	FL3 negative control	CD11b FITC	CD45 PE	(omission)[b]	Annexin V APC
	FL4 negative control	CD11b FITC	CD45 PE	7-AAD	(omission)[b]
	Negative control	Rat IgG$_{2b}$ FITC	Rat IgG$_{2b}$ PE	(omission)[b]	(omission)[b]
BrdU	Positive control	(FITC or GFP)[c]	CD45 PE	CD11b PerCP-Cy5.5	BrdU APC
	CD45^{high}/CD45^{dim} control[a]	(FITC or GFP)	CD45 PE	CD11b PerCP-Cy5.5	BrdU APC
	FL1 negative control[c]	(Ig or GFP- cells)	CD45 PE	CD11b PerCP-Cy5.5	BrdU APC
	FL2 negative control	(FITC or GFP)	Rat IgG$_{2b}$ PE	CD11b PerCP-Cy5.5	BrdU APC
	FL3 negative control	(FITC or GFP)	CD45 PE	Rat IgG$_{2b}$ PerCP-Cy5.5	BrdU APC
	FL4 negative control	(FITC or GFP)	CD45 PE	CD11b PerCP-Cy5.5	Mouse IgG$_1$ APC
	FL4 negative control[d]	(FITC or GFP)	CD45 PE	CD11b PerCP-Cy5.5	BrdU APC
	Negative control[d]	(Ig or GFP- cells)	Rat IgG$_{2b}$ PE	Rat IgG$_{2b}$ PerCP-Cy5.5	BrdU APC

[a]Use brain cells spiked with cells from spleen (or lymph nodes). Under some circumstances it may be ideal to run an entire set of staining controls with a mixture of brain/spleen cells

[b]The negative control should be omission of 7-AAD or Annexin V APC, rather than an Ig control

[c]Inclusion of a FITC-labeled activation marker or GFP is optional. FITC-labeled Ig of the relevant isotype or GFP$^-$ cells should be included as a negative control, respectively

[d]Uses cells from a mouse that was not injected with BrdU (antigen omission). Cells are labeled as normal with anti-BrdU APC antibody

5. Decant supernatant and vortex.
6. Fix and permeabilize the cells by adding 100 μl of Cytofix/Cytoperm™ Buffer.
7. Vortex, and incubate cells for 30 min at room temperature, under foil (*see* **Note 23**).
8. Wash with 1 ml 1× Perm/Wash™ Buffer. Spin at 4 °C at 400×*g* for 10 min, then decant supernatant (*see* **Note 24**), and vortex.
9. Resuspend cells in 100 μl of Cytoperm™ Permeabilization Buffer Plus and incubate for 10 min at 4 °C, under foil.

10. Wash with 1 ml 1× Perm/Wash™ Buffer. Spin at 4 °C at 400 × *g* for 10 min, then decant the supernatant and vortex.
11. Resuspend cells in 100 μl Cytofix/Cytoperm™ Buffer for 5 min at room temperature, under foil.
12. Wash with 1 ml 1× Perm/Wash™ Buffer. Spin at 4 °C at 400 × *g* for 10 min, then decant supernatant and vortex.
13. Resuspend cells with 100 μl diluted DNAse I (diluted to 300 μg/ml in DPBS, giving 30 μg per tube) and incubate at 37 °C for 1 h.
14. Wash with 1 ml 1× Perm/Wash™ Buffer. Spin at 4 °C at 400 × *g* for 10 min, then decant supernatant and vortex.
15. Stain for incorporated BrdU by adding 50 μl of 1× Perm/Wash™ Buffer containing anti-BrdU APC (1:50) (or control mouse IgG_1) and incubate for 20 min at room temperature.
16. Wash with 1 ml 1× Perm/Wash Buffer. Spin at 4 °C at 400 × *g* for 10 min, then decant supernatant and vortex.
17. Resuspend cells in staining buffer (~300 μl).
18. Acquire cells using a flow cytometer within 3 days.

3.8 Sample Acquisition

1. Adjust the Forward Scatter (FSC) and Side Scatter (SSC) parameters on linear scales (*see* **Note 25**).
2. Set a FSC threshold, to exclude debris and reduce file size (*see* **Notes 17** and **25**).
3. Adjust the fluorescence parameters so that the negative populations fall between 10^1 and 10^2 on logarithmic axes (Table 1).
4. Set up compensation using singly stained controls (Table 1).
5. Measure cell fluorescence by flow cytometry.
6. Acquire one to three million total events for samples and staining controls (*see* **Note 26**, Table 2).

3.9 Analysis: Identifying CD45⁺CD11b⁺ Cells

1. Identify cells of interest based on expression of surface markers (*see* **Note 27**). Draw a series of gates (SSC vs. CD11b, SSC vs. CD45, FSC vs. CD11b, and FSC vs. CD45), based on the fluorescence levels of isotype-stained controls (Table 2, Fig. 1).
2. Apply a combination of all gates and draw CD45/CD11b plots, with CD45 on the *y*-axis and CD11b on the *x*-axis. Only macrophages and microglia (all $CD45^+CD11b^+$ cells) should be visible (Fig. 1).
3. Add quadrants to separate $CD45^{high}$ and $CD45^{dim}$ cells by comparing brain cells to brain samples spiked with spleen cells (Table 2, Fig. 1) (*see* **Note 28**).

3.10 Analysis: Annexin V/Apoptosis

1. After gating on $CD11b^+/CD45^+$ cells, exclude dead cells by gating only on viable ($7\text{-}AAD^-$) cells, by comparing to the control sample where 7-AAD is omitted (Table 2, Fig. 2) (*see* **Note 29**).

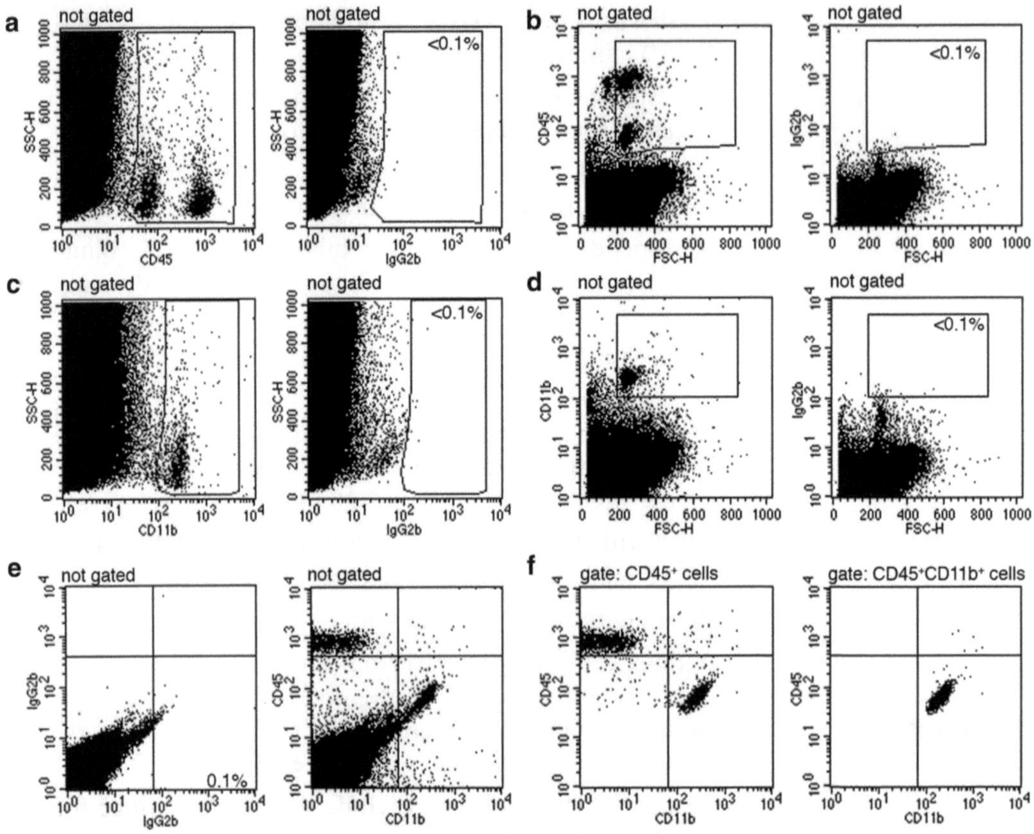

Fig. 1 Example of gating strategy and controls for analysis of CD45/CD11b flow cytometry profiles. (**a**) SSC/CD45 gate depicted on SSC/CD45 or SSC/IgG_{2b} flow cytometry profiles from brain homogenate spiked with spleen cells. (**b**) FSC/CD45 gate shown on CD45/FSC or IgG_{2b}/FSC flow cytometry profiles from brain homogenate spiked with spleen cells. (**c**) SSC/CD11b gate drawn on SSC/CD11b or SSC/IgG_{2b} flow cytometry profiles from brain homogenate spiked with spleen cells. (**d**) FSC/CD11b gate shown on CD11b/FSC or IgG_{2b}/FSC flow cytometry profiles from brain homogenate spiked with spleen cells. (**e**) Brain/spleen cells stained with isotype controls or antibodies against CD45 and CD11b antibodies, shown as CD45/CD11b plots. Quadrants discriminate CD45^{high} and CD45^{dim} cells. Events shown in (**a**)–(**e**) are not gated and represent several million total events collected. (**f**) CD45/CD11b profiles from brain/spleen mixture (*left*, gated on CD45$^+$ cells as in (**a**) and (**b**)) or neocortex from a 4-month-old mouse (*right*, gated on CD45$^+$CD11b$^+$ cells as in (**a**)–(**d**)). Quadrants discriminate CD45^{high} and CD45^{dim} cells

2. Determine the percentage of Annexin V$^+$ cells, by setting quadrants based on controls lacking Annexin V (Fig. 2) (*see* **Note 30**).

3.11 Analysis: BrdU/Proliferation

1. After gating on CD11b$^+$/CD45$^+$cells, set up quadrants to determine the percentage of BrdU$^+$ cells, versus controls lacking BrdU and isotype-stained controls (Table 2, Fig. 3) (*see* **Note 30**).

3.12 Application

1. Measure microglial proliferation and apoptosis in the experimental system of interest (Fig. 4).

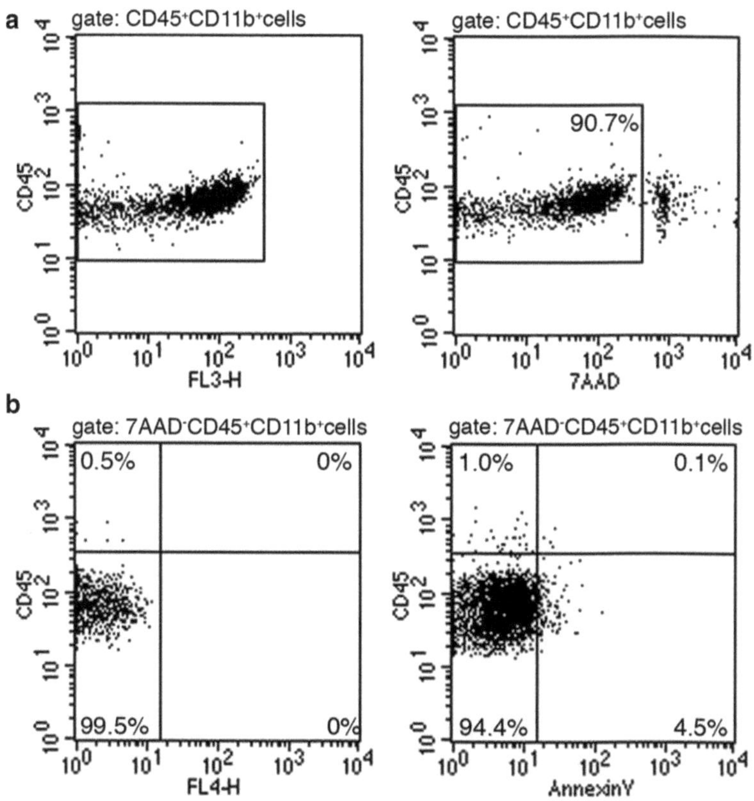

Fig. 2 Example of controls for analysis of apoptotic microglia/macrophages. (**a**) Live cells from the neocortex of an adult mouse are gated by excluding 7-AAD⁺ cells on CD11b⁺CD45⁺ gated CD45/7-AAD flow cytometry profiles (*right*), based on fluorescence levels of the control lacking 7-AAD (*left*). (**b**) Flow cytometry profiles gated on viable (7-AAD⁻) CD11b⁺CD45⁺ cells show fluorescence with Annexin V labeling (*right*) or binding buffer alone (*left*). Quadrants are drawn based on the fluorescence levels of the control lacking Annexin V, and discriminate CD45^{high} and CD45^{dim} cells as described in Fig. 1

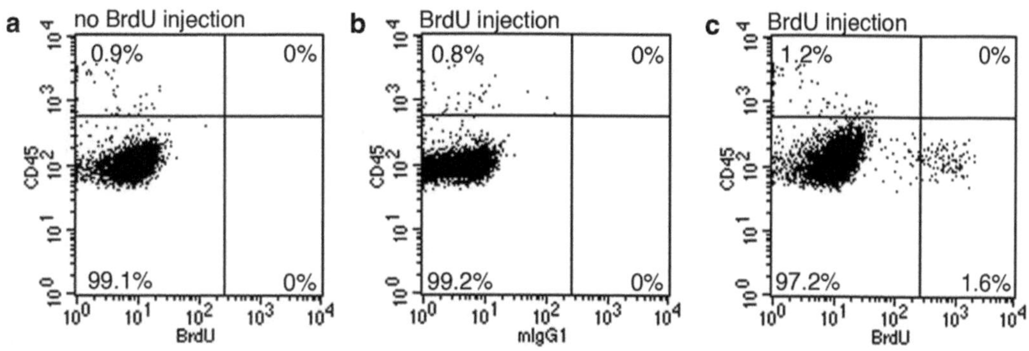

Fig. 3 Example of controls for analysis of BrdU⁺ microglia and macrophages. CD45/BrdU flow cytometry profiles, gated on CD11b⁺CD45⁺ cells, from the neocortex of either a naïve mouse without BrdU injection (**a**) or a mouse injected with BrdU (**c**), as well as a CD45/mIgG1 plot showing background staining in a BrdU-injected mouse (**b**). Quadrants are drawn based on the fluorescence levels of the control mice for BrdU, and discriminate CD45^{high} and CD45^{dim} cells using the strategy depicted in Fig. 1

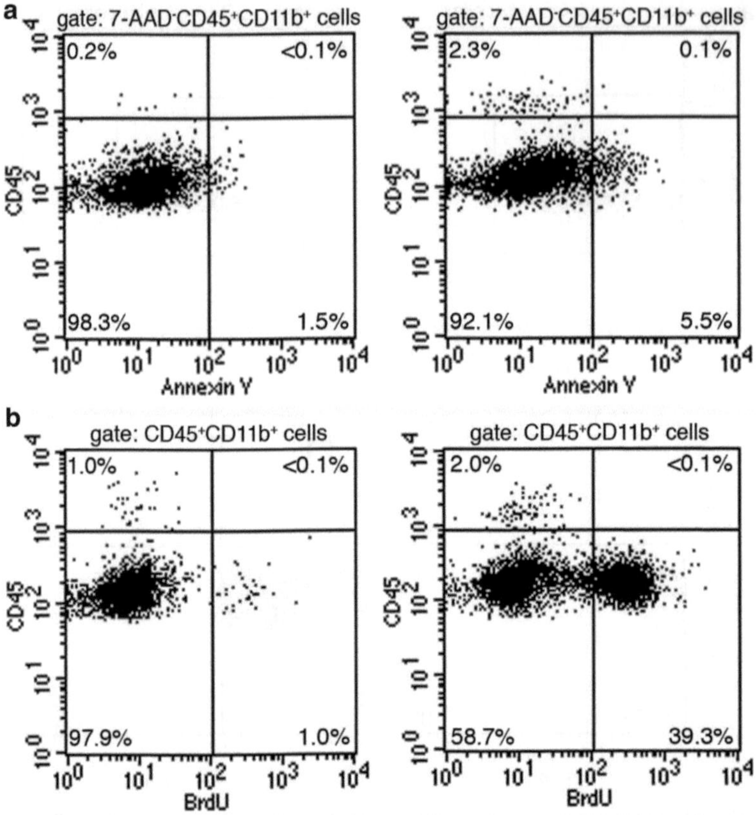

Fig. 4 Example of analysis of Annexin V+ and BrdU+ microglia and macrophages in an experimental setting. Microglial apoptosis (**a**) and proliferation (**b**) are increased in the ipsilateral hippocampus of mice 3 days after perforant pathway axonal lesion (*right*), compared to unlesioned contralateral hippocampus (*left*). Axonal lesions were carried out as previously described [8]

4 Notes

1. Each mouse is analyzed individually. At least one mouse should serve as a "no BrdU" control that is not injected with BrdU.
2. Aliquots of diluted BrdU can be stored at −20 °C, though any BrdU that precipitates upon freezing should be in solution before use.
3. Tissue is usually collected into HBSS, although RPMI may be preferred for analysis of apoptotic cells.
4. We usually add sodium azide to staining reagents at a final concentration of 0.1 %, except for analysis of Annexin V staining.
5. DNAse I stock solution may only be refrozen once before it loses activity. Additional DNAse I can be purchased from Sigma (D-4513), if necessary.
6. Many variations of the BrdU protocol are possible. We have obtained similar results by administering BrdU i.p. over 3 × 8 h intervals [8–10], single i.p. injections 1 h prior to perfusion

[11], or daily i.p. injections of a lower dose [19]. BrdU can also be administered via drinking water [29]. In addition, analysis of incorporated BrdU can be delayed by several days, to analyze the fate of previously proliferating cells [29, 30].

7. The purpose of perfusion is to eliminate contamination of the CNS by circulating blood cells. To ensure a good perfusion, do not perfuse the mouse too quickly or cause major bleeding elsewhere. Signs for good perfusion are that the forepaws turn white and that the liver pales, turning yellow. When finished, the brain should be white, not pink.
8. The meninges should be removed before analysis. Often the meninges are detached when the skull is removed; however, it may be necessary to do this with forceps.
9. Only a small amount of tissue is required for each stain for flow cytometry. If appropriate to the experimental setup, brains can be split into different hemispheres or regions, which may then be collected for different purposes (i.e., immunohistochemistry, RNA analyses). Furthermore, only a fraction of each brain region is required per staining for flow cytometry. For example, only ½ of a hippocampus [8, 10, 31, 32], entorhinal cortex [24, 31], or corpus callosum [19] should be stained. Use 1/6th to 1/8th of the neocortex from a single hemisphere per stain [20, 21].
10. To isolate hippocampi, use a scalpel blade to make an incision between the two cerebral hemispheres. At the front of the brain, cut all the way down to the base. Cut down 1–2 mm at the back of the cerebral hemispheres, just through the neocortex. Using the blunted side of the tip of the scalpel blade, lift the cortex away from the midbrain on each side. If necessary make gentle cuts with the scalpel to fully push the neocortex away (until flat), exposing the hippocampus. Make small incisions at the fimbria, and gently roll the hippocampi backwards, towards the entorhinal cortex, and remove. Though it is difficult to see after perfusion, the choroid plexus usually sticks to the hippocampi; it should be removed before analysis. This can be done by gently running the blunt side of the scalpel over the isolated hippocampi.
11. To isolate neocortex, first remove the hippocampus (*see* **Note 10**). Gently scoop out the striatum from the cortex, still lying at each side of the brain, then dissociate the cortex from the midbrain.
12. Alternatively, cells isolated from lymph nodes can be used to distinguish CD45^{high} cells from CD45^{dim} cells. We often collect inguinal lymph nodes for this purpose. They should be homogenized as indicated below; unlike spleen, no lysis of red blood cells is required.
13. These are live cells that need to be processed shortly after tissues are removed from mice. Analysis of apoptosis should be done

on the same day. Cells for BrdU analysis can be stored after the fixation step. These procedures are lengthy, and it is recommended to prepare as much as possible the day before and to begin with only a few samples until a routine has been established and techniques have been optimized.

14. This should be done as gently as possible, to prevent high levels of cellular autofluorescence.

15. This protocol can be modified for the analysis of cytokine production, by incubating cells with 1 μg/ml Brefeldin A for 4–6 h at 37 °C and 5 %CO_2 [20, 21], before proceeding with blocking and antibody staining.

16. Remove any clumps of dead cells that appear.

17. It is possible to estimate absolute cell numbers, as previously described [8, 10, 19–21, 31], if one knows what fraction of the brain region has been sampled. Thus it is important to split the cells as accurately as possible. A second sampling step occurs during sample acquisition at the flow cytometer. To account for this, we usually measure the volume of the cell suspension, before and after acquisition on the flow cytometer, to determine what fraction has been used. Alternatively, beads can be used to measure this [33]—however, in this case, the FSC threshold must be removed so that the beads remain visible, making file sizes much larger.

18. For compensation controls, to set up the flow cytometer, a series of cells stained with each antibody/dye individually is required, in addition to unstained controls. Staining controls include positive controls, Ig controls, and $CD45^{high}/CD45^{dim}$ controls. A low level of proliferation and apoptosis occurs in unmanipulated mice [8]. Including a series of controls whereby only one antibody/dye is replaced with Ig control/omitted allows determination of background staining in specific cell populations. Including a mix of brain and spleen cells, for both sets of stains, allows for consistent discrimination between $CD45^{high}/CD45^{dim}$ cells. Adding 5 μl of diluted spleen cells is sufficient for generating spleen cell-spiked brain control samples.

19. Unless washing with Perm/Wash™ Buffer (*see* **Note 24**), spins can be shortened to 6 min, if pressed for time.

20. Diluting to 0.5 μg/ml for anti-CD45 PE and to 1 μg/ml for anti-CD11b (FITC or PerCP-Cy™5.5) is typically sufficient to stain the number of cells described in this protocol (*see* **Note 9**). However, the concentration of each antibody should be titrated by the investigator to obtain optimal labeling.

21. Note that only 5 μl of Annexin V or 5 μl of 7-AAD should be added for FL3 and FL4 negative controls (Table 2), respectively. Stagger the addition of Annexin V/7-AAD if many samples are to be run.

22. It is possible to include an additional antibody, for example, a FITC-labeled activation marker, to determine whether its expression is related to cellular proliferation. We have successfully detected cellular expression of GFP on FL1 in combination with PE, PerCP-Cy5.5, and APC [8].
23. It is possible to pause the procedure here. Wash the cells with staining buffer, spin at 4 °C at 400 × *g* for 10 min, decant, and vortex. Cover with foil and store at 4 °C. Immediately prior to resuming the procedure, vortex the cells and add 1–2 ml of Perm/Wash™ Buffer for 1 h before continuing with **step 9**.
24. Use care when decanting after rinsing with Perm/Wash™ Buffer; the pellets are looser than when washed with staining buffer.
25. We usually center the microglial population at approximately 250 on the FSC and SSC scales, and set the FSC threshold to 80. It may help to draw a SSC versus CD11b gate and then to "back gate" to visualize CD11b$^+$ cells on FSC/SSC plots, since debris and dead cells have not been removed from the cell suspension. Note that microglial FSC and SSC become increased in the inflamed CNS [10, 27, 28].
26. A large number of total events should be collected, especially for analysis of rare events [10], since some events are debris, not actual cells.
27. Other cell types can be analyzed using this method by staining with different antibodies, including T cells [10, 19, 21, 31, 32, 34], Gr1$^+$ neutrophils [20, 21], CD11c$^+$ cells [19], and astrocytes [10].
28. It can help to visualize the brain/spleen plot after only gating on SSC versus CD45 and FSC versus CD45, since the CD11b$^-$ lymphocyte population in the spleen provides a very consistent level of CD45 expression, which is useful for discriminating CD45^{high}/CD45^{dim} cells on CD45/CD11b plots across multiple experiments (Fig. 1).
29. Alternatively, 7-AAD$^+$ cells can be gated directly and excluded from further analysis.
30. By setting quadrants based on the brain/spleen mix (Table 2), it is possible to calculate separately for macrophages and microglia (Fig. 2).

Acknowledgements

This work was supported by The Augustinus Foundation; Aase og Ejnar Danielsens Fond; The Carlsberg Foundation, Grosserer M. Brogaard og Hustrus Mindefond, Odense; Katrine og Vigo Skovgaards Fond; Fonden til Lægevidenskabens Fremme;

The Lundbeck Foundation; The Novo Nordisk Foundation; Overlægerådets Legatudvalg; the Danish Alzheimer's Society; and the Danish Medical Research Council.

References

1. Wirenfeldt M, Babcock AA, Vinters HV (2011) Microglia—insights into immune system structure, function, and reactivity in the central nervous system. Histol Histopathol 26(4):519–530
2. Hanisch UK, Kettenmann H (2007) Microglia: active sensor and versatile effector cells in the normal and pathologic brain. Nat Neurosci 10(11):1387–1394. doi:10.1038/nn1997
3. Davalos D, Grutzendler J, Yang G, Kim JV, Zuo Y, Jung S, Littman DR, Dustin ML, Gan WB (2005) ATP mediates rapid microglial response to local brain injury in vivo. Nat Neurosci 8(6):752–758. doi:10.1038/nn1472
4. Nimmerjahn A, Kirchhoff F, Helmchen F (2005) Resting microglial cells are highly dynamic surveillants of brain parenchyma in vivo. Science 308(5726):1314–1318. doi:10.1126/science.1110647
5. Ladeby R, Wirenfeldt M, Garcia-Ovejero D, Fenger C, Dissing-Olesen L, Dalmau I, Finsen B (2005) Microglial cell population dynamics in the injured adult central nervous system. Brain Res Brain Res Rev 48(2):196–206
6. Graeber MB, Tetzlaff W, Streit WJ, Kreutzberg GW (1988) Microglial cells but not astrocytes undergo mitosis following rat facial nerve axotomy. Neurosci Lett 85(3):317–321
7. Streit WJ, Kreutzberg GW (1988) Response of endogenous glial cells to motor neuron degeneration induced by toxic ricin. J Comp Neurol 268(2):248–263. doi:10.1002/cne.902680209
8. Wirenfeldt M, Dissing-Olesen L, Babcock AA, Nielsen M, Meldgaard M, Zimmer J, Azcoitia I, Leslie RG, Dagnaes-Hansen F, Finsen B (2007) Population control of resident and immigrant microglia by mitosis and apoptosis. Am J Pathol 171(2):617–631
9. Ladeby R, Wirenfeldt M, Dalmau I, Gregersen R, Garcia-Ovejero D, Babcock A, Owens T, Finsen B (2005) Proliferating resident microglia express the stem cell antigen CD34 in response to acute neural injury. Glia 50(2):121–131
10. Babcock AA, Wirenfeldt M, Holm T, Nielsen HH, Dissing-Olesen L, Toft-Hansen H, Millward JM, Landmann R, Rivest S, Finsen B, Owens T (2006) Toll-like receptor 2 signaling in response to brain injury: an innate bridge to neuroinflammation. J Neurosci 26(49):12826–12837. doi:10.1523/JNEUROSCI.4937-05.2006
11. Dissing-Olesen L, Ladeby R, Nielsen HH, Toft-Hansen H, Dalmau I, Finsen B (2007) Axonal lesion-induced microglial proliferation and microglial cluster formation in the mouse. Neuroscience 149(1):112–122. doi:10.1016/j.neuroscience.2007.06.037
12. Hailer NP, Grampp A, Nitsch R (1999) Proliferation of microglia and astrocytes in the dentate gyrus following entorhinal cortex lesion: a quantitative bromodeoxyuridine-labelling study. Eur J Neurosci 11(9):3359–3364
13. Jones LL, Banati RB, Graeber MB, Bonfanti L, Raivich G, Kreutzberg GW (1997) Population control of microglia: does apoptosis play a role? J Neurocytol 26(11):755–770
14. Lawson LJ, Perry VH, Gordon S (1992) Turnover of resident microglia in the normal adult mouse brain. Neuroscience 48(2):405–415
15. Gratzner HG, Leif RC (1981) An immunofluorescence method for monitoring DNA synthesis by flow cytometry. Cytometry 1(6):385–393. doi:10.1002/cyto.990010606
16. Gratzner HG (1982) Monoclonal antibody to 5-bromo- and 5-iododeoxyuridine: a new reagent for detection of DNA replication. Science 218(4571):474–475
17. Martin SJ, Reutelingsperger CP, McGahon AJ, Rader JA, van Schie RC, LaFace DM, Green DR (1995) Early redistribution of plasma membrane phosphatidylserine is a general feature of apoptosis regardless of the initiating stimulus: inhibition by overexpression of Bcl-2 and Abl. J Exp Med 182(5):1545–1556
18. Koopman G, Reutelingsperger CP, Kuijten GA, Keehnen RM, Pals ST, van Oers MH (1994) Annexin V for flow cytometric detection of phosphatidylserine expression on B cells undergoing apoptosis. Blood 84(5):1415–1420
19. Remington LT, Babcock AA, Zehntner SP, Owens T (2007) Microglial recruitment, activation, and proliferation in response to primary demyelination. Am J Pathol 170(5):1713–1724
20. Lambertsen KL, Clausen BH, Babcock AA, Gregersen R, Fenger C, Nielsen HH, Haugaard LS, Wirenfeldt M, Nielsen M, Dagnaes-Hansen F, Bluethmann H, Faergeman NJ, Meldgaard M, Deierborg T, Finsen B (2009) Microglia protect neurons against ischemia by synthesis of tumor necrosis

factor. J Neurosci 29(5):1319–1330. doi:10.1523/JNEUROSCI.5505-08.2009
21. Clausen BH, Lambertsen KL, Babcock AA, Holm TH, Dagnaes-Hansen F, Finsen B (2008) Interleukin-1beta and tumor necrosis factor-alpha are expressed by different subsets of microglia and macrophages after ischemic stroke in mice. J Neuroinflammation 5(1):46
22. Wirenfeldt M, Babcock AA, Ladeby R, Lambertsen KL, Dagnaes-Hansen F, Leslie RG, Owens T, Finsen B (2005) Reactive microgliosis engages distinct responses by microglial subpopulations after minor central nervous system injury. J Neurosci Res 82(4): 507–514
23. Sedgwick JD, Schwender S, Imrich H, Dorries R, Butcher GW, ter Meulen V (1991) Isolation and direct characterization of resident microglial cells from the normal and inflamed central nervous system. Proc Natl Acad Sci USA 88(16):7438–7442
24. Toft-Hansen H, Babcock AA, Millward JM, Owens T (2007) Downregulation of membrane type-matrix metalloproteinases in the inflamed or injured central nervous system. J Neuroinflammation 4:24. doi:10.1186/1742-2094-4-24
25. Brambilla R, Persaud T, Hu X, Karmally S, Shestopalov VI, Dvoriantchikova G, Ivanov D, Nathanson L, Barnum SR, Bethea JR (2009) Transgenic inhibition of astroglial NF-kappa B improves functional outcome in experimental autoimmune encephalomyelitis by suppressing chronic central nervous system inflammation. J Immunol 182(5):2628–2640. doi:10.4049/jimmunol.0802954
26. Renno T, Krakowski M, Piccirillo C, Lin JY, Owens T (1995) TNF-alpha expression by resident microglia and infiltrating leukocytes in the central nervous system of mice with experimental allergic encephalomyelitis. Regulation by Th1 cytokines. J Immunol 154(2):944–953
27. Sedgwick JD, Ford AL, Foulcher E, Airriess R (1998) Central nervous system microglial cell activation and proliferation follows direct interaction with tissue-infiltrating T cell blasts. J Immunol 160(11):5320–5330
28. Cao L, Palmer CD, Malon JT, De Leo JA (2009) Critical role of microglial CD40 in the maintenance of mechanical hypersensitivity in a murine model of neuropathic pain. Eur J Immunol 39(12):3562–3569. doi:10.1002/eji.200939657
29. Tough DF, Sprent J (1996) Viruses and T cell turnover: evidence for bystander proliferation. Immunol Rev 150:129–142
30. Nielsen HH, Toft-Hansen H, Lambertsen KL, Owens T, Finsen B (2011) Stimulation of adult oligodendrogenesis by myelin-specific T cells. Am J Pathol 179(4):2028–2041. doi:10.1016/j.ajpath.2011.06.006
31. Babcock AA, Toft-Hansen H, Owens T (2008) Signaling through MyD88 regulates leukocyte recruitment after brain injury. J Immunol 181(9):6481–6490
32. Babcock AA, Kuziel WA, Rivest S, Owens T (2003) Chemokine expression by glial cells directs leukocytes to sites of axonal injury in the CNS. J Neurosci 23(21):7922–7930
33. Toft-Hansen H, Fuchtbauer L, Owens T (2011) Inhibition of reactive astrocytosis in established experimental autoimmune encephalomyelitis favors infiltration by myeloid cells over T cells and enhances severity of disease. Glia 59(1):166–176. doi:10.1002/glia.21088
34. Khorooshi R, Babcock AA, Owens T (2008) NF-kappaB-driven STAT2 and CCL2 expression in astrocytes in response to brain injury. J Immunol 181(10):7284–7291

Chapter 16

Fluorescence Imaging of Intracellular Ca^{2+}, Na^{+}, and H^{+} in Cultured Microglia

Tom Schilling and Claudia Eder

Abstract

The behavior of microglial cells involves the activity of a variety of ion channels and ion transporters, which are implicated in the regulation of ion concentrations, membrane potential, and cell volume of microglia. Fluorescence imaging has been proven to be an elegant method to study ion concentration changes in intact microglial cells under physiological and pathophysiological conditions. The development of highly specific ion indicators has made it possible to detect changes in intracellular Ca^{2+}, Na^{+}, and H^{+} concentrations of microglial cells as a result of ion channel or ion transporter activity. Fluorescence signals of isolated dye-loaded microglial cells can be detected via a CCD camera equipped to a conventional microscope. This chapter summarizes protocols of loading of microglial cells with small-molecule ion indicators as well as protocols optimal for measurement and analysis of intracellular Ca^{2+}, Na^{+}, and H^{+} concentrations in microglia in vitro.

Key words Fluorescence imaging, CCD camera, Ion channels, Ion transporters, Microglia, Monochromator, Fura-2, SBFI, BCECF

1 Introduction

Fluorescence imaging provides a powerful method to investigate changes in the intracellular milieu of intact microglial cells. Due to the development of specific fluorescent ion indicators, it is now possible to precisely measure concentrations of intracellular ion concentrations and their changes in response to microglial stimulation. Thus, in addition to the patch-clamp technique, fluorescence imaging provides a useful tool to investigate ion channel activity in microglial cells. Furthermore, in fluorescence imaging experiments, the activity of electroneutral ion transporters can be assessed, which is impossible using the patch-clamp technique. For example, fluorescence imaging experiments on microglia can be used to determine the resting intracellular ion concentration of a microglial cell and to provide detailed spatiotemporal information about ion fluxes through microglial ion channels and ion transporters and

Bertrand Joseph and José Luis Venero (eds.), *Microglia: Methods and Protocols*, Methods in Molecular Biology, vol. 1041, DOI 10.1007/978-1-62703-520-0_16, © Springer Science+Business Media New York 2013

exchangers, about Ca^{2+} release from internal stores, and about H^+/pH changes in phagolysosomes and others.

For ion imaging, small-molecule ion indicators or genetically encoded indicators can be used. Ratiometric indicators are of great advantage as they allow accurate estimation of intracellular ion concentrations, since uneven dye loading, dye leakage, photobleaching, and changes in cell volume can be corrected. In general, ion indicators shift their absorption or emission spectrum or change their fluorescence intensity after binding to a specific ion. The resulting fluorescence can be detected by a camera, which is equipped to a microscope and connected with a computer, while fluorescence intensities of computer-generated images can be analyzed subsequently. Fluorescence imaging can be performed using conventional microscopy or confocal microscopy. Conventional microscopy can be used to detect ion changes in single cells or cell monolayers, whereas confocal microscopy allows imaging of cells in tissue as it provides an optical sectioning effect to reject out-of-focus fluorescence. In this chapter, we will introduce equipment and protocols required for fluorescence imaging of Ca^{2+}, Na^+, and H^+ signals in cultured microglia using a CCD camera equipped to a conventional microscope, which is suitable for ion measurements in isolated microglial cells in vitro.

2 Materials

2.1 Equipment for Fluorescence Imaging Experiments

Resolution of experimental data acquired with fluorescence imaging techniques is highly dependent on the quality of the equipment used. Essential parts are:

1. Microscope with an objective of at least 40× magnification.
2. Light source (laser, monochromator, mercury burner, or xenon lamp).
3. Filter sets (*see* Subheading 2.3).
4. CCD camera.
5. Acquisition and analysis software.
6. Vibration isolation table and cage.
7. Recording chamber for microscope stage.
8. Superfusion system.

For analysis of single cells, inverted microscopes are more valuable than upright microscopes, which are used for fluorescence imaging experiments on microglial cells in situ. The microscope should be equipped with an objective lens of at least 40× magnification (*see* **Note 1**). The microscope stage requires a recording chamber connected to a superfusion system for rapid and efficient solution exchange. For excitation purposes, a laser, a monochromator (*see* **Note 2**), or a fluorescent light source, such as a mercury or a xenon burner, can be used. The fluorescent probe is usually excited by light, generated according to a protocol of an acquisition software. Pictures of the emitted light signals are taken by an imaging software-driven CCD camera. This camera should be chosen according to emission spectra of the dyes; most cameras for fluorescence imaging systems are able to detect all commercially available dyes. For investigation of microglial ion channels and transporter activity, spatial resolution of the camera is more important than temporal resolution. A vibration isolation table, on which the microscope is mounted, minimizes mechanically induced artifacts during the recording. A cage increases signal to background ratio and minimizes optical artifacts by other light sources.

2.2 Cells

Cultured microglial (*see* **Note 3**) cells should be plated on sterile glass cover slips. It is important to test in advance, whether these cover slips are nonfluorescent for the excitation wavelength of the fluorescent dye used. Cell density for glass cover slips, fitting into a 24-well plate, should not exceed 100,000 cells per well to have a sufficient amount of nonfluorescent area, i.e., an area free of cells, for background subtraction.

2.3 Filter Sets

2.3.1 Measurement of Intracellular Calcium Concentrations

To detect changes of intracellular Ca^{2+} concentrations of microglial cells, Fura-2-acetoxymethyl ester (Fura-2AM) [1,2], Indo-1 [3], and Fluo-3 [4] have been used. In this chapter the application of Fura-2AM (*see* **Note 4**) will be described. With a K_d value of 224 nM, Fura-2AM is ideal to measure changes near the resting calcium concentration of 100 nM (*see* **Note 5**).

If no monochromator is available, the microscope should be equipped with band-pass filters of 340 nm and 380 nm wavelengths. To detect volume changes using Fura-2AM without a monochromator, an additional band-pass filter of 360 nm is required. Additionally, a 400 nm dichroic mirror and a 420 nm longpass emission filter (*see* **Note 6**) should be installed.

2.3.2 Measurement of Intracellular Sodium Concentrations

Changes in intracellular sodium concentration of microglial cells have been described using sodium-binding benzofuran-isophthalate acetoxymethyl ester (SBFI-AM) [5]. SBFI-AM (*see* **Note 7**) is a ratiometric dye with a K_d value of 11.3 mM. No additional filters are needed, if the microscope is equipped for calcium measurements with Fura-2AM, i.e., either a monochromator or 340 nm and 380 nm band-pass filters as well as a 400 nm dichroic mirror and a 420 nm longpass emission filter should be installed.

2.3.3 Intracellular pH Measurements

In microglial cells, intracellular pH (pH_i) measurements have been performed with 2′,7′-bis-(2-carboxyethyl)-5-(and-6)-carboxyfluorescein, acetoxymethyl ester (BCECF-AM) [6]. BCECF-AM has a pK_a of 6.98 (*see* **Note 8**) and is therefore a good choice to detect changes in pH_i. Like Fura-2AM and SBFI-AM, it has the advantages of ratiometric dyes and is cell membrane permeable. For dye excitation, the microscope should be equipped either with a 440 nm and a 490 nm band-pass filter or a monochromator. In addition, a 505 nm dichroic mirror and a 530 nm barrier filter should be installed.

2.4 Solutions

2.4.1 Measurement of Intracellular Calcium Concentrations

Experiments are performed using solution E_1 (*see* **Note 9**) and calcium-free solution E_2.

- Solution E_1: 130 mM NaCl, 5 mM KCl, 10 mM 4-(2-hydroxyethyl)piperazine-1-ethanesulfonic acid N-(2-hydroxyethyl)piperazine-N′-(2-ethanesulfonic acid) (HEPES), 10 mM D-glucose, 2 mM $CaCl_2$, and 1 mM $MgCl_2$ (pH = 7.4; adjusted with NaOH).
- Calcium-free solution E_2: 130 mM NaCl, 5 mM KCl, 10 mM HEPES, 10 mM D-glucose, 1 mM ethylene glycol-bis (2-aminoethylether)-N,N,N′,N′-tetraacetic acid (EGTA), and 4 mM $MgCl_2$ (pH = 7.4; adjusted with NaOH).
- To investigate intracellular calcium stores, prepare a 10 mM stock solution of the calcium ATPase inhibitor thapsigargin (*see* **Note 10**) in DMSO.
- To calibrate Fura-2AM fluorescence signals, calibration solutions E_3 and E_4 should be prepared.
- Calibration solution E_3: 130 mM NaCl, 5 mM KCl, 10 mM HEPES, 10 mM D-glucose, 2 mM EGTA, and 5 mM $MgCl_2$ (pH = 7.4; adjusted with NaOH).
- Calibration solution E_4: 130 mM NaCl, 5 mM KCl, 10 mM HEPES, 10 mM $CaCl_2$, and 5 mM $MgCl_2$ (pH = 7.4; adjusted with NaOH).

Reconstitute ionomycin and thapsigargin as a 10 mM stock solution in DMSO.

2.4.2 Measurement of Intracellular Sodium Concentrations

Experiments are performed using solution E_1. Prepare additionally sodium-free solution E_5.

- Solution E_1: 130 mM NaCl, 5 mM KCl, 10 mM HEPES, 10 mM D-glucose, 2 mM $CaCl_2$, and 1 mM $MgCl_2$ (pH = 7.4; adjusted with NaOH).
- Sodium-free solution E_5: 130 mM N-methyl-D-glucamine (NMG)-Cl, 5 mM KCl, 10 mM HEPES, 10 mM D-glucose, 2 mM $CaCl_2$, and 1 mM $MgCl_2$ (pH = 7.4; adjusted with NMGOH).

For calibration of SBFI-AM signals, stock solutions of gramicidin (10 mM in ethanol), monensin (10 mM in ethanol), and ouabain (20 mM in H_2O) should be available (*see* **Note 11**).

2.4.3 Intracellular pH Measurements

Solution E_1 (*see* **Note 12**) is used for experiments. Experimental data have to be calibrated after each experiment using solutions E_6, E_7, and E_8.

- Solution E_1: 130 mM NaCl, 5 mM KCl, 10 mM HEPES, 10 mM D-glucose, 2 mM $CaCl_2$, and 1 mM $MgCl_2$ (pH = 7.4; adjusted with NaOH).
- Solution E_6: 130 mM KCl, 10 mM HEPES, 10 mM D-glucose, 2 mM $CaCl_2$, and 1 mM $MgCl_2$ (pH = 7.6; adjusted with KOH).
- Solution E_7: 130 mM KCl, 10 mM HEPES, 10 mM D-glucose, 2 mM $CaCl_2$, and 1 mM $MgCl_2$ (pH = 7.0; adjusted with KOH).
- Solution E_8: 130 mM KCl, 10 mM 2-(N-morpholino)ethanesulfonic acid hydrate, 4-morpholineethanesulfonic acid (MES), 10 mM D-glucose, 2 mM $CaCl_2$, and 1 mM $MgCl_2$ (pH = 6.4; adjusted with KOH).

Furthermore, dissolve nigericin in ethanol as a 10 mM stock solution.

2.5 Data Analysis

Data can be analyzed with most versions of commercially distributed fluorescence imaging software. A second software license is advisable to be able to perform experiments while analyzing previous experiments on another computer at the same time. As an open-source alternative, the NIH offers the platform-independent program ImageJ (*see* **Note 13**), which is able to perform most image processing tasks and can be accustomed with plug-ins to one's needs.

3 Methods

3.1 Ca^{2+} Imaging

3.1.1 Detection of Intracellular Ca^{2+} Concentration

1. Prepare a 3 mM stock solution (*see* **Note 14**) of Fura-2 AM in DMSO (*see* **Note 15**).
2. Dilute this stock solution in 2 ml solution E_1 to a final concentration of 3 μM Fura-2AM. Do not expose solution to light.
3. Place one glass cover slip with microglial cells in Fura-2AM solution for 30 min at room temperature.
4. Wash microglial cells on glass cover slip in solution E_1 for 30 min at room temperature.
5. Fill recording chamber with solution E_1. Transfer glass cover slip with dye-loaded microglial cells into recording chamber of microscope.
6. Take a bright-field picture of a viewing field with a sufficient number of microglial cells (*see* **Note 16**).
7. Take a sample picture of the same cells with excitation light at 380 nm. Increase exposure time until no further increase in

emitted fluorescence of microglial cells can be achieved. Adjust the protocol of the imaging software for this exposure time.

8. Set the protocol of the imaging software to take one picture with excitation at 340 nm and one at 380 nm every 20 s (*see* **Note 17**). To monitor the experiment, calculate a ratiometric picture in the protocol editor by dividing the emission signals at 340 and 380 nm.
9. Start the protocol. Wait for a baseline level of Ca^{2+} signals for at least 15 data points.
10. Superfuse cells with solution E_1 as a control experiment. No change in Ca^{2+} signals should be observed (*see* **Note 18**).
11. Superfuse cells with activators dissolved in solution E_1. Observe changes in ratiometric calcium signals (*see* **Note 19**). Ca^{2+} signals of microglial cells should reach baseline levels in at least ten data points before using another stimulus.

3.1.2 Detection of Ca^{2+} Influx from Extracellular Space

1. Follow **steps 1–10** of previous Subheading.
2. Superfuse cells with solution E_1 containing the desired activator and observe increases in intracellular calcium concentrations (*see* **Note 20**). Record at least ten data points after calcium signals reached baseline levels before proceeding with the next steps.
3. Superfuse cells with calcium-free solution E_2. A small decrease of fluorescence intensities should be observed. Wait for a stable baseline of Ca^{2+} signals under these conditions (*see* **Note 21**).
4. Stimulate microglial cells with E_2 containing the desired activator. If the fluorescence ratio signal does not change under these conditions, the activator induces calcium influx from the extracellular space. If a Ca^{2+} signal is still detected (*see* **Note 22**), the activator elicits intracellular calcium increases from intracellular Ca^{2+} stores (Subheading 3.1.3).
5. Superfuse cells with solution E_1. A moderate increase of the ratiometric signal (*see* **Note 23**) should be observed.
6. Repeat **step 2** to confirm that there is no inherent decline of Ca^{2+} signals over time.

3.1.3 Ca^{2+} Release from Intracellular Stores

1. Follow **steps 1–9** of Subheading 3.1.1. Change the protocol to take a fluorescence picture every 5 s.
2. Superfuse microglial cells with solution E_2. Observe a decrease in the ratiometric signal.
3. Exchange solution with E_2 containing 500 nM thapsigargin (*see* **Note 24**). An increase in fluorescence intensity signals should be detected, indicating depletion of intracellular Ca^{2+} stores. Wait for a stable baseline after this transient rise in intracellular calcium concentrations.

4. Superfuse microglial cells with the desired activator dissolved in E_2. No change of fluorescence signals under these conditions is a sign that the activator leads to Ca^{2+} release from intracellular calcium stores.

3.1.4 Calibration of Intracellular Ca^{2+} Signals

1. The system should always be recalibrated, when any part of the optical pathway (i.e., monochromator, filter sets, objective lenses, camera, recording chamber, glass cover slips) has been changed (*see* **Note 25**).
2. Follow **steps 1–9** in Subheading 3.1.1.
3. Superfuse cells with solution E_3 containing 10 μM ionomycin and 1 μM thapsigargin. Wait for a stable ratiometric fluorescence signal.
4. Superfuse cells with solution E_4 containing 10 μM ionomycin. Wait until the intracellular fluorescence signals reach steady state.
5. Repeat this calibration procedure at least two more times with new cells on glass cover slips.

3.1.5 Data Analysis

1. In a first step, the calibration experiments should be analyzed.
2. Open the bright-field picture with your image analysis software.
3. Place several regions of interest (ROI) in background areas, in which no cells are visible.
4. Determine the mean background fluorescence intensity in the 340 nm fluorescence picture. Subtract this value from the stack of fluorescence pictures excited at 340 nm.
5. Repeat the procedure with fluorescence pictures excited at 380 nm.
6. Place ROIs in the bright-field picture in every visible cell.
7. Determine the fluorescence intensity F for each cell at 340 nm and 380 nm excitation when exposed to E_3 ($F340_{min}$, $F380_{min}$) and E_4 ($F340_{max}$, $F380_{max}$).
8. Calculate for each cell $R_{min} = F340_{min}/F380_{min}$, $R_{max} = F340_{max}/F380_{max}$, and $\beta = F380_{min}/F380_{max}$.
9. Calculate the mean values for R_{min}, R_{max}, and β from all experiments.
10. To determine changes in the intracellular calcium concentration, repeat **steps 2–5** for an experiment.
11. In the bright-field picture, put a ROI into each cell or into identified subcellular structures.
12. Measure fluorescence intensities F340 and F380 for all ROIs over time in background-corrected pictures (*see* **Note 26**). Calculate for each ROI the ratiometric fluorescence signal $R = F340/F380$ over time.

13. Calculate changes in intracellular calcium concentrations $[Ca^{2+}]_i$ for each ROI according to Eq. (1) [7]:

$$[Ca^{2+}]_i = K_d\beta(R - R_{min})/(R_{max} - R) \quad (1)$$

Use values for R_{min}, R_{max}, and β as derived in **step 9** and a K_d of 224 nM for Fura-2AM.

14. Measure the increase in intracellular calcium concentration for each cell and analyze its dependence on extracellular calcium and on release of calcium from intracellular stores. The following parameters can be used to characterize calcium signals: maximal increase in calcium concentration, time to increase of calcium concentration, time to maximal calcium concentration, and decay time of fluorescence signals. If the calcium concentration declines to baseline levels, the area under the curve can be measured.

3.2 Na^+ Imaging

3.2.1 Detection of Intracellular Na^+ Changes

1. Prepare a 2 mM stock solution (*see* **Note 27**) of SBFI-AM in equal volumes of DMSO and pluronic acid.
2. Dilute SBFI-AM stock solution in 2 ml solution E_1 to a final concentration of 10 μM. Prevent exposure to bright light.
3. Place microglial cells on glass cover slip in SBFI-AM solution for 45–60 min at room temperature.
4. Wash dye-loaded microglial cells in solution E_1 for 30 min at room temperature.
5. Mount glass cover slip with dye-loaded microglial cells in recording chamber filled with solution E_1.
6. Check fluorescence intensity of SBFI-AM in microglial cells by exciting the intracellular dye at 380 nm (*see* **Note 28**).
7. Choose a viewing field with several microglial cells. Take a bright-field picture (*see* **Note 29**).
8. Take a sample picture of the same cells excited at 380 nm. Increase the exposure time to maximize the emitted fluorescence signal. Use this exposure time in a protocol of the imaging software.
9. Adjust the protocol of the imaging software to take every 20 s a picture at 340 nm and at 380 nm. To observe the experiment, create in the protocol editor a picture that divides the 340 nm signal by the 380 nm signal.
10. Start the protocol. Wait for a stable baseline level of the ratiometric signal.
11. Check the superfusion system by exchanging solution E_1 in the recording chamber. Ratiometric Na^+ signals should not alter.
12. Superfuse cells with activators dissolved in solution E_1. Monitor changes in ratiometric SBFI-AM signals of microglial cells.

Ratiometric signals should reach baseline levels in at least ten data points before using another stimulus.

13. Repeat experiments using solution E_5 to ensure that changes in SBFI fluorescence are dependent on extracellular sodium (*see* **Note 30**).

3.2.2 Calibration of SBFI Fluorescence Signals

1. Whenever an optical element like objectives or the type of glass cover slips is changed, the system should be recalibrated.
2. Follow **steps 1–10** in Subheading 3.2.1.
3. Superfuse cells with solution E_5 containing 5 μM gramicidin, 10 μM ouabain, and 10 μM monensin. Wait until a steady state of the intracellular fluorescence signal has been achieved.
4. Exchange the solution in the recording chamber with solution E_1 containing 5 μM gramicidin, 10 μM ouabain, and 10 μM monensin. Wait for a stable ratiometric fluorescence signal.
5. Repeat this calibration at least two more times with new cells.

3.2.3 Data Analysis

Adapt the description from Subheading 3.1.5. To calculate intracellular sodium concentrations, a K_d value of 11.3 mM for SBFI-AM can be assumed (*see* **Note 31**).

3.3 Intracellular pH Imaging

3.3.1 Measurement of pH_i Changes

1. Prepare a 2 mM stock solution (*see* **Note 32**) of BCECF-AM with DMSO as the solvent. Do not expose BCECF-AM solution to bright light.
2. Dilute this stock solution in 2 ml E_1 to achieve a 2 μM BCECF-AM solution.
3. Transfer a glass cover slip with microglial cells into this BCECF-AM solution.
4. After 10 min, wash dye-loaded microglial cells in solution E_1 for another 10 min at room temperature.
5. Fill recording chamber of the microscope with solution E_1 (*see* **Note 33**). Mount glass cover slip with BCECF-AM-loaded microglial cells in recording chamber.
6. Take a bright-field picture of a viewing field. Take care that a sufficient number of microglial cells are visible (*see* **Note 34**).
7. Take a sample picture of the same cells with excitation light at 490 nm. Change exposure time to a value, where the emitted fluorescence intensity has a half-maximal value. Adjust the protocol of the imaging software for this exposure time.
8. Set the protocol of the imaging software to take one picture with excitation at 490 nm and one at 440 nm every 20 s. To monitor the experiment, calculate a ratiometric picture in the protocol editor by dividing the emission signals at 490 and 440 nm excitation wavelengths.

9. Start the protocol. Wait for a baseline level of pH_i signals for at least 15 data points.
10. Superfuse cells with solution E_1 as a control experiment. No change in pH_i signals should be observed (*see* **Note 35**).
11. Superfuse cells with activators dissolved in solution E_1. Observe pH_i changes in microglial cells (*see* **Note 36**). Signals should reach steady state in at least ten data points before using another stimulus.
12. Calibrate BCECF-AM fluorescence signals (*see* **Note 37**) at the end of each experiment. Superfuse cells with solution E_1 containing 10 μM nigericin. Wait for a stable fluorescence signal in at least ten data points.
13. Superfuse cells subsequently with extracellular solutions E_6, E_7, and E_8 containing 10 μM nigericin. After each solution, a stable ratiometric signal should be observed.

3.3.2 Data Analysis

1. Open the bright-field picture with your image analysis software.
2. Place several regions of interest (ROI) in background areas, in which no cells are visible.
3. Determine the mean background fluorescence intensity in the 490 nm fluorescence picture. Subtract this value from the stack of fluorescence pictures excited at 490 nm.
4. Repeat the procedure with fluorescence pictures excited at 440 nm.
5. Create ROIs in the bright-field picture in every visible cell or in identified subcellular compartments.
6. Determine the fluorescence intensity F for each cell at 490 nm and 440 nm excitation from background subtracted pictures (*see* **Note 38**).
7. Calculate for each cell the ratiometric fluorescence signal $R = F490/F440$ over time.
8. Determine the ratio for each cell in the presence of extracellular solution E_6 (pH 7.6), E_7 (pH 7.0), and E_8 (pH 6.4).
9. Perform a linear regression fit of these three data points between pH_i and fluorescence ratio for each cell.
10. Convert fluorescence ratios for each cell into pH_i values according to this linear regression curve.

3.4 Determination of Cell Volume Changes (See Note 39)

1. Follow **steps 1–6** in Subheading 3.1.1.
2. Take a sample picture of the same viewing field with excitation light at 360 nm. Change exposure time to optimize the fluorescence signals emitted by microglial cells. Adjust exposure time in protocol of the imaging software accordingly.
3. Start a protocol in imaging software for excitation at 360 nm every 20 s.

4. Superfuse microglial cells with solution E_1 containing desired activators or a solution of increased/decreased osmolarity. Observe changes in the emitted fluorescence signal (*see* **Note 40**).
5. For cell volume changes, analyze signals in background-corrected fluorescence images excited at 360 nm. An increase of the emitted fluorescence signal reflects an increase in Fura-2AM concentration and therefore cell shrinkage. Accordingly, a decrease of the Fura-2AM signal at 360 nm is detected during cell swelling.

4 Notes

1. Objective lenses with 40× magnification are usually a good compromise between cellular resolution and cell number per viewing field. 20× magnification might be sufficient for detection of cytoplasmic fluorescence changes. If subcellular compartments are investigated, 63× magnification needs to be considered. Inverted microscopes should preferentially be equipped with water immersion objectives.

   Fluorophores like Fura-2AM or SBFI-AM, which are excited in the UV range, need special objectives for optimal UV light transmission.
2. Monochromators have a variety of advantages. They can produce excitation light of a specific wavelength in a continuous spectrum. Monochromators do not need excitation filters in addition to dichroic and band-pass emission filters. Most importantly, monochromators are able to change excitation wavelengths fast and without mechanical impact on the observed specimen. This is especially important if simultaneously patch-clamp recordings are performed.
3. Preparation of primary microglial cultures is described in ref. [8]. Alternatively, a microglial cell line like BV-2 can be used.
4. Fura-2AM has several advantages. It is a ratiometric dye and can therefore be calibrated in situ (*see* Subheading 3.1.4). Additionally, photobleaching and dye leakage can easily be detected and corrected for. Fura-2AM also emits a bright signal, which allows to load microglial cells at a concentration, where the dye does not act as a significant calcium buffer. The major disadvantage of Fura-2AM is its excitation wavelength in the energy-rich ultraviolet range, which can lead to chemical alterations of cellular proteins.
5. If intracellular calcium concentrations rise during the experiment into the micromolar range, Fura-6F with a K_d value of 5.5 μM might be a better choice.
6. If simultaneously a second fluorophore with an emission spectrum similar to Fura-2AM is used, a 510 nm band-pass emission

filter might be necessary to separate both signals. Compared to the suggested longpass filter, this band-pass filter reduces the fluorescence intensity detected by the CCD camera.

7. Disadvantages of SBFI-AM are the excitation in the UV range and the high K_d value. The former disadvantage can be overcome by using the non-ratiometric dyes sodium green or CoroNa green; both dyes have an even higher K_d value than SBFI-AM.

8. This pK_a is not optimal for measurements of pH_i changes in acidic compartments like lysosomes or mitochondria. A targeted expression of pH_i-sensitive GFP variants might be indicated [9].

9. Alternatively, the following bicarbonate-buffered solution, fully oxygenated at 37 °C with carbogen (95 % O_2, 5 % CO_2), can be used: 129 mM NaCl, 3 mM KCl, 1.25 mM NaH_2PO_4, 10 mM D-glucose, 1.6 mM $CaCl_2$, 1.8 $MgSO_4$, and 21 mM $NaHCO_3$, 21. The calcium-free as well as the calibration solutions need to be adjusted accordingly. This bicarbonate-buffered solution should not be used to load fluorescent dyes into microglial cells.

10. Thapsigargin binds irreversibly to calcium ATPases; a reversible inhibitor is cyclopiazonic acid. A 10 mM stock solution of cyclopiazonic acid should be prepared in DMSO.

11. Stock solutions of gramicidin and ouabain should be ultrasonicated or vigorously vortexed before use.

12. To detect HCO_3^--dependent processes (e.g., Cl^-/HCO_3^- exchanger or Na^+/HCO_3^- co-transporter), use the following bicarbonate-buffered solution fully oxygenated at 37 °C with 95 % O_2 and 5 % CO_2: 129 mM NaCl, 3 mM KCl, 1.25 mM NaH_2PO_4, 10 mM D-glucose, 1.6 mM $CaCl_2$, 1.8 $MgSO_4$, and 21 mM $NaHCO_3$. To load BCECF-AM into microglial cells, HEPES-buffered solution E_1 should be used.

13. A free download of ImageJ and its plug-ins is available under http://rsbweb.nih.gov/ij/.

14. Stock solutions should be freshly prepared and kept in dark.

15. If an excessive amount of Fura-2AM crystals is visible during the experiments, the dye is not sufficiently dissolved. In this case, a mixture of 50 % DMSO and 50 % pluronic acid should be used to prepare the stock solution.

16. Take care that several parts of the picture are not covered by dye-loaded cells. These areas are essential for background subtraction (*see* Subheading 3.1.5).

17. Some Ca^{2+} signals are faster and need a higher sample rate. In most cases it is adequate to take a picture every 5 s.

18. If superfusion with extracellular solution alone evokes calcium signals, microglial cells may detect mechanical stress.

Superfusion should be rearranged to prevent this. Another common problem is a superfusion-induced movement of the whole glass cover slip. Additional weights made of inert material like platinum can be placed on the cover slip to prevent movement of the observed cells.

19. An increase of the intracellular Ca^{2+} concentration is characterized by a decrease of the fluorescence intensity at 380 nm and a simultaneous increase of the fluorescence intensity at 340 nm.
20. Some receptors show desensitization, which could influence the second calcium response. It should be tested with solution E_1 whether repetitive stimulation evokes stable responses.
21. A prolonged (>10 min) exposure to calcium-free solutions should be avoided because it can deplete intracellular calcium stores.
22. Stock solution of activators should be prepared in a calcium-free solution. Otherwise small calcium signals might be evoked in calcium-free solution E_2, which do not reflect calcium release from intracellular stores.
23. This probably reflects in microglial cells the activity of calcium release-activated calcium (CRAC) channels [10].
24. If a reversible inhibition of the calcium ATPase is necessary, 10 μM cyclopiazonic acid can be used.
25. This calibration method is not applicable for non-ratiometric dyes like Fluo-3.
26. Fluorescence signals of all analyzed cells should be checked thoroughly for dye leakage and photobleaching. This is indicated by a rapid and simultaneous decline in all fluorescence signals in one or more cells. As a consequence the values of fluorescence intensities are very low, causing an excessive noise signal in the ratiometric analysis.
27. Stock solutions should be freshly prepared and kept in dark.
28. In case of an insufficient fluorescence signal, ultrasonicate or vortex SBFI-AM solutions or increase dye-loading time.
29. Take care that several parts of the picture are not covered by dye-loaded cells. These areas are essential for background subtraction during data analysis (*see* Subheading 3.2.3).
30. Changes in SBFI-AM fluorescence signals do not necessarily reflect increases in intracellular Na^+ concentrations. Alterations in intracellular pH, in the intracellular potassium concentration or direct chemical alterations of SBFI-AM can influence the dye's emission signal.
31. The K_d value of SBFI-AM can change depending on factors like pH and potassium concentration. In reference [11] is described, how the K_d value of SBFI-AM can be determined in situ.

32. Stock solutions should be freshly prepared and kept in dark.
33. To detect HCO_3^--dependent processes (e.g., Cl^-/HCO_3^- exchanger or Na^+/HCO_3^- co-transporter), use the following bicarbonate-buffered solution fully oxygenated at 37 °C with 95 % O_2 and 5 % CO_2: 129 mM NaCl, 3 mM KCl, 1.25 mM NaH_2PO_4, 10 mM D-glucose, 1.6 mM $CaCl_2$, 1.8 $MgSO_4$, and 21 mM $NaHCO_3$. To load BCECF-AM into microglial cells, HEPES-buffered solution E_1 should be used.
34. Several parts of the picture should not be covered by dye-loaded cells. These areas are essential for background subtraction during data analysis (*see* Subheading 3.3.2).
35. If superfusion with extracellular solution evokes changes in ratiometric signals, microglial cells might be exposed to mechanical stress and superfusion should be rearranged.
36. A change of the intracellular pH concentration is characterized by an alteration of the fluorescence intensity at 490 nm with no apparent change of the fluorescence intensity at the isosbestic point at 440 nm.
37. A calibration method similar to the one described by [7] for Fura-2AM signals can be found for BCECF-AM in ref. [12].
38. Fluorescence signals of all analyzed cells should be checked thoroughly for dye leakage and photobleaching. This is indicated by a decline in both fluorescence signals in one or more cells. These cells should be excluded from further analysis.
39. Volume changes are detected as changes in Fura-2AM concentration, measured at 360 nm excitation wavelength. This wavelength is the isosbestic point for Fura-2 AM, at which the fluorescence intensity is independent of the intracellular calcium concentration.
40. This is not a ratiometric signal; the dye is only excited at 360 nm wavelength.

Acknowledgements

C.E. is supported by European Union FP7 collaborative grant TargetBraIn (No. 279017).

References

1. Walz W, Ilschner S, Ohlemeyer C et al (1993) Extracellular ATP activates a cation conductance and a K^+ conductance in cultured microglial cells from mouse brain. J Neurosci 13(10): 4403–4411
2. Schilling T, Repp H, Richter H et al (2002) Lysophospholipids induce membrane hyperpolarization in microglia by activation of IKCa1 Ca^{2+}-dependent K^+ channels. Neuroscience 109(4):827–835

3. Taupenot L, Ciesielskitreska J, Ulrich G et al (1996) Chromogranin a triggers a phenotypic transformation and the generation of nitric oxide in brain microglial cells. Neuroscience 72(2):377–389
4. Nolte C, Möller T, Walter T et al (1996) Complement 5a controls motility of murine microglial cells in vitro via activation of an inhibitory G-protein and the rearrangement of the actin cytoskeleton. Neuroscience 73(4): 1091–1107
5. Schilling T, Eder C (2004) A novel physiological mechanism of glycine-induced immunomodulation: Na^{+}-coupled amino acid transporter currents in cultured brain macrophages. J Physiol 559(Pt 1):35–40
6. Faff L, Ohlemeyer C, Kettenmann H (1996) Intracellular pH regulation in cultured microglial cells from mouse brain. J Neurosci Res 46(3):294–304
7. Grynkiewicz G, Poenie M, Tsien R (1985) A new generation of Ca^{2+} indicators with greatly improved fluorescence properties. J Biol Chem 260(6):3440–3450
8. Eder C, Klee R, Heinemann U (1997) Pharmacological properties of Ca^{2+}-activated K^{+} currents of ramified murine brain macrophages. Naunyn Schmiedebergs Arch Pharmacol 356(2):233–239
9. Pinton P, Rimessi A, Romagnoli A et al (2007) Biosensors for the detection of calcium and pH. Methods Cell Biol 80(06):297–325
10. Nörenberg W, Cordes A, Blöhbaum G et al (1997) Coexistence of purino- and pyrimidinoceptors on activated rat microglial cells. Br J Pharmacol 121(6):1087–1098
11. Diarra A, Sheldon C, Church J (2001) In situ calibration and $[H^{+}]$ sensitivity of the fluorescent Na^{+} indicator SBFI. Am J Physiol Cell Physiol 280(6):C1623–C1633
12. James-Kracke MR (1992) Quick and accurate method to convert BCECF fluorescence to pH_i: calibration in three different types of cell preparations. J Cell Physiol 151(3):596–603

Chapter 17

Patch Clamp Protocols to Study Ion Channel Activity in Microglia

Tom Schilling and Claudia Eder

Abstract

Microglia express a variety of ion channels, which can be distinguished based on their ion selectivity into K^+, H^+, Na^+, Ca^{2+}, nonselective cation, and Cl^- channels. With respect to their activation mode, voltage-, Ca^{2+}-, calcium release-, G protein-, swelling-, and stretch-activated ion channels have been described in microglia. The best method to study the activity of microglial ion channels is the patch clamp technique. The activity of microglial ion channels under physiological conditions is best explored using the perforated patch clamp technique, which allows recordings of membrane potential or ion currents, while the intracellular milieu of the cells remains intact. In whole-cell patch clamp recordings, application of specific voltage protocols with defined intra- and extracellular solutions allows precise identification of a certain ion channel type in microglia as well as the investigation of the channel's biophysical and pharmacological properties. This chapter summarizes patch clamp protocols optimal for recording and analysis of microglial ion channel activity in vitro and in situ.

Key words Microglia, Ion channel, Patch clamp technique, Whole-cell recordings, Perforated patch recordings, Voltage clamp, Current clamp, Voltage ramp, Voltage step

1 Introduction

Microglial cells express a variety of distinct ion channels, which play pivotal roles in regulating microglial functions, such as proliferation, migration, cytokine production, and reactive oxygen species generation (for reviews *see* refs. 1–4). Unlike many other cell types, microglial cells do not constitutively express all ion channel types. Expression of microglial ion channels strongly depends on the functional state of the cells and on the cells' microenvironment. Furthermore, differences in the ion channel expression pattern have been found between cultured microglial cells and microglia of brain tissue, as well as between different species. Since microglial ion channels provide potential therapeutic targets for neurological and psychiatric diseases [3], it is of particular interest to identify

Bertrand Joseph and José Luis Venero (eds.), *Microglia: Methods and Protocols*, Methods in Molecular Biology, vol. 1041, DOI 10.1007/978-1-62703-520-0_17, © Springer Science+Business Media New York 2013

ion channels expressed in microglia under certain physiological and pathophysiological conditions in the brain.

The patch clamp technique has been proven to be the best method to study ion channel activity in excitable and non-excitable cells, including microglia [5]. Here, we focus on protocols used for whole-cell and perforated patch clamp recordings, since these recording modes provide excellent approaches to study microglial ion channel activity under physiological conditions. Whole-cell recordings are used to assess ion channel activity of the entire cell membrane. Due to the precise definition of intracellular and extracellular solutions, it is possible to investigate certain ion channel types in isolation. The perforated patch clamp method has been developed, because intracellular second messenger systems can be dialyzed out during whole-cell recordings leading to current rundown. In perforated patch clamp recordings, pore-forming molecules, such as nystatin, amphotericin, or gramicidin, allow electrical access to the cell but prevent washout of larger intracellular constituents. Therefore, whole-cell ion currents or membrane potential changes can be measured, while the cell interior remains intact (*see* **Note 1**).

In this chapter, we list intracellular and extracellular solutions and summarize voltage protocols optimal for measurements of currents through known microglial K^+, Na^+, H^+, nonselective cation, Ca^{2+}, and Cl^- channels.

2 Materials

2.1 Equipment for Patch Clamp Experiments

The patch clamp rig should at least consist of the following components: vibration isolation table, microscope (inverted microscope for in vitro recordings; upright microscope for in situ or in vitro recordings) with an appropriate recording chamber, micromanipulators (for positioning of patch pipette and superfusion pipette), and patch clamp amplifier and corresponding software for data acquisition and analysis. For superfusion of cells with ion channel activators or inhibitors, the use of a multibarrel microperfusion pipette is recommended in addition to a system allowing complete exchange of bath solutions (*see* **Note 2**).

In addition, a puller is required to prepare patch pipettes prior to patch clamp experiments. Patch pipettes are made from borosilicate glass capillaries (*see* Subheading 3.2). AgCl-coated silver wires are used as patch and reference electrodes (*see* Subheading 3.3).

2.2 Cells

Use cultured primary microglial cells (*see* **Note 3**) or a microglia cell line (e.g., BV-2 cells) for in vitro studies. Prepare acute or cultured organotypic brain slices (*see* **Note 4**) to study microglial cells in situ.

2.3 Solutions

Prepare and use extracellular solution E_1 for storage of brain slices and for in situ recordings of microglia in brain slices:

- Extracellular solution E_1: 129 mM NaCl, 3 mM KCl, 1.8 mM $MgSO_4$, 21 mM $NaHCO_3$, 1.25 mM NaH_2PO_4, 1.6 mM $CaCl_2$, 10 mM D-glucose (oxygenated with 95 % O_2, 5 % CO_2; pH = 7.4) (*see* **Note 5**). For studies on the effects of polyvalent cations, such as La^{3+}, Gd^{3+}, Ba^{2+}, or Zn^{2+}, replace $MgSO_4$ by $MgCl_2$, and omit NaH_2PO_4 from this extracellular solution.

Prepare extracellular solutions E_2–E_6 for whole-cell recordings of microglial cells in vitro, and use intracellular solutions I_1–I_7 (*see* **Note 6**) for both in vitro and in situ recordings of microglia (*see* **Note 7**).

2.3.1 Solutions for Measurements of Voltage-Activated K^+ or Na^+ Currents

- Intracellular solution I_1: 120 mM KCl, 5 mM NaCl, 2 mM $MgCl_2$, 1 mM $CaCl_2$, 10 mM N-[2-hydroxyethyl]piperazine-N′-[2-ethanesulfonic acid] (HEPES), 11 mM ethylenebis(oxonitrilo)tetraacetate (EGTA) (pH = 7.3; adjusted with KOH).

To investigate voltage-activated K^+ or Na^+ channels, use extracellular solution E_1 for in situ recordings or extracellular solution E_2 for in vitro measurements:

- Extracellular solution E_2: 130 mM NaCl, 5 mM KCl, 2 mM $CaCl_2$, 1 mM $MgCl_2$, 10 mM HEPES, 10 mM D-glucose (pH = 7.4; adjusted with NaOH).

2.3.2 Solutions for Measurements of Ca^{2+}-Activated K^+ Currents

Ca^{2+}-activated K^+ currents can be measured using intracellular solution I_2 containing a low concentration of calcium buffer. Increases in the intracellular Ca^{2+} concentration should be elicited either by stimulating Ca^{2+} influx from the extracellular space or by inducing Ca^{2+} release from intracellular stores.

- Low Ca^{2+}-buffered intracellular solution I_2: 120 mM KCl, 2 mM $MgCl_2$, 10 mM HEPES, 0.1 mM 1,2-Bis(2-aminophenoxy)ethane-N,N,N′,N′-tetraacetic acid (BAPTA) (pH = 7.3; adjusted with KOH).

Alternatively, Ca^{2+}-activated K^+ currents can be measured using intracellular solutions at which the free intracellular Ca^{2+} concentration is clamped to values sufficient to directly activate Ca^{2+}-activated K^+ channels. Using intracellular solution I_3, an intracellular free Ca^{2+} concentration of 1 μM is obtained (*see* **Note 8**).

- High Ca^{2+}-containing intracellular solution I_3: 120 mM KCl, 2 mM $MgCl_2$, 5 mM BAPTA, 4.43 mM $CaCl_2$, 10 mM HEPES (pH = 7.3; adjusted with KOH).

Free Ca^{2+} concentrations of solutions can be calculated using the program WINMAXC (*see* **Note 9**).

Use extracellular solution E_1 for in situ measurements or extracellular solution E_2 for in vitro experiments.

2.3.3 Solutions for Measurements of G Protein-Activated K^+ Currents

To measure G protein-activated K^+ currents, use intracellular solution I_1 and add 50 μM GTPγS.

Use extracellular solution E_1 for in situ measurements or extracellular solution E_2 for in vitro experiments.

2.3.4 Solutions for Measurements of H^+ Currents

Intracellular and extracellular solutions are based on tetramethylammonium (TMA) hydroxide and methanesulfonic ($MeSO_3$) acid.

- Intracellular solution I_4: 80 mM $TMAMeSO_3$, 100 mM 2-(N-morpholino)ethanesulfonic acid (MES), 2 mM $MgCl_2$, 1 mM BAPTA (pH 6.0; adjusted with TMAOH). When the intracellular solution needs to be clamped to pH 7.0, use 100 mM HEPES instead of MES.
- Extracellular solution E_3: 80 mM $TMAMeSO_3$, 2 mM $CaCl_2$, 100 mM HEPES, 1 mM EGTA (pH = 7.5 or 7.0). Use 100 mM MES instead of HEPES for studies at pH = 5.0, 6.0, or 6.5 and use 100 mM tricine instead of HEPES for studies at pH = 8.0.

2.3.5 Solutions for Measurements of Nonselective Cation (TRP) Channel Currents

The extracellular and intracellular solution should be defined in a way that they differ in their reversal potentials for nonselective cation (TRP) and anion (Cl^-) currents. Use extracellular solution E_1 for in situ measurements or extracellular solution E_2 for in vitro experiments. Prepare intracellular solution I_5 (*see* **Note 10**).

- Intracellular solution I_5: 20 mM NaCl, 100 mM Na-gluconate, 2 mM $MgCl_2$, 1 mM $CaCl_2$, 10 mM HEPES, 11 mM EGTA, 2 mM Na_2ATP, 0.5 mM Na_2GTP (pH = 7.3; adjusted with NaOH).

Add 10 μM 4α-Phorbol 12,13-didecanoate (4α-PDD) to the extracellular solution, to activate TRPV4 channels. To activate TRPM2 channels, add 100 μM ADP-ribose to the intracellular solution. Increase the free Ca^{2+} concentration of the intracellular solution to 40 μM to measure TRPM4 channel activity. To activate TRPM7 channels, omit Mg^{2+} and ATP from the intracellular solution.

2.3.6 Solutions for Measurements of Calcium Release-Activated Ca^{2+} (CRAC) Channels

Prepare extracellular solution E_4 (*see* **Note 11**). Use E_4 with intracellular solution I_6 and add 10 μM inositol 1,4,5-trisphosphate ($InsP_3$) to intracellular solution I_6.

- Extracellular solution E_4: 130 mM NaCl, 10 mM $CaCl_2$, 1 mM $MgCl_2$, 10 mM HEPES, 10 mM D-glucose (pH = 7.4; adjusted with NaOH).

- Intracellular solution I_6: 120 mM NaCl, 2 mM $MgCl_2$, 1 mM $CaCl_2$, 10 mM HEPES, 11 mM EGTA, 2 mM Na_2ATP, 0.5 mM Na_2GTP (pH = 7.3; adjusted with NaOH).

2.3.7 Solutions for Measurements of Cl^- Currents

Solutions for Cl^- current measurements should differ in their reversal potentials for nonselective cation (TRP) and anion (Cl^-) currents. Use intracellular solution I_5 with extracellular solution E_1 for in situ measurements or extracellular solution E_2 for in vitro experiments (*see* **Note 10**). Add 100 μM $LaCl_3$ to the extracellular solution to block voltage-gated H^+ currents, if necessary.

To study swelling-activated Cl^- currents, prepare intracellular solution I_6 and use this with isoosmolar extracellular solution E_5 and hypoosmolar solution E_6 (*see* **Note 12**):

- Isoosmolar (300 mosmol/kg) solution E_5 containing 50 mM NaCl, 170 mM D-mannitol, 2 mM $CaCl_2$, 1 mM $MgCl_2$, 10 mM HEPES, 10 mM D-glucose (pH = 7.4; adjusted with NaOH).
- Hypoosmolar (200 mosmol/kg) solution E_6 containing 50 mM NaCl, 70 mM D-mannitol, 2 mM $CaCl_2$, 1 mM $MgCl_2$, 10 mM HEPES, 10 mM D-glucose (pH = 7.4; adjusted with NaOH).

2.3.8 Solutions for Perforated Patch Clamp Recordings

Use extracellular solution E_1 for in situ measurements or extracellular solution E_2 for in vitro experiments.

- Prepare solution I_7 containing 120 mM K_2SO_4, 16 mM KCl, 5 mM $MgSO_4$, 10 mM HEPES (pH = 7.3; adjusted with KOH).
- Prepare a stock solution of nystatin in DMSO (25 mg/ml), which should be stored in a freezer. Dilute this stock subsequently in solution I_7 to a final concentration of 150–250 μg/ml (*see* **Note 13**). Sonicate this solution for 10 min prior to its use for backfilling the patch pipette. The final concentration of nystatin used for perforated patch recordings needs to be adjusted carefully (*see* **Note 14**). Amphotericin (200 μg/ml) or gramicidin (5 μg/ml) can be used instead of nystatin (*see* **Note 15**).

3 Methods

3.1 Preparation of Cells for In Vitro or In Situ Patch Clamp Experiments

To study ion channel activity of isolated microglial cells in vitro, either primary cultured microglial cells (*see* **Note 3**) or a microglial cell line can be used. Depending on the recording chambers used, microglial cells should be plated either into plastic dishes or onto glass coverslips at least 1 day prior to patch clamp recordings. As microglial cells adhere well to glass or plastic surfaces, coating of recording chambers is not required.

To study ion channel activity of microglial cells in situ, use either acutely prepared or cultured organotypic brain slices (*see* **Note 4**). For investigations of microglial cells in brain tissue, use transgenic mice, in which microglial cells have been labeled with a genetically encoded fluorescent dye, e.g., Iba1/GFP-labeled microglia or CX3CR1/GFP-labeled microglia.

If animals with GFP-labeled microglia are not available, stain microglial cells in brain slices prior to experiments with an isolectin-B4-conjugated Alexa dye (e.g., Alexa 488-IB_4). For microglial staining, keep individual brain slices in a chamber with oxygenated solution E_1 containing 10 μg/ml Alexa-IB_4 (*see* **Note 16**). Prepare the final Alexa dye-containing solution from a stock solution of 1 mg/ml Alexa-IB_4, which should also be prepared in solution E_1. Prior to each patch clamp experiment, incubate one brain slice in the dark for 20 min in Alexa-IB_4-containing solution E_1 at room temperature. After the incubation period, transfer the brain slice to the recording chamber under the microscope and wash the slice with oxygenated solution E_1 for 20 min before starting patch clamp recordings (*see* **Note 17**).

3.2 Preparation of Patch Pipettes

Patch electrodes consist of a patch pipette filled with an intracellular solution that has contact to the patch clamp amplifier via an AgCl-coated silver wire. Patch pipettes are prepared from borosilicate glass capillaries, which are commercially available in varying diameters and lengths (*see* **Note 18**). The appropriate capillary diameter depends on the pipette holder of the patch clamp amplifier in use.

Usually pullers form patch pipettes in two or more steps. The first step defines the geometry of the tip (*see* **Note 19**). The last step determines the pipette resistance. Increasing the temperature in this step leads to a smaller diameter of the tip orifice, i.e., a higher pipette resistance.

Prepare patch electrodes of 3–5 MΩ (*see* **Note 20**) immediately before their use. Store patch electrodes in a box to protect tips from dust exposure.

Polish pipette tips for a higher seal resistance, if necessary. Care should be taken that this polishing procedure does not substantially increase the pipette resistance above 5 MΩ. Coat pipette tips with insulating polymers like Sylgard 184, if a decreased noise level is desired (*see* **Note 21**).

3.3 Preparation of Reference Electrodes

A reference electrode (also called bath electrode) should connect the ground of the patch clamp amplifier with the extracellular solution in the recording chamber. These electrodes are usually silver wires with a AgCl coating (*see* **Note 22**). They are commercially distributed in different varieties.

A better reference electrode, especially when solutions contain low chloride concentrations, is an agar bridge. They are commercially available but can also be prepared by heating a 3 M KCl solution containing 2–5 % agar. Fill a glass capillary (*see* **Note 23**) with this mixture and introduce a Ag/AgCl wire in one end of the glass tube. Fixate this wire with heat-shrinkable tubing. After 1 day remove excess agar outside the glass tube and discard tubes containing air bubbles. Store agar bridges in 3 M KCl solution between experiments.

3.4 Patch Clamp Recordings: Establishment of a Giga-Seal and of a Perforated Patch or Whole-Cell Configuration

It is recommended to perform in situ patch clamp recordings at a temperature of 34–36 °C (*see* **Note 24**), while in vitro patch clamp recordings can also be carried out at room temperature (20–23 °C).

1. Maintain cells in extracellular solution E_1 for in situ experiments or in extracellular solution E_2 for in vitro experiments.
2. Bring the reference Ag/AgCl electrode in contact with the extracellular solution in the recording chamber.
3. Fill a patch pipette with the appropriate intracellular solution. Attach the pipette to the pipette holder (*see* **Note 25**). Make sure that the electrode wire is in contact with the intracellular solution and no air bubbles are trapped in the patch pipette.
4. Choose a microglial cell from which recordings should be performed (*see* **Note 26**).
5. Apply a test pulse at an appropriate frequency (e.g., 10 mV for 10 ms every second) to visualize changes in pipette resistance, when approaching the cell. The pipette resistance should be 3–5 MΩ (*see* **Notes 20** and **27**).
6. Bring the patch pipette in close contact with the cell surface, while applying constantly a positive pressure on the pipette. The recording pipette is in close proximity of the cell membrane, when the pipette resistance increases by a few MΩ (*see* **Note 28**). Apply gentle suction to the patch pipette to increase the resistance between the patch pipette and the cell membrane (*see* **Note 29**) until it reaches values of more than 1 GΩ, i.e., a giga-seal (*see* **Note 30**).
7. After formation of the giga-seal, break the membrane within the pipette tip by fast application of additional suction to reach the whole-cell configuration.
8. For perforated patch clamp recordings, the patch pipette tip should be dipped briefly in solution I_7 (*see* **Note 31**) and then backfilled with solution I_7 to which the pore-forming agent, i.e., nystatin, amphotericin, or gramicidin, has been added. After formation of the giga-seal, small pores are formed in the membrane patch within the pipette. To visualize this pore-forming process, apply a test pulse similar to the one used during giga-seal formation. The pores lower the series resistance allowing voltage or current clamp recordings of the entire cell membrane. Optimally this pore-forming process lasts for about 10–30 min, and the series resistance reaches a

stable value of less than 20 MΩ. Make sure that the input resistance, i.e., the giga-seal, remains stable during the pore-forming period.

9. After establishing electrical access to the inside of the cell, series resistance and membrane capacitance (*see* **Note 32**) should be compensated for. Some amplifier models are able to automatically compensate for series resistance and membrane capacitance. The series resistance in the whole-cell configuration should not exceed two times the value of the pipette resistance (*see* **Note 33**), while the input resistance should still be in the GΩ range (*see* **Note 34**).
10. The superfusion pipette should be positioned very close (at a distance of a few μm) to the recorded cell to permit a rapid and complete exchange of solutions outside the cell (*see* **Note 35**).

3.5 Patch Clamp Recordings: Detection of Changes in Membrane Potential in the Current Clamp Mode

1. Maintain cells in extracellular solution E_1 for in situ experiments or in extracellular solution E_2 for in vitro experiments.
2. Switch the amplifier to current clamp mode, while no holding current should be applied. This allows immediate determination of the resting membrane potential of a cell.
3. To investigate changes in membrane potential during drug application, record membrane potential continuously before, during, and after application of certain drugs. All current clamp recordings should ideally be performed in the perforated patch configuration to allow recordings under physiological conditions from intact cells and to prevent washout of intracellular components.

3.6 Patch Clamp Recordings: Detection of Ion Currents in the Voltage Clamp Mode

1. Maintain cells in extracellular solution E_1 for in situ experiments or in extracellular solution E_2 for in vitro experiments.
2. Switch the amplifier to voltage clamp mode to measure ion currents in microglia.
3. Clamp cells to a holding potential, which is close to the resting membrane potential of the cell, does not activate the investigated ion channel and allows full activation of the ion channel of interest (*see* **Note 36**).
4. To elicit and visualize ion channel activity, apply either a voltage ramp or a voltage step protocol (*see* **Note 37**). Voltage ramp protocols allow determination of current activation threshold and observation of voltage-dependent behavior. Voltage-step protocols allow assessment of current kinetics, i.e., time-dependent activation and inactivation behavior of the currents at a certain voltage. When using voltage-step protocols, it is recommended to record a leak protocol after each measurement. For this, step from the holding potential to a potential, where no channel activity is detected, e.g., from −60 to −70 mV. The current measured in this protocol is the

leak current. To minimize noise, average ten traces of this leak current recording.

5. Ion substitution experiments can be performed to identify the activated channel. Omit the suspected charge carrier from the solutions and observe, if the current can still be elicited. Care should be taken that the substituted ion is not conducted by the ion channel under investigation (*see* **Note 38**).
6. Use tail current protocols to determine current reversal potentials (*see* **Note 39**). Tail current measurements are voltage-step protocols, consisting of two consecutive voltage steps. The first step is set to a fixed voltage value that fully activates the channel under investigation (*see* **Note 40**). The second step measures tail currents by varying the voltage around the expected equilibrium potential. The tail currents, i.e., the currents observed in this second voltage step, should switch from outward to inward currents or vice versa. The voltage, where no current flows, is the reversal (i.e., equilibrium) potential for this ion channel.
7. Apply known ion channel activators and inhibitors and observe their effect on the measured current. Usually a combination of different inhibitors and measurement of their K_d value is necessary to conclusively identify an ion channel type.

3.6.1 Patch Clamp Recordings: Detection of K⁺ Currents

1. Maintain cells in extracellular solution E_1 for in situ experiments or in extracellular solution E_2 for in vitro experiments.
2. Perform recordings in the perforated patch clamp or whole-cell mode. For whole-cell recordings, fill the patch pipette with intracellular solution I_1 for recordings of inward-rectifying or outward-rectifying voltage-activated K^+ currents, with intracellular solution I_2 or I_3 for recordings of Ca^{2+}-activated K^+ currents, or with intracellular solution I_1 containing 50 μM GTPγS for recordings of G protein-activated K^+ currents.
3. Clamp cells to a holding potential of −60 mV (*see* **Note 41**).
4. Apply a voltage ramp from −150 mV to +30 mV for 400 ms.
5. Apply 200 ms lasting voltage steps to potentials between −150 mV and +30 mV in 10 mV increments, while an interval of 20 s between voltage pulses should be allowed (*see* **Note 42**).
6. To block Kir2.1 inward-rectifying voltage-activated potassium channels (*see* **Note 43**), use 100 μM Ba^{2+}.
7. To identify Kv1.3 outward-rectifying voltage-activated K^+ channels (*see* **Note 44**), test the effect of 100 nM margatoxin or 50 nM agitoxin.
8. To identify KCa3.1 (*see* **Note 45**) Ca^{2+}-activated K^+ channels, test the effect of 100 nM charybdotoxin (*see* **Note 46**) and 10 μM TRAM-34 (*see* **Note 47**).

9. To identify KCa1.1 (*see* **Note 48**) Ca^{2+}-activated K^+ channels, use 10 μM paxillin or 100 nM iberiotoxin.
10. Apply 100 nM apamin or 5 nM tamapin, to identify KCa2.3 (*see* **Note 49**) Ca^{2+}-activated K^+ channels.
11. Add 2 mM GDP-βS to the intracellular solution to inhibit G protein-activated K^+ channels (*see* **Note 50**).
12. To identify ATP-sensitive K^+ channels (K_{ATP}), test the effect of 10 μM glibenclamide or 500 μM tolbutamide.

3.6.2 Patch Clamp Recordings: Detection of Na^+ Currents

1. Maintain cells in extracellular solution E_1 for in situ experiments or in extracellular solution E_2 for in vitro experiments.
2. Perform recordings in the perforated patch clamp or whole-cell mode. For whole-cell recordings, fill the patch pipette with intracellular solution I_1 (*see* **Note 51**).
3. Clamp cells to a holding potential of −80 mV.
4. Apply 50 ms lasting voltage steps to potentials between −80 mV and +30 mV in 10 mV increments at an interval of 10 s (*see* **Note 52**).
5. Test whether 1 μM tetrodotoxin (TTX) (*see* **Note 53**) induces Na^+ current inhibition.

3.6.3 Patch Clamp Recordings: Detection of H^+ Currents

1. Maintain cells in extracellular solution E_1 for in situ experiments or in extracellular solution E_3 for in vitro experiments.
2. Perform recordings in the perforated patch clamp or whole-cell mode. For whole-cell recordings, fill the patch pipette with intracellular solution I_4.
3. Clamp cells to a holding potential of −60 mV (*see* **Note 54**).
4. Apply voltage steps for a duration of 4 s in 20 mV increments from −60 mV to +80 mV at 60 s intervals (*see* **Notes 55** and **56**).
5. Change pH of intracellular and/or extracellular solutions and repeat the voltage-step protocol in order to test whether the H^+ current activation threshold changes as predicted (*see* **Note 57**).
6. Test whether 100 μM Zn^{2+} or 100 μM La^{3+} induces H^+ current inhibition.

3.6.4 Patch Clamp Recordings: Detection of Nonselective Cation (TRP) Channel Currents

1. Maintain cells in extracellular solution E_1 for in situ experiments or in extracellular solution E_2 for in vitro experiments.
2. Perform recordings in the perforated patch clamp or whole-cell mode. For whole-cell recordings, fill the patch pipette with intracellular solution I_5.
3. Clamp cells to a holding potential of 0 mV.
4. Apply voltage ramps from −90 mV to +90 mV for 400 ms every 20 s.

5. Alternatively, apply 200 ms lasting voltage steps to potentials between −90 mV and +90 mV in 10 mV increments at an interval of 10 s (*see* **Note 58**).
6. Test first the effects of 50 μM Gd^{3+} or 10 μM ruthenium red, which inhibit most TRP channels.
7. Test for TRPV1 channels by applying 10 μM capsazepine or 100 nM 5′-iodo-resiniferatoxin.
8. Test 50 μM 2-APB or 100 μM flufenamic acid, which inhibit most TRPC channels (*see* **Note 59**).
9. Test 10 μM clotrimazole (*see* **Note 60**) to inhibit TRPM2 channels.

3.6.5 Patch Clamp Recordings: Detection of Calcium Release-Activated Ca^{2+} (CRAC) Currents

1. Maintain cells in extracellular solution E_4.
2. Perform recordings in the whole-cell mode. Fill the patch pipette with intracellular solution I_6 to which 10 μM $InsP_3$ has been added.
3. Clamp cells to a holding potential of 0 mV.
4. Apply voltage ramps from −90 mV to +90 mV for 400 ms every 20 s (*see* **Note 61**).
5. Apply 200 ms lasting voltage steps to potentials between −90 mV and +90 mV in 10 mV increments at an interval of 10 s.
6. Test the inhibitory effect of 100 μM 2-APB or 10 μM Gd^{3+} on CRAC currents.

3.6.6 Patch Clamp Recordings: Detection of Cl^- Currents

1. Maintain cells in extracellular solution E_1 for in situ experiments or in extracellular solution E_2 for in vitro experiments.
2. Perform recordings in the perforated patch clamp or whole-cell mode. For whole-cell recordings, fill the patch pipette with intracellular solution I_5.
3. Clamp cells to a holding potential of −40 mV.
4. Apply voltage ramps from −90 mV to +90 mV for 400 ms every 10 s.
5. Alternatively, apply 200 ms lasting voltage steps to potentials between −90 mV and +90 mV in 10 mV increments at an interval of 10 s (*see* **Note 62**).
6. To activate swelling-induced Cl^- currents, maintain cells in isoosmolar extracellular solution E_5 and use intracellular solution I_6.
7. Clamp cells to a holding potential of 0 mV and use voltage ramp or voltage-step protocols as described before.
8. Superfuse cells with hypoosmolar solution E_6.
9. Test the effect of 1 mM DIDS or 100 μM NPPB on Cl^- currents.

3.7 Data Analyses

3.7.1 Analysis of Current Clamp Recordings

Determine the mean membrane potential of microglial cells, if experiments have been performed in the current clamp mode. These should be preferentially perforated patch clamp experiments. Compare this value with the mean membrane potential of cells in experiments with certain ionic substitutions. Measure membrane potentials of microglial cells before, during, and after application of ion channel activators/inhibitors. Determine after drug application the delay of changes in membrane potentials and the time to reach a new stable membrane potential.

3.7.2 Analysis of Voltage Clamp Recordings: Voltage-Activated Currents

Determine the activation threshold of voltage-activated channels (e.g., voltage-activated K^+, Na^+, or H^+ channels) from voltage ramp or voltage-step protocols. If voltage-step protocols have been used, subtract leak currents (*see* **Note 63**) from the recorded ion currents.

Use voltage-step protocols to observe whether ion currents show time-dependent (i.e., changes in current over time in one voltage step) or time-independent (i.e., no current change over time) behavior. Fit current traces with exponential functions (inward-rectifying voltage-activated K^+ channels, voltage-activated H^+ channels) or according to a n^4j model of a Hodgkin-Huxley kinetics (outward-rectifying voltage-activated K^+ channels). Determine the activation and inactivation time constants at different voltages.

Measure the reversal potential (*see* **Note 64**) of the investigated ion current. The best way to determine the reversal potential is to measure the voltage at which no tail currents flow (*see* Tail current protocols in Subheadings 3.6). Using voltage ramps, this is the membrane potential at which the current trace intersects with the zero line. The reversal potential gives a hint to the selectivity of the investigated channel (*see* **Note 65**).

Determine the current density (*see* **Note 66**), which is dependent on the number of expressed channel proteins and their single channel conductance (*see* **Note 67**). Measure the current amplitude at a given potential and subtract the leak current (*see* **Note 63**) from this value. Divide the leak-subtracted current amplitude by cell capacitance (*see* Subheading 3.4) as a measure of cell surface area to calculate the current density.

Quantify the effect of ion substitutions. Measure the change in current density, reversal potential, activation threshold, and time-dependent behavior before and after certain ion substitutions.

Quantify the effect of ion channel inhibitors. Measure (if possible, from leak-subtracted current amplitudes) the percentage of current inhibition. Calculate IC50 values of inhibitors from a concentration-response curve. For this calculation, measurements with at least five different inhibitor concentrations are necessary. Fit all data points according to a Hill equation to determine the IC50 value. Determine whether current inhibition is voltage dependent or voltage independent. Examine whether the inhibitor changes activation threshold or reversal potential of the current. Determine whether current inhibition is fully, partially, or not reversible.

3.7.3 Analysis of Voltage Clamp Recordings: Ligand-Activated Currents

Extract for ligand-activated (e.g., Ca^{2+}-, G protein-, or calcium release-activated) channels the current induced by these agents by subtracting the current trace before drug application from the one during drug application. The same principle applies to swelling-/stretch-activated (e.g., Cl^-) channels. Establish whether the current exhibits a linear (i.e., a voltage independent) or nonlinear (i.e., a voltage dependent) I-V relationship. Examine voltage-step protocols for time-dependent or time-independent behavior of ion currents.

Determine the reversal potential of ion currents either from tail current protocols or from voltage ramp protocols.

Calculate current densities from leak-subtracted amplitudes using voltage-step protocols. Alternatively, use voltage ramp protocols to calculate the conductance and normalize this value to cell capacitance.

Quantify changes in current density/normalized conductance, reversal potential, and time- and voltage-dependent behavior before and after certain ion substitutions.

Determine the effect of ion channel inhibitors. Calculate the percentage of current or conductance inhibition and determine the IC50 values of inhibitors using a Hill equation. Examine whether inhibition of currents is voltage dependent or independent. Compare reversal potentials before, during, and after application of ion channel inhibitors. Determine whether current inhibition is reversible.

4 Notes

1. A more detailed description on how to investigate biophysical and pharmacological properties of ion channels using the patch clamp technique can be found in [5].
2. Bath application of drugs is possible, but care should be taken that the measured ion channel or receptor does not desensitize when the extracellular solution is slowly exchanged. Electrophoretical or pressure-induced drug application is another possibility to consider.
3. A method for isolation of microglial cells and their subsequent culture is described in [6].
4. Protocols for preparation of hippocampal brain slices can be found in [7] for acutely isolated brain slices and in [8] for organotypic cultures.

5. A 10× concentrated stock of extracellular solution E_1 can be prepared, if the $NaHCO_3$ stock solution is stored separately. Extracellular solution E_1 should be prepared freshly every day from these stock solutions.
6. Add 2 mM Na_2ATP and 0.5 mM Na_2GTP to the intracellular solution if current rundown is observed during whole-cell recordings.
7. Check osmolarity of all solutions to prevent swelling or shrinkage of cells. It is recommended that the extracellular solution has a slightly higher osmolarity than the intracellular solution.
8. To measure KCa1.1 (BK) channels, it is advisable to increase the free calcium concentration of intracellular solution I_3 to 10 μM.
9. A Windows version of this program can be downloaded from http://www.stanford.edu/~cpatton/downloads.htm.
10. The reversal potential for extracellular solution E_1 or E_2 versus intracellular solution I_5 is +5 mV for nonselective cation channels and −40 mV for Cl^- channels.
11. Alternatively, omit KCl from extracellular solution E_1 and raise the $CaCl_2$ concentration to 10 mM.
12. The reversal potential for extracellular solution E_5 or E_6 versus intracellular solution I_6 is −20 mV for nonselective cation channels and +20 mV for Cl^- channels.
13. Do not store the final nystatin-containing pipette solution for more than 1 h at room temperature. It is recommended to prepare for each new cell under investigation a fresh nystatin-containing solution from stock.
14. The final concentration of the pore-forming substances nystatin, amphotericin, or gramicidin used for perforated patch recordings may vary between cell preparations depending on their membrane properties. A concentration should be chosen, at which the series resistance decreases within 10–30 min continuously, reaching stable values below 20 MΩ. The use of concentrations above the optimum results in either losing the giga-seal or in rupture of the cell membrane under the pipette leading to a whole-cell configuration. No or insufficient changes in series resistance occur over a period of more than 30 min recording time, if the concentration of the agent is too low.
15. Amphotericin or gramicidin does not have any substantial advantage compared to nystatin. They have similar limitations regarding their chemical stability.
16. The stock and final solution of fluorescent dyes should not be exposed to bright light and kept in the dark prior and during the experiment.
17. For patch clamp recordings of microglial cells in situ, we recommend to add a fluorescent dye with a different excitation

wavelength (e.g., Alexa647) to the intracellular solution. This approach provides a good control whether an extracellularly labeled microglial cell is identical to the cell stained intracellularly during the patch clamp recording. Add the Alexa dye at a final concentration of 2 μM to the intracellular solution used. Prepare the final solution from a frozen stock solution of 1 mM Alexa dye, which can be prepared in intracellular solution or water.

18. The silver chloride coating of the electrode wire is easily scraped off by glass capillaries. Therefore, it is recommended to fire polish both ends of the capillaries before pulling the patch pipette.
19. A short conical tip facilitates diffusion of the intracellular solution into the cell. This is of advantage for perforated patch recordings as well as for measurements of Ca^{2+}-activated, G protein-activated, and CRAC channels.
20. A pipette resistance of up to 10 MΩ is tolerable, if the intracellular solution mainly contains large molecules as charge carriers (e.g., to measure proton channel activity, *see* Subheading 2.3.4).
21. Neither tip polishing nor tip coating are usually required for patch clamp recordings of microglial cells with an exemption being single channel recordings. Try first to renew the filament of the microelectrode puller, if the seal resistance is constantly insufficient.
22. All Ag/AgCl electrodes, i.e., pipette and reference electrodes, should be stored in bleach overnight.
23. Alternatively, disposable pipette tips with a trimmed opening can be used instead of a small glass tube.
24. Due to the bicarbonate buffer, the pH value of extracellular solution E_1 is temperature dependent.
25. During this procedure the grounded table should be touched to avoid electrostatic discharges, which can destroy the patch clamp amplifier.
26. IB_4-conjugated Alexa dyes label microglial cells as well as endothelial cells. Due to their different morphology, microglia can be easily distinguished from endothelial cells, which delineate blood vessels.
27. A much higher resistance can be caused by air bubbles in the patch pipette, lack of contact between electrode wire and intracellular solution, no contact between reference electrode and extracellular solution, or a very small pipette tip opening.
28. Alternatively, it is possible to observe cell membrane dimpling, when the recording pipette is close enough to the cell's surface.
29. This membrane resistance is also called input resistance. It is mainly determined by the quality of the seal and the conductivity of all ion channels.

30. Check that patch pipettes do not drift during the recording if you experience problems with this step. Make sure that vibrations from the surroundings are not transmitted to the patch pipette. Use only freshly pulled pipettes; do not store patch pipettes for more than 1 day. Polish pipette tips using a microforge, if necessary.
31. Do not omit this step even if you use glass capillaries with filaments. Tip filling with solution I_7 not containing pore-forming molecules is required to achieve giga-seal formation.
32. The membrane capacitance correlates with the cell surface area and can therefore be used to normalize currents.
33. A high series resistance, especially in combination with a low input resistance, causes a voltage drop, i.e., the command voltage from the amplifier is not fully detected by ion channels in the cell membrane. Additionally, diffusion of intracellular solution from the patch pipette into the cell is limited.
34. A decreased input resistance can be caused by losing the tight seal between pipette and cell membrane. An input resistance in the MΩ range is also detectable, when channels are already open, e.g., a high calcium concentration in the pipette gates calcium-activated potassium channels.
35. The flow of the superfusion pipette should be tested with control solution before the actual patch clamp recording is performed. Sometimes the flow from the superfusion pipette does not exchange the solution around the patched cell; the superfusion pipette needs to be repositioned in this case. Another common problem is blockage of the superfusion system by air bubbles. Test the superfusion pipette flow at a safe distance from the patched cell. It should also be tested with drug-free control solution that ion channel activity is not altered during solution exchange.
36. If possible, activation of ion channels other than the investigated voltage-activated ion channel should be avoided. Otherwise a permanent ion influx or efflux is induced with subsequent cell volume changes.
37. Adjust the sample frequency to a value that the relevant voltage step or voltage ramp records at least 2,000 data points. Typical sample frequencies are 5–10 kHz for K^+ channel measurements, 50–100 kHz for Na^+ channel measurements, or 0.5–1 kHz for H^+ channel measurements.
38. Typical substitutions are N-methyl-D-glucamine (NMG) or TMA for monovalent cations and gluconate for Cl^- ions. Omission of Ca^{2+} or Mg^{2+} should be accompanied by inclusion of chelators like BAPTA or EGTA.

39. A reversal potential of a channel is also called equilibrium potential and allows to determine which ions are conducted by the investigated channel.
40. Care should be taken that the current, elicited in this first step, does not change substantially during the whole measurement. If changes occur, change the voltage value in this first step or allow more time between steps.
41. For voltage-activated potassium channel measurements, the membrane potential should be −60 mV. Outward-rectifying Kv1.3 K^+ channels are partially inactivated at more positive holding potentials; inward-rectifying Kir2.1 K^+ channels activate at more negative holding potentials.
42. Some ion channels in microglia, e.g., voltage-activated Kv1.3 K^+ channels, exhibit strong use dependence. Therefore, during voltage-step protocols, sufficient time for recovery should be allowed between individual voltage steps.
43. An example for an inward-rectifying K^+ current in microglia can be found in [1], Fig. 1.
44. A typical outward-rectifying voltage-activated K+ current is shown in [1], Fig. 2.
45. This channel is also known as KCNN4, IKCa1, IK1, and SK4. A representative KCa3.1 current is depicted in [1], Fig. 3.
46. Charybdotoxin also inhibits Kv1.3 and KCa1.1 channels.
47. TRAM-34 does not only inhibit Ca^{2+}-activated K^+ channels but also nonselective cation channels in microglia [9].
48. This KCa1.1 channel has also the following names: KCNMA1, BK, Slo, and maxiK. A representative current recording for this channel is shown in [7], Fig. 4.
49. KCa2.3 channels are also called KCNN3 and SK3. A representative current recording is shown in Fig. 1 of [10].
50. Figure 6 in [11] shows an example of a G protein-activated K^+ current.
51. Substitute K^+ ions in all solutions with Cs^+, if large K^+ currents interfere with the measurement of Na^+ currents.
52. A typical Na^+ current recording in microglia is shown in Fig. 2 of [12].
53. Microglial cells predominantly express TTX-sensitive Nav1.6 and Nav1.1 sodium channels but also TTX-resistant Nav1.5 subunits [13].
54. The holding potential should be set below the activation threshold $V_{threshold}$ of voltage-activated H^+ channels, which can be calculated from $V_{threshold} = 20\ mV - 40\ mV(pH_o - pH_i)$ [14].

55. For investigation of voltage-activated H^+ channels, time between individual voltage steps should be at least 30 s to minimize intracellular proton depletion and accumulation of protons outside the cell membrane [15].
56. Figure 5 in [1] shows a typical recording of a voltage-activated H^+ current.
57. The activation threshold of voltage-activated H^+ channels can be calculated using the equation $V_{threshold} = 20\ mV - 40\ mV\ (pH_o - pH_i)$ [14].
58. Nonselective cation currents are shown in Fig. 1 of [16]. A typical example of a TRPV1 current is shown in [17] Fig. 1, a TRPV4 current in [18] Fig. 2, a TRPM2 current in [19] Fig. 3, a TRPM4 current in [20] Fig. 4, and a TRPM7 current in [21] Fig. 1.
59. In contrast to other TRPC channels, TRPC6-mediated currents are potentiated by flufenamic acid.
60. Clotrimazole also inhibits Ca^{2+}-activated KCa3.1 K^+ channels
61. A representative example of a CRAC current is shown in [22], Fig. 7.
62. Figure 6 in [1] shows a typical recording of a microglial stretch-activated Cl^- current.
63. Leak currents have a linear I-V relationship. Therefore, the leak current for a voltage step of 20 mV is two times the leak current for a 10 mV voltage step. This leak subtraction method cannot be applied to ion channels, which are not voltage activated (e.g., the calcium-activated channel KCa3.1).
64. The reversal potential (or equilibrium potential) is often called Nernst potential in the literature. They are distinct, when more than one ion channel with a large permeability is open or the ion channel is permeable for more than one ionic species. In other words, if only one channel type is open and this channel is selective for one ion, reversal and Nernst potential are identical. The Nernst potential for one ion can be determined using the following online calculator: http://www.physiologyweb.com/calculators/nernst_potential_calculator.html
65. For example, the activated channel is most likely a potassium channel, if the reversal potential is near the Nernst potential for potassium. The reversal potential is difficult to interpret, when Nernst potentials for several ions do not differ significantly. Often the Nernst potential for Cl^- is 0 mV, which is also the reversal potential for nonselective cation/TRP channels. This is not the case for the solutions described in Subheadings 2.3.5 and 2.3.7—the reversal potentials for Cl^- and cation currents differ significantly.

66. Microglial activation often results in changes of cell size and morphology. Current density normalizes the current for cell surface area and is therefore a better marker for ion channel expression than current amplitude.
67. Both channel number and single channel conductance can change depending on intra- or extracellular activators/modulators.

Acknowledgements

C.E. is supported by European Union FP7 collaborative grant TargetBraIn (No. 279017).

References

1. Eder C (1998) Ion channels in microglia (brain macrophages). Am J Physiol 275(2 Pt 1):C327–C342
2. Eder C (2005) Regulation of microglial behavior by ion channel activity. J Neurosci Res 81(3):314–321
3. Eder C (2010) Ion channels in monocytes and microglia/brain macrophages: promising therapeutic targets for neurological diseases. J Neuroimmunol 224(1–2):51–55
4. Kettenmann H, Hanisch U-K, Noda M et al (2011) Physiology of microglia. Physiol Rev 91(2):461–553
5. Hille B (2001) Ion channels of excitable membranes, 3rd edn. Sinauer Associates, Sunderland, MA
6. Eder C, Klee R, Heinemann U (1997) Pharmacological properties of Ca^{2+}-activated K^{+} currents of ramified murine brain macrophages. Naunyn Schmiedebergs Arch Pharmacol 356(2):233–239
7. Schilling T, Eder C (2007) Ion channel expression in resting and activated microglia of hippocampal slices from juvenile mice. Brain Res 1186:21–28
8. Stoppini L, Buchs P-A, Muller D (1991) A simple method for organotypic cultures of nervous tissue. J Neurosci Methods 37(2):173–182
9. Schilling T, Eder C (2007) TRAM-34 inhibits nonselective cation channels. Pflugers Arch 454(4):559–563
10. Liu B-S, Ferreira R, Lively S et al (2013) Microglial SK3 and SK4 currents and activation state are modulated by the neuroprotective drug, riluzole. J Neuroimmune Pharmacol 8(1):227–237
11. Schilling T, Repp H, Richter H et al (2002) Lysophospholipids induce membrane hyperpolarization in microglia by activation of IKCa1 Ca(2+)-dependent K(+) channels. Neuroscience 109(4):827–835
12. Nörenberg W, Illes P, Gebicke-Haerter PJ (1994) Sodium channel in isolated human brain macrophages (microglia). Glia 10(3): 165–172
13. Black JA, Waxman SG (2011) Sodium channels and microglial function. Exp Neurol 234(2):302–315
14. Cherny VV, Markin VS, DeCoursey T (1995) The voltage-activated hydrogen ion conductance in rat alveolar epithelial cells is determined by the pH gradient. J Gen Physiol 105(6):861–896
15. DeCoursey TE (2003) Voltage-gated proton channels and other proton transfer pathways. Physiol Rev 83(2):475–579
16. Schilling T, Eder C (2009) Non-selective cation channel activity is required for lysophosphatidylcholine-induced monocyte migration. J Cell Physiol 221(2):325–334
17. Schilling T, Eder C (2009) Importance of the non-selective cation channel TRPV1 for microglial reactive oxygen species generation. J Neuroimmunol 216(1–2):118–121
18. Konno M, Shirakawa H, Iida S et al (2012) Stimulation of transient receptor potential vanilloid 4 channel suppresses abnormal activation of microglia induced by lipopolysaccharide. Glia 60(5):761–770
19. Kraft R, Grimm C, Grosse K et al (2004) Hydrogen peroxide and ADP-ribose induce TRPM2-mediated calcium influx and cation

currents in microglia. Am J Physiol Cell Physiol 286(1):C129–C137
20. Beck A, Penner R, Fleig A (2008) Lipopolysaccharide-induced down-regulation of Ca^{2+} release-activated Ca^{2+} currents (I_{CRAC}) but not Ca^{2+}-activated TRPM4-like currents (I_{CAN}) in cultured mouse microglial cells. J Physiol 586(2):427–439
21. Jiang X, Newell EW, Schlichter LC (2003) Regulation of a TRPM7-like current in rat brain microglia. J Biol Chem 278(44): 42867–42876
22. Nörenberg W, Cordes A, Blöhbaum G et al (1997) Coexistence of purino- and pyrimidinoceptors on activated rat microglial cells. Br J Pharmacol 121(6):1087–1098

Part VI

Analysis of Microglia Polarization

Chapter 18

Studying M1 and M2 States in Adult Microglia

Sadanand M. Gaikwad and Michael T. Heneka

Abstract

Microglial cell function receives increasing interest. To date, the majority of experiments are performed by using immortalized microglia-like cells or primary microglia prepared from pre- or postnatal rodent brain. As those may not adequately reflect the microglial biology in the adult brain, this protocol advocates a procedure which allows for the isolation, purification, and subsequent analysis of microglial cells. Once isolated, the principal state of activation, M1 or M2, can be determined in adult microglia using fluorescence-activated cell sorting, quantitative PCR, and/or Western blotting. Likewise, adult microglia generated by this protocol can be used for functional analysis through cell cultivation for a limited time.

Key words Adult microglia, M1/M2 phenotype, Neurodegeneration, Cytokine, Inflammatory gene, Alzheimer, Neuroinflammation

1 Introduction

The notion that the innate immune system is widely implicated in a variety of physiological and pathological processes in the brain, including synaptic pruning, debris clearance by phagocytosis, and trophic factor supply to surrounding bystander cells, has instigated research on microglial cells, the major cell type representing the brain's innate immune defense in recent years. Until now, however, the majority of studies on microglia have been performed using immortalized microglia-like cells and primary microglia derived from pre- or postnatal rat or mouse brain preparations. These cells, however, may only in part reflect the biology of microglial cells as they may lack substantial information which may only be acquired over time, once brain development has been finalized and the morphology and function of microglia has been shaped by neurotransmitters, endocrine factors, and their specific neuronal and astroglial environment. Based on this, one can assume that the experimental results obtained from in vitro studies using immortalized microglia-like cells or primary microglia should be further backed up by the analysis of adult microglia. Since not all of these studies can be

Bertrand Joseph and José Luis Venero (eds.), *Microglia: Methods and Protocols*, Methods in Molecular Biology, vol. 1041, DOI 10.1007/978-1-62703-520-0_18, © Springer Science+Business Media New York 2013

done in the living animal, e.g., by the use of two-photon in vivo laser scanning microscopy of microglia-expressing fluorophores of interest, protocols are required which allow for the analysis of adult microglia in the context of a specific physiological state or disease.

Microglial activation in response to physiological or pathophysiological stimuli will determine their morphology and function, and thus microglia can be polarized into various directions of which two, the M1 and M2 phenotypes, just have begun to be investigated and described. It is therefore likely that with an increasing knowledge about microglia functions in the brain, further activation states will be discovered. The M1 phenotype usually refers to microglial cells stimulated by interferon gamma and/or tumor necrosis factor alpha which expresses proinflammatory cytokines, while M2 cells produce anti-inflammatory cytokines, including interleukin 10 and interleukin 4, and are thought to dampen the immune response and instead promote tissue repair, similar to the concept of TH1/TH2 T cell polarization [1].

Neurodegenerative processes inevitably lead to the activation of microglial cells, which may acutely serve a beneficial purpose, but when chronically activated further promote and exacerbate neurodegeneration [2]. Using the example of Alzheimer's disease, we below describe a procedure how adult microglia can be isolated from a related transgenic mouse model and then studied to delineate their precise role in this devastating neurological disorder. A brief description of the mouse strains analyzed is followed by a listing of all materials and tools needed for the subsequent steps. In brief, the described methodology includes the murine anesthesia, surgery, and intracardial perfusion which is performed to clear the brain's vasculature from circulating immune cells and immune mediators. Thereafter the gentle dissociation of the brain is described, followed by centrifugation and separation procedures to clear the microglia from myelin and debris. Then the protocol pictures fluorescence-activated cell sorting (FACS), which through the use of adequate markers allows for the identification and isolation of adult microglia from the brain. Given the technical possibilities modern FACS provides, subsequent steps of analysis may include the purification of RNA and analysis of the transcriptome by qPCR or similar methods. Alternatively and depending on the amount of cells, these methods may also be used for the analysis of microglia by Western blot, immunocytochemistry, and others. A further option is to use the isolated cells and to cultivate those further. This procedure allows to perform a functional analysis, e.g., using assays which allow the assessment of migration or phagocytosis (not described). Importantly, the described methods of microglial isolation and analysis can be adopted for other models of neurodegenerative disease including but not restricted to the analysis of microglia in the substantia nigra of MPTP-treated rodents or for models of experimental allergic encephalitis.

2 Materials

2.1 Animals and Ages

1. APP/PS1 transgenic animals (# 005864, The Jackson Laboratory, [3]) and wild-type (WT) animals (The Jackson Laboratory) both were from C57/Bl6 genetic background.
2. Animal ages during experiments were 6–10 months.
3. Mice were housed under standard conditions at 22 °C, a 12h light–dark cycle with free access to food and water.
4. Animal care and handling was performed according to the declaration of Helsinki and approved by the local ethical committee.
5. The following animal groups were analyzed: WT and APP/PS1.

2.2 Intracardial Perfusion of Mice

Tools needed:

1. Blunt-ended scissors, with approximately 2.5- to 3-in. blade length.
2. Blunt-ended forceps, approximately 6 in., iridectomy scissors, hemostats [2].
3. Metal spatula, scalpel, a small, single-use needle (6) (*see* **Note 1**).
4. Pointed scissors, with approximately 1-in. blade length.
5. Container for the mouse with a corked surface (*see* **Note 1**).

Solutions:

1. Ice-cold 1× PBS including approximately 10 U/ml heparin (a 25 kU vial will make 2.5 L of PBS+Hep) (*see* **Note 2**).
2. 70 % Ethanol and dH_2O.

Perfusion setup:

1. Perfusion pump (which maintains a flow no higher than 0.5 ml/min for up to 3–5 min) (*see* **Note 3**).
2. Extension tubing 100 cm.
3. Butterfly needles, 23G, with 12-in. tubing.

Anesthesia:

1. Isoflurane (2-chloro-2-(difluoromethoxy)-1,1,1-trifluoroethane) (*see* **Note 4**).

2.3 Adult Mouse Brain Microglia Isolation

1. Medium B: 45 ml of 1× Hank's buffered salt solution (HBSS) and 5 ml of fetal calf serum (FCS) 10 % mixed by vortex and Collagenase Type IV (0.2 mg/ml, working activity of 770 U/mg) (*see* **Note 5**).
2. To inactivate the enzymatic reaction, 20 ml of 20 % fetal calf serum (FCS) in 1× HBSS.

3. 100 % Percoll, mixture of 100 % Percoll and 10× HBSS was prepared in ratio of 9:1 (45 ml of 100 % Percoll and 5 ml of ice-cold 10× HBSS).
4. 75 % Percoll, for one brain 7.5 ml of 100 % was mixed with 2.5 ml of ice-cold PBS (1×).
5. 25 % Percoll, for one brain 2.5 ml of 100 % Percoll was mixed with 7.5 ml of ice-cold PBS (1×) (*see* **Note 6**).
6. 18G, 23G needles with 20 ml syringes, cell strainer with 70 μm nylon mesh (*see* **Note 7**).

2.4 Cell Staining and Sorting

1. CD16/32 antibody (*Fc* receptor blocking antibody).
2. Anti-mouse CD11b- and CD45-conjugated antibody (*see* **Note 8**).
3. Eppendorf tubes (1.5 ml) and microcentrifuge tubes—centrifuge.
4. FACS tubes.
5. Cytometry and cell sorting facility (*see* **Note 9**).

2.5 Real-Time PCR Components

1. mRNA Isolation Kit.
2. cDNA Preparation Kit.
3. PCR primers or fluorescent probe and real-time PCR machine (*see* **Note 10**).

2.6 Immunoblotting

Prepare all solutions using ultrapure water (prepared by purifying deionized water to attain a sensitivity of 18 MΩ cm at 25 °C) and analytical grade reagents. Prepare and store all reagents at room temperature (unless otherwise indicated). Diligently follow all waste disposal regulations when disposing waste materials. We do not add sodium azide to the reagents.

2.7 Solutions and Reagents

1. To prepare 1× phosphate-buffered saline (PBS), add 8 g (137 mM) of NaCl, 0.2 g (2.7 mM) of KCl, 1.44 g (10 mM) of Na_2HPO_4, and 0.24 g (10 mM) of KH_2PO_4 to 800 ml of dH_2O. Then adjust the pH to 7.4 with HCl (*see* **Note 16**) and adjust the volume to 1liter with the addition of dH_2O and autoclave to sterilize.
2. To prepare 1 ml volume of 2× SDS lysis buffer: take 200 μl (100 mM) of Tris–HCl prepared in pH 6.8 then 200 μl (4 %) of SDS, 100 μl (10 %) of glycerol, and 20 μl (2 %) of 2-mercaptoEtOH mixed and store at 4 °C until use (*see* **Note 11**).
3. Just before use add protease inhibitors to the lysis buffer 3 μl Pic and 3 μl (200 mM of) PMS (*see* **Note 12**).
4. To prepare 10 % APS, 1 g ammonium persulfate is mixed with 10 ml dH_2O. Mix well.

5. 1× transfer buffer was prepared by adding 3.03 g of Tris–HCl, 14.4 g of glycine, 200 ml of methanol, and 1 ml of SDS (10 %) (*see* **Note 13**).
6. To prepare 10× PAGE buffer, 144 g of glycine and 30 g of Tris base are filled to 1 L with dH_2O and mixed.
7. Running buffer: 100 ml of 10× SDS-PAGE buffers is mixed with 5 ml of 20 % SDS and filled up to 1 L with dH_2O and mixed (*see* **Note 14**).
8. Blotting buffer is prepared by mixing 100 ml of 10× PAGE buffer and 200 ml of methanol and filling up to 1 L with dH_2O and mixed.
9. Tank buffer 10× is prepared by mixing 15 g of Tris–HCl, 72 g glycine, and 50 ml of SDS (10 %) and filling up to by adding 500 ml of dH_2O.
10. Wash buffer: Use 1× TBST is prepared by adding 1 L of TBS 1× and 1 ml of Tween20.
11. 100 ml of 5 % nonfat dry milk is prepared in 1× TBST by adding 5 g of milk powder (5 %), 10 ml of 10× TBS (20 mM Tris–HCl pH8, 100 mM NaCl), and 50 µl Tween20 (0.05 %).
12. Blocking buffer: Use 1× TBST with 5 % nonfat dry milk; for 150 ml, add 7.5 g nonfat dry milk to 150 ml 1× TBST and mix well (*see* **Note 15**), then store at 4 °C.
13. Bovine serum albumin (BSA).
14. Primary antibody dilution buffer: Use 1× TBST with 5 % BSA or 5 % nonfat dry milk as indicated on the primary antibody datasheet; for 20 ml, add 1.0 g BSA or nonfat dry milk to 20 ml 1× TBST and mix well (*see* **Note 15**) store at 4 °C.
15. HRP Western Blot Detection System: Secondary antibody conjugated to horseradish peroxidase (HRP), Anti-Biotin HRP-linked antibody, chemiluminescent reagent, and peroxide.
16. Prestained Protein Marker, Broad Range (premixed format).
17. Blotting membrane: This protocol has been optimized for nitrocellulose membranes (recommended). PVDF membranes may also be used. A pore size of 0.2 µm is generally recommended.

3 Methods

3.1 Intracardial Perfusion and Surgery

1. Place tubing into ice-cold 1× PBS buffer. Connect the extension tube through the perfusion pump and the other end with butterfly needles, 23G, with 12-in. tubing.
2. Turn on the pump to flow PBS through the extension tube to wash the extension tube and butterfly needles along with

tubing. Make sure that there are no bubbles in any of the tubes (*see* **Note 3**).

3. Place the mouse in a container that has a small amount of isoflurane (or equivalent) (*see* **Note 4**).
4. Once the mouse no longer shows a response to the painful stimulus, wet the abdomen with 70 % ethanol. Place the mouse on the corked surface with the abdomen facing up, fixated using small needles (*see* **Note 1**).
5. Grab the skin with a blunt-ended forceps at the level of the diaphragm, and cut to expose the liver. Cut laterally and then up, cutting through the ribs. Lift flap and continue cutting until the heart is easy to access. Pin the loose flap of the skin out of the way and free the heart by tearing any connective tissue with the forceps (*see* **Note 16**).
6. Secure the chest flap to the corked surface with a 20G needle (*see* **Note 17**).
7. Place the butterfly needle into the left ventricle, and turn on the perfusion pump to flow no faster than 0.5 ml/min of ice-cold 1× PBS BUFFER. Then immediately cut the right atrium (*see* **Note 18**).
8. Continue the buffer flow until the liver has become a light coffee color. This should be noticeable within 4–5 min.
9. Turn off pump. If perfusing more mice, lift the tube out of buffer for a couple of seconds and switch the pump-to-pump buffer. Allow the pump to flow until any air bubbles have been expelled. This ensures that only buffer is in the line. Turn off pump.
10. Decapitate mouse at the level of the shoulders. Cut through the scalp with a razor blade to expose the skull. Using a small sharp-ended scissors and a scalpel, scrape off the skin and muscle at the back of the brain.
11. Carefully expose the brain by scraping through the skull bone (*see* **Note 19**). After most of the skull bone is removed, with the help of metal spatula gently take out the brain and place it into ice-cold 1×HBSS medium.

3.2 Adult Microglia Isolation

1. Remove the mouse brain and determine the weight of tissue in 1 ml of ice-cold 1× HBSS-contained 35 mm dish to make sure the 0.4 g limit per brain is not exceeded.
2. With the help of a new scalpel, separate the two hemispheres, most of the visible meninges must be carefully removed under a microscope.
3. Transfer the brain of one animal into one small Petri dish containing 5 ml of Medium B (enzymatic solution) and finely chop the brain with a scalpel and incubate it for 90 min at 37 °C, 5 % CO_2, and continuously stir (Fig. 1) (*see* **Note 20**).

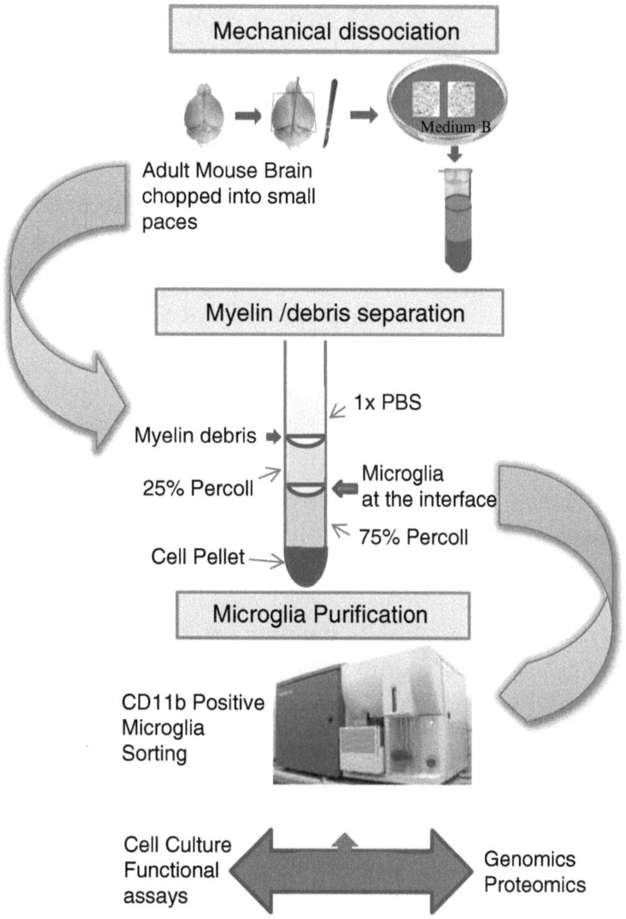

Fig. 1 Protocol steps to isolate microglia from an adult mouse brain are outlined and illustrated in this schematic procedure workflow

4. After this incubation transfer the tissue to a 50 ml falcon tube and inactivate enzymatic reaction using 20 ml of 20 % fetal calf serum (FCS) in 1× HBSS, and centrifuge at 200 × *g* for 10 min at 4 °C.
5. Gently mix the tissue suspension with a 20 ml syringe and 18G needle 10–12 times and then repeat it with a 23G needle (*see* **Note 21**). Remove all clumps and make single-cell suspensions, filtering them through the cell strainer with 70 μm nylon mesh (*see* **Note 7**) into a new 50 ml conical tube. Centrifuge at 155 × *g* for 10 min at 4 °C.
6. A discontinuous gradient density centrifugation step using Percoll is then performed: Resuspend the pellet with 1 ml of 75 % Percoll and then add the remaining 9 ml of 75 % Percoll.
7. Very carefully add 10 ml of 25 % Percoll by using a serological 5 ml pipette, avoid mixing the phases! On top of this very slowly add 10 ml pure 1× PBS and centrifuge at 600 × *g* for

25 min at room temperature with slow acceleration and no brake (*see* **Note 22**).

8. Gently discard the material at the interphase between 25 % Percoll and 1× PBS, which contains myelin and cell debris (Fig. 1).
9. Next, gently collect the lightly visible interphase ring between 75 % and 25 %, which contains a mixed glial cell population, into a new 50 ml flacon tube (around 10 ml from the ring) (Fig. 1).
10. Add threefold ice-cold 1× PBS (around 30 ml) to the same 50 ml flacon tube.
11. Perform a final centrifugation at 200 × *g* for 10 min at 4 °C with slow brake.
12. Gently discard the supernatant, keeping in mind that approximately 100 μl may remain in the falcon tube. Add additional ice-cold 900 μl 1×HBSS and count the cell number for further surface marker staining.

3.3 CD11b Positive Microglia Staining and Sorting

FACS staining protocol: Dual-color method used APC and PE conjugates.

1. After the cell count, distribute equal numbers of the cells into the appropriate numbers of 1.5 ml Eppendorf tubes (e.g., CD11b, CD45, CD11b+CD45, unstained) (*see* **Note 23**).
2. Centrifuge at 9,391 × *g* for 1 min and discard the HBSS-contained supernatant.
3. Gently resuspend the cell pellet in 500 μl of 1× HBSS buffer.
4. Centrifuge at 9,391 × *g* for 1 min and discard the 1× HBSS buffer.
5. Gently resuspend the cell pellet in 200 μl of 1× HBSS buffer and add 2 μl of CD16/32 Fc receptor blocking antibody (*~0.5–1 μg/100 μl*) and incubate for 20 min on ice.
6. Add 1 μl of fluorochrome-conjugated primary antibodies respectively to each tube except to the unstained control (*~0.5–1 μg/100 μl*) and incubate for 30 min on ice (*see* **Note 8**).
7. Add 400 μl of 1× HBSS buffer (just to avoid stress during centrifugation), centrifuge at 9,391 × *g* for 1 min, and discard the 1× HBSS buffer.
8. Wash the cells one more time by gently resuspending the cell pellet in 500 μl of 1× HBSS buffer and centrifuging at 9,391 × *g* for 1 min to completely remove unbound fluorochrome-conjugated antibodies.
9. Finally, gently resuspend the cell pellet in 500 μl of 1× HBSS buffer and filter through 70 μm nylon mesh into the FACS tubes. CD11b-high and CD45-low positive cells are sorted and collected by a flow cytometric cell sorter like the BD FACSAria III (Fig. 1).

Table 1
List of the M1 and M2 gene markers

Microglia phenotype	Gene target
Neutral	Iba1 C1qa Slc13a3
M1	iNOS IL-1, IL-6 TNFα, Mcp-1 Cxcl-2 Tspo
M2	IL-10 Ym1 Arg-I Fizz TGFβ Lgals3

3.4 Real-Time PCR

After sorting CD11b-high and CD45-low positive microglia cell, populations are further used for transcriptional quantification of inflammatory gene markers for M1 and M2. (Selected M1 and M2 markers are listed in Table 1.)

mRNA isolation and cDNA preparation are done according to respective datasheet (*see* **Note 24**). The primers' list is given in Table 2 (we use Taqman probe for further gene quantification) (*see* **Note** 7).

3.5 Immunoblotting

After sorting CD11b-high and CD45-low positive microglia cell, populations are further used for quantification of inflammatory protein markers for M1 and M2 translational regulation. (Selected M1 and M2 protein markers are listed in Table 3.)

Preparation of Cell Lysates and Protein Blotting

1. Sorted microglial cells are washed once with 1× PBS. Cell lysis is conducted by adding 2× SDS lysis buffer (at approximately 2×10^6 to 1×10^7 cells per ml).
2. Transfer extract to a microcentrifuge tube (1.5 ml Epi) and heat to 100 °C, heating at this temperature for 5 min, then sonicate with three to four bursts of 5–10 s each. Next cool it on ice.
3. Microcentrifuge the sample for 5 min at full speed (12,000 rpm).
4. Measure the protein concentration from each sample by a standard available method (*see* **Note 25**).
5. Load the same concentration of each sample (around 20 μl) onto SDS-PAGE gel (10 cm × 10 cm) and either 10 μl of prestained molecular weight markers for electrotransfer

Table 2
List of Taqman probes used for detection gene transcripts

Gene target	Taqman probe
Iba1	Mm00479862_g1
C1qa	Mm00432142_m1
Slc13a3	Mm00475280_m1
iNOS	Mm00440485_m1
IL-1, IL-6	Mm00434228_m1, Mm00446190_1
TNFα, Mcp-1	Mm00443258_m1, Mm00441242_m1
Cxcl-2	Mm00436450_m1
Tspo	Mm00437828_m1
IL-10	Mm00439616_m1
Ym1	Mm00657889_mH
Arg-I	Mm00475988_m1
Fizz	Mm00445109_m1
TGFβ	Mm00498255_m1
Lgals3	Mm00802901_m1

Table 3
List of the M1 and M2 protein markers

Phenotype	Protein
Neutral	Mac-1 F4/80 Isolectin B4
M1	iNOS IL-6, IL-1 TNFα MARCO MHC II TSPO
M2	CD163 CD206 Ym1 Arginase I FIZZ

verification or 10 μl of biotinylated protein ladder to determine molecular weights.

6. Finally, transfer the electrophoresed protein to nitrocellulose or PVDF membrane by electrotransfer.

Blocking Membrane and Antibody Incubation

Please make sure the sizes of the membranes listed below have a volume of 10 cm × 10 cm (100 cm^2).

1. Blocking membrane: After the electrotransfer of membrane, wash with 20 ml of 1× TBS for 5 min at room temperature.
2. Add 20 ml of blocking buffer and incubate the membrane for 1 h at room temperature.
3. Wash the membrane once for 5 min with 20 ml of 1× TBST buffer.
4. Incubate with primary antibody at the appropriate recommended antibody dilutions. A total 10 ml of primary antibody dilution buffer should be used at 4 °C overnight with gentle agitation.
5. The next day wash the membrane three times for 5 min with 20 ml of TBST buffer.
6. Prepare the HRP-linked antibody in a dilution of 1:1,000 in 20 ml of blocking buffer (or if you have it, Anti-biotin) and incubate for 1 h at room temperature with gentle agitation.
7. Wash the membrane three times for 5 min with 20 ml of TBST buffer and proceed to developing the image.

Developing Images

1. Take a new membrane incubation plastic tray and mix the Chemiluminescent Detection Substrates in a 1:1 ratio. The amount may vary depending on the size of the membrane, in each case 1 ml (for small membranes) or 2 ml (for large membranes).
2. Carefully transfer the membrane onto the plastic tray prepared in **step 1** above (carefully agitate the tray to and fro).
3. Now cautiously use a blunt-ended forceps to remove the membrane from the tray and place on the white board in the documentation system device set.
4. Put paraffin paper over the membrane; use a small roller to remove air bubbles.
5. Switch on the documentation unit on the Western blot.
6. Turn on the Epi white light to visualize the membrane and adjust the membrane to the proper orientation.
7. Turn off the Epi white light and adjust the illumination intensity and the duration of membrane exposure.
8. Expose to film, develop the image, and save it for further use.

4 Notes

1. The needle number is not specified. We use a 23G needle to fix the mouse on corked surface and secure the four paws to surface spreading them as wide as possible.
2. The use of heparin is not a must, but you need to be quick to perfuse if you do not want to use it.
3. Not all laboratories have a perfusion pump facility. Instead, at the one end of the extension tube, attach a 50 ml syringe filled with 1×PBS and at the other end attach the butterfly needle (23G) with a 12-in. tubing. Make sure that there are no bubbles in any of the tubes.
4. This is only to relax the mouse enough such that the mouse can easily die. If kept too long in the container, pinch the toes to judge its level of response to painful stimulus.
5. Other enzymes have been used for this protocol but we have found that these enzymes yield a good end product. Prepare freshly prior to the start of the experiment.
6. Prepared 75 % and 25 % Percoll can be stored on ice until use.
7. We used Taqman probe and based a gene expression assay, but one could order real-time PCR primers and study the expression pattern.
8. This preparation generates huge amounts of debris which cannot be removed by gradient centrifugation. Cell suspensions can be improved using 70 μm nylon mesh, or mesh of an even finer grade, as necessary.
9. Conjugated antibodies need to be selected carefully since the selected fluorochrome peak excitation and emission wavelengths should not overlap each other in double-stain FACS analysis.
10. As the cell sorting facility may be inaccessible, one could alternatively use the CD11b magnetic cell sorting.
11. SDS precipitates at 4 °C; therefore, the lysis buffer needs to be gently heated prior to use.
12. Any serine protease inhibitor may be used.
13. Dilute 100 ml of 10× native buffer (0.25 M Tris–HCl and 1.92 M glycine) to 800 ml with dH_2O and mix 200 ml of methanol. Avoid mixing methanol directly into the 10× buffer since it precipitates its ingredients. In this case the precipitate can be redissolved by the addition of 800 ml of dH_2O.
14. Another simple method to prepare running buffer is to prepare 10× native buffer, measure 30.3 g Tris–HCl and 144 g glycine, mix well, add makeup to 1 L with dH_2O, then add 10 ml of 10 % SDS. Always add SDS solution at last, since it makes bubbles.

15. Add 100 ml of 10× TBS to a 1 L volume graduated cylinder and filling it to about 800 ml with dH_2O. Transfer 50 g of skim milk powder into the cylinder and stir until dissolved. Fill to 1 L with dH_2O. Separate 500 ml for the blocking solution. To the remaining 500 ml solution, add 250 μl of Tween20, dissolve, and use it as diluents for the antibody.
16. Be sure to draw the scissors away from the organs when cutting to avoid damaging circulation more than necessary.
17. *Do not use scissors to free the heart*—this causes undesired bleeding.
18. This process needs to be done quickly before onset of blood clotting.
19. This needs to be done carefully as it can easily damage the hemisphere.
20. Some groups have shown that even a 2-h incubation time may be used.
21. To have single-cell suspensions, better to use first 18G needle and then 23G needle may improve the preparation and avoid tissue clumps.
22. This step must be done very slowly. When properly done, one should clearly see different layers.
23. From one mouse, approximately 700,000 cells can be obtained.
24. Any high-quality mRNA Isolation Kit may be used which provides high-quality mRNA from sorted cell populations. Any cDNA Isolation Kit may be used to prepare cDNA from isolated high-quality mRNA.
25. Any available standard method may be used to quantification of protein concentration.

Acknowledgement

This work was supported by a grant of the Deutsche Forschungsgemeinschaft to MTH (KFO177, TP8) and of the EU-FP7 program on neuroinflammation (INMIND).

References

1. Gordon S (2003) Alternative activation of macrophages. Nat Rev Immunol 3:23–35
2. Heneka MT, O'Banion MK, Terwel D et al (2010) Neuroinflammatory processes in Alzheimer's disease. J Neural Transm 117: 919–947
3. Jankowsky JL, Slunt HH, Ratovitski T et al (2001) Co-expression of multiple transgenes in mouse CNS: a comparison of strategies. Biomol Eng 17:157–165

Chapter 19

Isolating, Culturing, and Polarizing Primary Human Adult and Fetal Microglia

Bryce A. Durafourt, Craig S. Moore, Manon Blain, and Jack P. Antel

Abstract

Microglia are an important component of the innate immune system within the central nervous system (CNS). Isolation and in vitro culturing of microglia can provide insight towards the basic biology of these cells as well as their interactions with neurons, astrocytes, and oligodendrocytes. While studies of rodent microglia and microglial cell lines have provided a basis for our understanding of these cells, human adult microglia exhibit distinct properties when compared to rodent cells. Furthermore, the study of human fetal microglia provides a window into the developing CNS. This chapter describes the protocols used to isolate, purify, and culture both human adult and fetal microglia. Under basal culture conditions, human microglia survive for extended periods in the absence of growth factors, thus allowing their properties to be investigated under resting conditions. In addition, both human adult and fetal microglia can be used to study how they respond to different polarization conditions. As is the case with macrophages, it is also possible to polarize microglia towards the pro-inflammatory "M1" and the anti-inflammatory "M2" phenotypes, as described in this chapter.

Key words Microglia, Myeloid cells, Human, Isolation, Culture, Polarization, M1, M2

1 Introduction

In the scientific literature, much of our understanding as it pertains to the biology of microglia has relied on the use of primary rodent-derived cells and transformed human cell lines. While the use of these cells/cell lines has led to many discoveries that we predict will translate and apply to humans, these continue to require validation in primary human culture systems. Human microglia possess distinct properties, including their capacity to survive for prolonged periods under basal culture conditions. Access to both human fetal and adult central nervous system (CNS) tissue provides the unique and invaluable opportunity to isolate primary human neural cells and gain insight into their basic properties during development, and in the mature brain under physiologic and pathological conditions. Microglia represent an important member of the innate immune

Bertrand Joseph and José Luis Venero (eds.), *Microglia: Methods and Protocols*, Methods in Molecular Biology, vol. 1041,
DOI 10.1007/978-1-62703-520-0_19, © Springer Science+Business Media New York 2013

system within the CNS and play critical roles in influencing the adaptive immune response and biological activities of other CNS-resident cells (e.g., neurons, astrocytes, oligodendrocytes). In this chapter, we will discuss in detail the protocols that can be used to successfully isolate, purify, and culture human microglia from either fetal or adult human CNS tissue. In addition, we will also provide protocols that can be used to polarize microglia into either a classical, pro-inflammatory "M1" or alternative, anti-inflammatory "M2" phenotype. Polarization of microglia is performed using different disease-relevant molecules, reflecting the plasticity of these cells in vivo. In culture, cell polarization allows us to compare and contrast the molecular mechanisms that may help distinguish microglia from other myeloid cell types and provide rationale and insight for further in situ investigation in diseased tissue.

Based on whether CNS tissue is fetal- or adult-derived, in this chapter we will provide different cell isolation, purification, and culturing protocols. Human fetal samples are obtained at a pre-myelinating gestational age between 14 and 20 weeks; adult tissue is surgically derived and most often collected from tissue resections from either the frontal or temporal lobe of patients undergoing surgery for intractable epilepsy. The same isolation and culture techniques have also been applied to autopsy-derived material. The use of different cell isolation techniques is primarily due to the presence of myelin in the adult tissue, which must be successfully removed to prevent cell toxicity and unwanted cellular biology that could incur as a consequence of the presence of myelin during culture. It should be noted that compared to surgically resected material, CNS tissue derived at autopsy introduces both premorbid (such as preterminal conditions including ischemia and hypoxia) and postmortem (delay to obtaining tissue) variables. We acknowledge that variables in working with human tissue exceed those encountered using tissue obtained from rodents or other animals raised under strict laboratory conditions. Use of surgical tissue for research purposes must be coordinated with examination of tissue for clinical diagnostic purposes.

2 Materials

2.1 General Material

1. Gloves, lab coat.
2. Incubator (37 °C, 5 % CO_2).
3. Shaking water bath (37 °C).
4. Tissue culture hood.
5. Vacuum aspirator apparatus.
6. Swinging bucket centrifuge.
7. Disposal beaker containing ~50 ml bleach.

8. Parafilm.
9. Pipette aid and 5, 10, and 25 ml serological pipettes.
10. 10 μl and 200 μl micropipettes and tips.
11. Autoclaved 100 ml and 500 ml bottles.
12. Mashing filter (*see* **Note 6** on preparing the filter).
13. Plunger of 20 ml syringe.
14. Phosphate-buffered saline (PBS).
15. DNase.
16. Trypsin.
17. 50 ml conical tubes.
18. Fetal calf serum (FCS).
19. Poly-L-lysine (optional).
20. Tissue-coated culture flasks (T12.5, T25, T75, T175) and plates (24-well plates, 12-well plates, or 6-well plates as required).
21. Hemocytometer.

2.2 Adult Prep

1. Percoll®.
2. Minimum Essential Medium (MEM) containing 5 % FCS, 0.1 % glucose, 1 % penicillin–streptomycin, and 1 % glutamine.
3. Plastic 3 ml transfer pipette.
4. Pathology cassette and formalin.
5. Glass Pasteur pipettes and bulb.
6. 40 ml Nalgene centrifuge tubes.
7. Fixed rotor centrifuge.
8. Ethylenediaminetetraacetic acid (EDTA).

2.3 Fetal Prep

1. Petri dish (150 × 15 mm).
2. Scalpels.
3. Sterile forceps.
4. Dulbecco's Modified Eagle Medium (DMEM) containing 5 % FCS, 0.1 % glucose, 1 % penicillin-streptomycin, and 1 % glutamine.
5. 70 μm cell strainer.

2.4 Polarization

1. Macrophage colony-stimulating factor (M-CSF).
2. Granulocyte-macrophage colony-stimulating factor (GM-CSF).
3. Interleukin-4 (IL-4).
4. Interleukin-13 (IL-13).
5. Interferon gamma (IFN-γ).
6. Lipopolysaccharide (LPS; serotype 0127:B8).

3 Methods

These protocols must be carried out in a sterile tissue culture hood, using autoclaved solutions and autoclaved reusable or sterile disposable equipment. Universal precautions must be followed when processing human tissue; personnel must wear lab coats and gloves. Special care must also be taken especially when handling sharp instruments. Any reusable equipment must be left in a 10 % bleach solution for at least 24 h prior to being washed. Unless otherwise stated, all centrifugation steps should be carried out at room temperature, in a swinging bucket rotor, with brakes set to "high."

3.1 Isolation of Human Adult Microglia

1. Human brain tissue (*see* **Note 1**) is obtained in a Cavitron Ultrasonic Surgical Aspirator (CUSA®) bag, in the form of pieces less than 1 mm^3. Following resection from the CNS, process tissue as quickly as possible (ideally within 1 h).
2. Before starting the procedure, turn on a shaking water bath to 37 °C.
3. Carefully open the CUSA® bag in the tissue culture hood and pour contents into 50 ml conical tubes. A maximum of 14 tubes should be used; if there is still liquid in the CUSA® bag, keep the bag for use in the next step.
4. Allow the tissue to settle at the bottom of the tubes, and then pour out ¾ of the liquid. Pour any remaining CUSA® bag contents into the conical tubes. Repeat until the CUSA® bag is empty, at which point the CUSA® bag should be rinsed with 50–100 ml PBS, taking care not to leave any tissue sticking to the inside of the CUSA® bag.
5. Allow the tissue to settle and pour off ¾ of the liquid as in **Step 4** above, then add PBS in order to wash and remove as much blood as possible. Repeat this step until the liquid becomes almost clear, usually after two to three washes.
6. Pool the tissue to one tube by first allowing tissue to settle and pouring off ¾ of the liquid as above. Once it has settled, note the total amount of tissue (*see* **Note 2**).
7. Use a transfer pipette to remove 1 ml of tissue and place it into a pathology tissue cassette. Place the cassette into a specimen container containing formalin and return this specimen to the pathology department (*see* **Note 3**).
8. Remove as many blood clots as possible using a Pasteur pipette (*see* **Note 4**).
9. Pour the cleaned tissue into a 100 ml bottle and rinse the conical tube with ~5 ml PBS.
10. Add 5 ml 0.5 % trypsin and 5 ml DNase (25 μg/ml). If total tissue volume is >15 ml, add an additional 5 ml each of trypsin and DNase.

11. Add PBS to a final volume of 68 ml.
12. Tightly cap the bottle, and wrap the mouth of the bottle and cap with parafilm (*see* **Note 5**) and shake at 180 rpm, 37 °C for 30 min.
13. During the shaking period, set up the required materials for the following steps. Set up mashing filter (*see* **Note 6**) in the neck of a 500 ml bottle. Remove the plunger of a 20 ml syringe and place on top of the mashing filter.
14. Following the 30 min shaking incubation, add 10 ml of FCS to inactivate the trypsin.
15. Determine the maximum volume of PBS to be used for the mashing steps (*see* **Note** 7).
16. Pipette 5–10 ml of tissue into the filter, add PBS, and mash using the syringe plunger. When adding PBS, do not exceed a maximum volume of 50 ml per tube calculated above. If the prep is very large and this volume exceeds 500 ml, move the filter setup to a new 500 ml bottle when the first bottle is full.
17. Continue adding 5–10 ml of tissue into the filter and mashing with PBS (*see* **Note 8**).
18. Once all tissue has been mashed, wash the bottle that contained the tissue with 20 ml of PBS, ensuring all pieces of tissue have been collected and added to the filter. Continue mashing with PBS until tissue no longer passes through the filter. It is normal that some clumps of myelin debris may not pass through the filter.
19. Divide the contents of the bottle equally into the number of tubes determined in **Step 15** above. Fill to 50 ml with PBS and spin 300 × *g*, 10 min.
20. During this spin, add 9 ml of Percoll® to each sterile Nalgene centrifuge tube. Prepare a 50 ml conical tube with PBS for use in the next step.
21. Pour out the entire supernatant of each tube and pour in ~20 ml PBS. Do not add >21 ml PBS as this is the volume used in the next step. Adding <21 ml at this step ensures that no tissue is wasted.
22. Resuspend the tissue by pipetting up and down. Process one conical tube at a time. Transfer first 10 ml from the conical tube into a Percoll® tube. Next, transfer 11 ml, using the tube prepared with PBS to supplement to the required volume.
23. Weigh the centrifuge tubes in order to match for balance. The maximum difference in weights between paired tubes should not exceed 0.05 g. Add PBS if required to bring weights within acceptable range (adding 10 μl for each 0.01 g difference).
24. Spin 31,000 × *g*, 4 °C, 30 min, with no brakes, in a fixed rotor centrifuge.

25. During this spin, set up an aspirator with 10 ml of bleach, and prepare 50 ml conical tubes to match the number of Nalgene tubes used above.
26. Carefully remove the Nalgene tubes from the centrifuge. The top layer contains myelin, the middle layer contains glial cells, and the bottom layer contains red blood cells.
27. Using a vacuum aspirator, remove the top layer of myelin (*see* **Note 9**).
28. Set a pipette aid to "slow" and transfer the cell layer (~15–18 ml) to a 50 ml conical tube (*see* **Note 10**).
29. Fill each conical tube with PBS to a final volume of 50 ml. Invert tubes to mix.
30. Spin $900 \times g$, 10 min.
31. During this spin, warm MEM containing 5 % FCS, 0.1 % glucose, 1 % penicillin-streptomycin, and 1 % glutamine to 37 °C.
32. Carefully pour out the supernatant as much as possible.
33. Add 7 ml of medium to one tube, resuspend the pellet, and transfer to the next tube. Continue this process until all pellets within each tube have been resuspended.
34. Use 10 ml of medium to wash the empty tubes, ensuring all cells are collected in the final tube.
35. Spin $300 \times g$, 10 min, using low brakes.
36. Resuspend with 2 ml MEM 5 % FCS and remove a small aliquot (10 μl) to count.
37. Plate at 2×10^6 cells per ml in a T12.5 plastic flask (max 3 ml) or 6 ml per T25 flask if the prep is larger (i.e., $>6 \times 10^6$ cells).
38. Allow the microglia 24 h to adhere to the flask. Floating cells (including oligodendrocytes and progenitors) should be removed at this point (*see* **Note 11**). Wash flask 2× with 3 ml medium (MEM 5 % FCS). Do not touch the pipette to the surface of the flask to avoid disturbing adherent microglia.
39. Add fresh MEM 5 % FCS (3 ml for T12.5 flask, 6 ml for T25 flask).
40. Allow microglia to recover from the isolation process for 3–5 days prior to trypsinization and plating.
41. Set up a vacuum aspirator, and prepare a 50 ml conical tube containing 5 ml of warm medium (MEM 5 % FCS).
42. Prepare 10 ml trypsin solution (0.05 % trypsin, 2 mM EDTA).
43. Aspirate medium and rinse with 3–5 ml (3 ml for T12.5, 5 ml for T25) of PBS.
44. Aspirate the PBS and replace with 3–5 ml (3 ml for T12.5, 5 ml for T25) of trypsin solution.

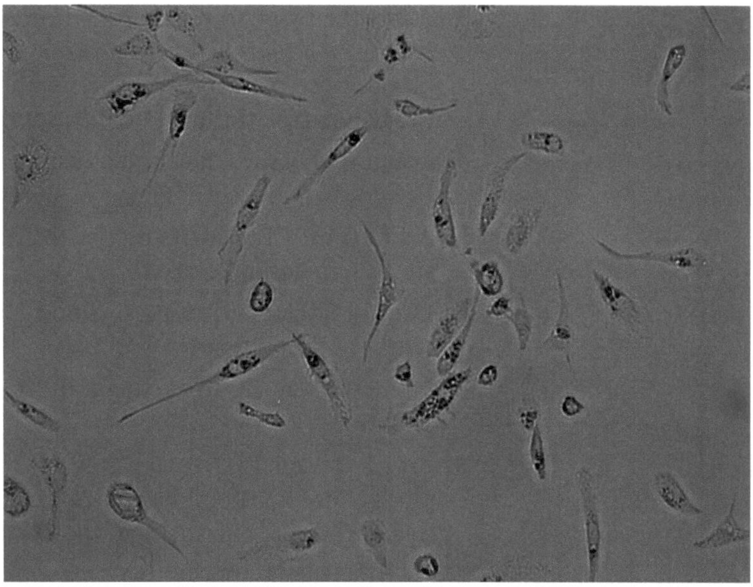

Fig. 1 Human adult microglia (25×). Cells were isolated according to the described protocol and grown in MEM containing 5 % FCS, 0.1 % glucose, 1 % penicillin–streptomycin, and 1 % glutamine. In culture, adult microglia cells adopt a bipolar morphology and exhibit numerous vesicles visible under light microscopy

45. After 10 s, aspirate the trypsin, leaving only a thin film of liquid covering the surface of the flask.
46. Place flask in the incubator for 5 min or until cells have lifted (*see* **Note 12**).
47. Hit the flask three times on a counter surface (quickly and forcefully) to dislodge the cells. Verify under a microscope that the cells have been dislodged.
48. Quickly return to the hood, pipette 3–5 ml medium into the flask (expel liquid to further detach cells), and transfer cells to the tube prepared with medium. Rinse the flask 2× with medium, transferring each time to the tube.
49. Spin 300 ×*g*, 10 min.
50. Resuspend in 1–2 ml of medium and remove a small aliquot (10 μl) to count.
51. Plate the microglia at desired density, usually 100,000 cells/ml, in either a 24-well plate (1 ml/well) or a 6-well plate (3–4 ml/well) (*see* **Note 13**). *See* Fig. 1 for the appearance of typical human adult microglia after 4–5 days in culture.

3.2 Isolation of Human Fetal Microglia

1. Before starting the procedure, warm the shaking water bath to 37 °C.
2. Upon receipt of fetal brain tissue (*see* **Note 14**), transfer the tissue to a 50 ml conical tube.

3. Allow the tissue to settle at the bottom of tube, and then pour off excess liquid. If the liquid contains a significant amount of blood, rinse two to three times with PBS (each time allowing the tissue to settle and carefully pouring off ¾ of the liquid).
4. Pour off ¾ of the liquid and transfer the tissue to a 150 × 15 mm Petri dish.
5. Add 10 ml PBS onto the tissue and remove the meninges (if present) using sterile forceps.
6. Use the forceps to remove blood clots, taking care to avoid removing brain tissue.
7. Use two scalpels to crosscut the brain into small pieces of 1–2 mm^3.
8. Add 5–10 ml PBS to the minced tissue and transfer with a pipette into a 100 ml bottle. Add another 10 ml PBS to wash the Petri dish and transfer to the bottle. Fill with PBS to a final volume of 40 ml.
9. Add 5 ml 0.5 % trypsin and 10 ml DNase (25 μg/ml). If volume of tissue exceeds 20 ml, add an additional 5 ml 0.5 % trypsin and 5 ml DNase (25 μg/ml).
10. Tightly cap the bottle, wrap the mouth of the bottle and cap with parafilm (*see* **Note 5**), and shake at 180 rpm, 37 °C for 15 min.
11. During this incubation step, prepare reagents for the following steps. Set up mashing filter (*see* **Note 6**) in the neck of a 500 ml bottle. Remove the plunger of a 20 ml syringe and place on top of the mashing filter.
12. At the end of the incubation, add 5 ml of FCS to inactivate the trypsin.
13. Transfer the dissociated brain pieces onto the filter by adding 5 ml of tissue at a time, adding PBS and mashing with the plunger.
14. Once all the tissue has been mashed, wash the bottle with PBS and transfer onto the filter. Continue to mash and wash with PBS until the mesh is as clear as possible. It is normal that some debris (such as blood clots) may be left on the mesh. The final volume in the bottle should be 150–400 ml, depending on the size of prep.
15. Distribute the liquid from the bottle into 1 conical tube per 50 ml volume (i.e., three to eight tubes total), and spin at 450 × *g*, 10 min.
16. During this spin, warm the medium (DMEM containing 5 % FCS, 0.1 % glucose, 1 % penicillin–streptomycin, 1 % glutamine) to 37 °C.
17. Carefully pour off the supernatant, being careful not to disturb the pellet; 10–15 ml of liquid should remain.
18. Resuspend the first pellet by pipetting 5 ml of medium.

19. Transfer this first volume to the second pellet, and resuspend. Continue until all the pellets have been resuspended and the tissue has been pooled to a single tube. Wash the empty tubes using another 5 ml of medium and transfer to the tube with the pooled tissue.
20. Take an aliquot of the tissue (50 μl) to count, noting that this sample may need to be diluted more than 1:10 if the prep is large.
21. Spin 300 × *g*, 10 min.
22. During this spin, count the cells, taking care to exclude the red blood cells from the count.
23. Pour off the supernatant and resuspend in 20 ml of medium. If cell suspension contains clumps (i.e., dead cells), the suspension should be passed through a 70 μM cell strainer.
24. Bring cell suspension to a concentration of 100×10^6 cells/ml in DMEM 5 % FCS (*see* **Note 15**).
25. Plate cells at 6×10^6 cells/ml in 40–50 ml per T175 flask or 15–20 ml per T75 flask (*see* **Note 16**).
26. Allow astrocytes and microglia to grow for 5–7 days in the flask. At this time, if the astrocyte bed is thick and many floating microglia are visible under the microscope, the microglia can be harvested (go to **Step 31**). If the astrocyte layer is not yet thick and microglia are still adherent, follow the steps below (**27–30**) to change the medium in the flask.
27. Pour the medium out of the flask into a 50 ml conical tube, carefully by inverting the flask, without disturbing the astrocyte layer.
28. Add 25 ml (T175 flask) or 10 ml (T75 flask) fresh medium to the flask, by inverting the flask and expelling liquid onto the surface opposite the cell layer (this avoids disturbing the astrocyte bed). Return the flask to the incubator for the duration of the following spin.
29. Spin the collected medium 300 × *g*, 10 min.
30. Pour off the supernatant and resuspend in 25 ml (T175 flask) or 10 ml (T75 flask) fresh medium. Add this medium back to the flask, again expelling onto the surface of the flask opposite the cell layer. Return the flask to the incubator for another 5–7 days.
31. When the astrocyte layer is thick and microglia are floating, collect the microglia by carefully pouring off the liquid from the flask into a 50 ml conical tube (*see* **Note 17**).
32. Spin the collected medium 300 × *g*, 10 min.
33. Resuspend in 1–2 ml of fresh medium and remove a small aliquot (10 μl) to count.

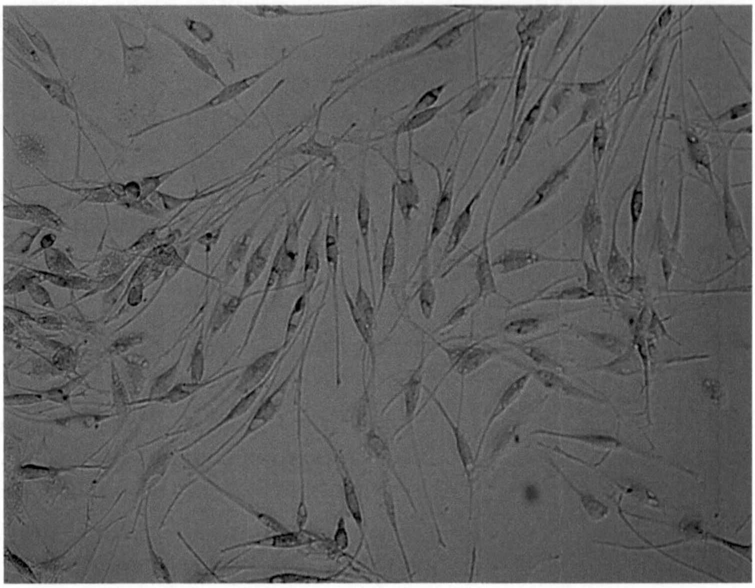

Fig. 2 Human fetal microglia (25×). Cells were isolated according to the described protocol and grown in DMEM containing 5 % FCS, 0.1 % glucose, 1 % penicillin–streptomycin, and 1 % glutamine. In culture, fetal microglia are mitotic with long, slender, bipolar processes

34. Plate the microglia at desired density, usually 100,000 cells/ml, in either a 24-well plate (1 ml/well) or a 6-well plate (3–4 ml/well) (*see* **Note 13**). *See* Fig. 2 for the appearance of typical human fetal microglia after 4–5 days in culture.

3.3 Polarization of Human Microglia

1. Plated microglia can be left untreated for multiple weeks in culture; medium should be changed every 5–7 days. To generate classically activated (M1) microglia, follow **Steps 2** and **3** below; to generate alternatively activated (M2) microglia, follow **Steps 4** and **5** below.
2. To generate M1 microglia, treat cells with GM-CSF (5 ng/ml) for 48 h, then remove half the medium and replace with fresh medium containing a full dose of GM-CSF (such that the final concentration of GM-CSF is again 5 ng/ml) for an additional 48–72 h.
3. Activate M1 cells with IFN-γ (20 ng/ml) for 1 h followed by 24–48 h treatment with LPS (serotype 0127:B8, 100 ng/ml).
4. To generate M2 microglia (*see* **Note 18**), treat the cells as per **Step 2** above, using M-CSF (25 ng/ml) instead of GM-CSF.
5. Activate M2 cells by treatment with IL-4 (20 ng/ml) and IL-13 (20 ng/ml) for 24 h. M2 cells must be primed with an additional dose of IL-4 (20 ng/ml) and IL-13 (20 ng/ml) (without changing the medium) for 24 h in order to achieve full M2 polarization.

6. Polarization can be confirmed by flow cytometry assessing the expression of surface markers associated with the M1 (CD80 and CCR7) and M2 (CD206 and CD209) phenotypes [1].

4 Notes

1. Tissue is obtained as a mixture of white and grey matter of temporal lobe brain tissue from patients undergoing surgery for non-tumor-related intractable epilepsy. Tissue is derived from outside of the suspected focal site. As the material in the CUSA® bag is in small pieces in PBS, tissue hypoxia likely occurs later than if tissue is obtained in block format. If tissue is received in block format, tissue should be manually chopped into small sections as quickly as possible prior to processing.
2. Recording the volume of tissue is required to determine the number of centrifuge tubes that will be used in subsequent steps. Tissue volume is also helpful in approximating cell yield (*see* Fig. 3).
3. This specimen is required in case the diagnosis is later questioned or if a tumor is suspected. In addition, sections can later be made from this stored tissue and be useful in the event that cells obtained from this specimen exhibit abnormal properties.
4. This process can be made easier by dividing the tissue into multiple tubes. Alternatively, the tissue may be spread in a large Petri dish. The objective is to make the tissue as clean as possible, but should not exceed 20 min to limit loss of viable cells.
5. Wrapping the bottle cap in parafilm prevents contamination from splashing of water in the shaking water bath.

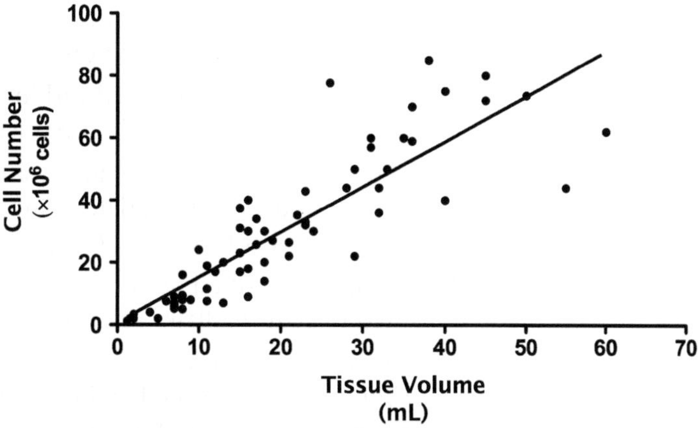

Fig. 3 Total cell yield versus starting tissue volume. Average total cell yield (including microglia and oligodendrocytes) is approximately 1.4×10^6 cells per ml of starting tissue

6. Prepare the mashing filter using a two-piece Buchner funnel and a 132 μm nylon mesh. Cut a square piece of nylon mesh 1–2 cm larger than the diameter of the funnel and place tightly between the two funnel pieces. Tape tightly with autoclave tape to ensure the mesh does not become detached during mashing. Autoclave the prepared mashing filter prior to use.
7. At this point, it is appropriate to determine the number of tubes that will be used for the Percoll® spin. As our fixed rotor centrifuge can accommodate a maximum of 14 tubes, we first determine if one or two spins will be necessary. Our guideline is to use 1–2 ml of tissue per Nalgene tube. For ≤14 ml of tissue, use 1 tube per ml, rounding down to an even number of tubes (to facilitate balancing the centrifuge). For 15–20 ml of tissue, use 14 tubes. For >20 ml of tissue, more than 14 tubes are required for best yield, and thus two spins are necessary. In this case, use 1 tube per ml, rounding down to an even number of tubes and using a maximum of 28 tubes.
8. Adding too much tissue without mashing or adding PBS can clog the mesh and make the process slower.
9. Remove as much myelin as possible, especially if also collecting other cell types, as free myelin is toxic to oligodendrocytes.
10. Care should be taken not to touch the red blood cell layer below; this process can be facilitated by slightly tilting the tube.
11. Methods of culturing oligodendrocytes from adult human brain tissue will not be discussed here; *see* [2].
12. After 5 min, verify under a microscope that some cells are floating. If cells are not yet floating, return flask to incubator for an additional 1–2 min. Leaving cells in trypsin for too long can be detrimental to the cells.
13. If using glass chamber slides, surfaces should be treated with poly-L-lysine prior to plating. Tissue culture-coated plates and flasks do not require additional coating prior to use.
14. Tissue should be processed within 24 h of dissection; a delay of 36 h or longer may result in reduced cell viability.
15. If other glial cells (progenitors, astrocytes, or neurons) are to be cultured, an appropriate volume of cell suspension should be removed at this point; for specific methods, *see* [3–5].
16. To improve adhesion of cells to the flasks, poly-L-lysine can be used to coat the flasks prior to plating.
17. The astrocyte layer is extremely fragile at this point, and care should be taken when handling the flask and when pouring off the medium. Astrocytes must not be dislodged during this step as this may result in astrocyte contamination of the microglial cultures.

18. Alternatively activated microglia generated using IL-4 and IL-13 are referred to as M2a cells. Other classes of M2 cells also exist, namely, M2b (generated by activation with immune complexes and Toll-like receptor ligands) and the "deactivated" M2c (generated using IL-10). The appropriate activation paradigm should be selected accordingly.

References

1. Durafourt BA, Moore CS, Zammit DA et al (2012) Comparison of polarization properties of human adult microglia and blood-derived macrophages. Glia 60(5): 717–727
2. Ruffini F, Arbour N, Blain M et al (2004) Distinctive properties of human adult brain-derived myelin progenitor cells. Am J Pathol 165(6):2167–2175
3. Cui QL, Fragoso G, Miron VE et al (2010) Response of human oligodendrocyte progenitors to growth factors and axon signals. J Neuropathol Exp Neurol 69(9):930–944
4. Darlington PJ, Goldman JS, Cui QL et al (2008) Widespread immunoreactivity for neuronal nuclei in cultured human and rodent astrocytes. J Neurochem 104(5):1201–1209
5. D'Souza S, Alinauskas K, McCrea E et al (1995) Differential susceptibility of human CNS-derived cell populations to TNF-dependent and independent immune-mediated injury. J Neurosci 15(11):7293–7300

Part VII

Co-culture Systems to Analysis Microglia Interactions with Other Cell Types

Chapter 20

Understanding Microglia–Neuron Cross Talk: Relevance of the Microglia–Neuron Cocultures

Fernando G. Correa, Miriam Hernangómez, and Carmen Guaza

Abstract

Microglia–neuron interaction is a complex process involving a plethora of ligands and receptors. The outcome of this intricate process will depend on the prevailing signals (i.e., whether the microglial cells will produce pro-inflammatory cytokines and/or phagocyte a dying neuron or whether it will produce neurotrophic factors and support neuronal growth, among other possible scenarios).

In order to study this complex process, several tools have been developed, ranging from in vivo models (*knockout* and *knock-in* mice, conditional transgenic mice, imaging techniques) to in vitro models (microglia–neuron cocultures, transwell cell cultures). Here we describe a protocol for primary microglia–neuron coculture. This coculture allows to combine neurons and microglial cells coming from *wild-type* and *KO* mice, making this coculture a useful method to study in vitro the interaction of different sets of ligand–receptor.

Key words Microglia, Neurons, Cocultures, Innate immunity, Neuroimmunoregulatory molecules, CD200, CD200R

1 Introduction

Microglia constitute the highly versatile resident macrophages in the central nervous system (CNS), comprising 5–20 % of the total glial cell population (Lawson et al.). As reviewed by Yang et al. [1], in 1919 a disciple of Ramón y Cajal, Pío del Río-Hortega recognized microglial cells as a different cell type from the other glial cells. He also characterized microglial transformation from a ramified phenotype into ameboid phagocytic macrophage-like cells in the stab wounds made in animal brains [2]. From these observations, del Río-Hortega concluded that microglia originated from peripheral mononuclear cells [2].

Interestingly, and after a long debate in which even the existence of microglial cells was questioned (reviewed in [3]), microglia has finally been established as a distinct glial cell population from a myelomonocity origin [4]. Recently, it has been demonstrated that

Bertrand Joseph and José Luis Venero (eds.), *Microglia: Methods and Protocols*, Methods in Molecular Biology, vol. 1041, DOI 10.1007/978-1-62703-520-0_20, © Springer Science+Business Media New York 2013

microglial cells are originated from a distinct myeloid precursor that migrates from the hematopoietic islands in the yolk sac to the developing brain parenchyma at embryonic day 8.5 [5].

Microglia are widely distributed in the adult CNS, but differences in their cellular density among different brain areas have been reported in mice [6, 7] and humans [8]. Thus, microglial cells are more abundant in the telencephalon than in the diencephalon or the mesencephalon, with the rhombencephalon containing the lowest amount of these cells [7] (reviewed in [9]). Not only there is an uneven distribution of microglia across the cerebral anatomical regions but between the gray and white matter (the latter containing higher microglial cell density than the former). Moreover, cumulative data show that microglial expression of tissue macrophage markers varies within the different brain regions as well as their response to different stimuli [10–12].

In the adult healthy brain, microglia has been described to be in "resting" or "quiescent" state with a ramified morphology. However, this term has limitations and can lead to the misleading idea that microglia are dormant cells awaiting a signal that will wake them up. On the contrary, recent elegant in vivo imaging experiments show "resting" microglia as highly dynamic cells, continuously branching thin processes that survey and sample their microenvironment [13, 14] in search for potential threats and danger signals. Thanks to this routine immunosurveillance of the CNS, microglia will remove apoptotic bodies and other potentially toxic cellular debris (myelin debris, amyloid deposits, protein aggregates, etc.) [15–17]. In order to accomplish this task, it is essential that microglial cells are capable of distinguishing between "self" and "nonself" signals (Fig. 1). Clearance of pathogens and toxic cell debris during infection or tissue damage is based on the recognition of "nonself" and "altered-self" patterns by microglia, but also astrocytes, oligodendrocytes, and neurons have been shown to be able to recognize those patterns [18, 19]. There is a plethora of the so-called "eat-me" signals expressed by pathogens and apoptotic or necrotic cells. Some of these signals are a heterogeneous group of molecules known as pathogen-associated molecular patterns (PAMPs) and are characterized by being highly conserved through evolution with little antigenic variability [15]. These PAMPs are constituents of the microbial structure which induce in the host a strong innate immune reaction directed towards the removal of the pathogen by phagocytosis [20]. The classical example of a PAMP is the lipopolysaccharide (LPS), component of the Gram-negative external membrane. Analogous to PAMPs, it has been proposed that cells undergoing programmed cell death express de novo apoptotic cell-associated molecular patterns (ACAMPs) [15, 21, 22]. These ACAMPs would play a key role in the embryonic process in which whole cell populations need to be cleared out without mounting an inflammatory response and

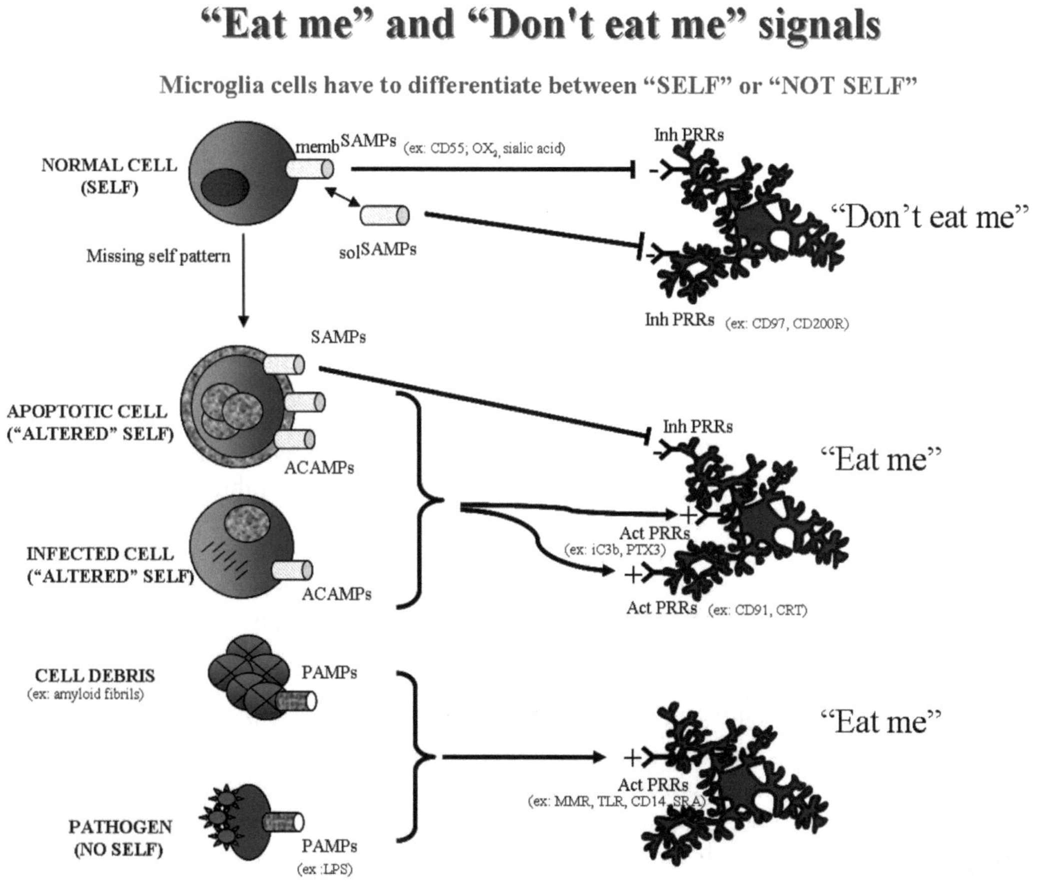

Fig. 1 Scheme showing how microglial cells differentiate "self" from "nonself" cues. Adapted from Medzhitov and Janeway [18]

minimizing the presence of cellular debris [23]. Some ACAMPs include oxidized low-density proteins, alteration of the membrane electrical charges, nucleic acids, and phosphatidylserine [19, 24]. Similarly, damaged or stressed tissues release/express the so-called danger-associated molecular patterns (DAMPs). Some DAMPs including heat shock proteins (HSP), adenosine, ATP, high-mobility group box chromosomal protein 1 (HMGB-1), galectins, and thioredoxin present adjuvant and pro-inflammatory activity [25]. Phagocytic cells recognize these PAMPs, ACAMPs, and/or DAMPs, which can be either membrane bound or soluble, through their pattern recognition receptors (PRRs) [15, 22, 26]. Some of the PRRs include toll-like receptors (TLR), scavenger and mannose receptors, CD14, CD36, complement receptors, phosphatidylserine receptor (PSR), and milk fat globulin (MFG-EGF8) (reviewed in [27]). Therefore, the activation of microglia, rather than an unspecific process, is highly dependent on the stimulus that originated it.

In addition to this plethora of signals that might elicit an innate immune activation directed to the elimination of the pathogen, the apoptotic cell, or the tissue debris, there is a complex set of interactions that silence and reshape microglial response [18]. For instance, electrically active neurons inhibit the interferon-gamma (IFNγ)-induced increase in MHC class II expression in microglia [28]. Some neurotransmitters have modulatory effects on microglial response, whereas others, like substance P, enhance the activation of microglia [29]. Neurotrophins secreted by healthy neurons such as nerve growth factor (NGF), brain-derived growth factor (BDNF), and to a lower extent neurotrophin-3 (NT-3) were able to reverse the induction of MHC class II molecules in microglia [30]. Interestingly, neurons express membrane-bound molecules and/or secrete soluble mediators that function as "don't-eat-me" signals, reshaping microglial response and inhibiting their phagocytic activity on the site of injury [19, 31]. These signals were originally grouped in a family of heterogeneous molecules called self-associated molecular patterns (SAMPs). It was suggested that these SAMPs could interact with novel inhibitory PRRs, negatively modulating the innate immune response and promoting tissue repair [19]. These SAMPs signals are present in almost all host "normal-self" cells but not in pathogens, and they are downregulated in apoptotic/necrotic cells (altered self). Thus, this "normal-self" identity can be lost. An example of a molecular marker of "normal self" is the presence of sialic acid in the terminal of glycoproteins and glycolipids, which can interact with different microglial receptors [32]. One of these inhibitory receptors belongs to a family known as siglecs [33]. Siglecs contain immunoreceptor tyrosine-based inhibitory motifs (ITIMs) in their cytoplasmic tail which downregulate phagocytosis in macrophages and microglia [33, 34]. Most microorganisms lack sialic acid in their glycolipids and/or glycoproteins, and in some cases, virally infected or transformed cells can show a deficient pattern or sialylation, which is a strong signal of "altered self" to promote phagocytosis. Also in apoptotic cells, a reduced expression of sialic acid has been shown [27].

Of particular interest is the physical interaction (by means of their membrane-bound proteins) between neurons and microglia. Thus, a group of SAMPs was renamed neuroimmune regulatory proteins (NIRegs) to highlight their role in modulating and reshaping an adverse immune response and skewing microglia towards a protective phenotype. Interestingly, many of these NIRegs contain an ITIM domain in their cytoplasmic tails which negatively regulated microglial response to insults.

As mentioned before, siglecs are members of a subgroup of the Ig superfamily that recognize the sialic acid residues on the periphery of cell surface glycolipids and glycoprotein. Siglecs contain an ITIM domain in their cytoplasmic tails which signal via the recruitment of tyrosine phosphatases such as SHP1 (reviewed in [35]).

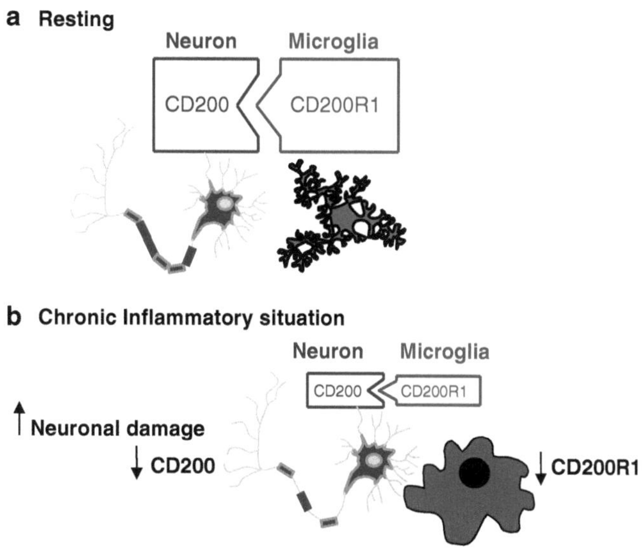

Fig. 2 (**a**) In a basal situation the cross talk between neuron and microglia via the CD200–CD200R interaction is correct and the inhibitory signal is working adequately. (**b**) In an inflammatory situation the cross talk between neuron and microglia through the CD200–CD200R interaction is altered. The reduced inhibitory input from CD200 causes a disturbed equilibrium which results in activation of microglia and neuronal damage

In this sense, when microglial siglecs interact with the neuronal glycocalyx, neurotoxicity is alleviated [36].

Another interesting NIReg protein is the pair CD200–CD200 receptor (CD200R). CD200 is a 41–47 KDa protein belonging to the immunoglobulin Ig supergene family characterized by two IgSF domains [37, 38]. This surface protein is highly conserved and within the CNS is mainly expressed by neurons and vascular endothelium [39]. In contrast, CD200R, which also contains two IgSF domains and a longer cytoplasmic tail with an ITIM domain [37], is chiefly expressed by myeloid cells and microglia [39, 40]. CD200–CD200R interaction plays a critical role in neuronal protection in the setting of inflammatory-mediated neurodegeneration (Fig. 2); in particular, this immunoregulatory system, when deficient, may contribute to chronic inflammation in multiple sclerosis, Alzheimer disease, or in the aging brain. Deletion of CD200 results in myeloid cells dysregulation and enhanced susceptibility to autoimmune response [40–42].

Another neuron–microglia ligand–receptor interaction that can modulate microglial responsiveness is the neuronal CD47 (also expressed by endothelium, astrocytes, and dendritic cells) [43] and the microglial signal-regulatory protein SIRPα ((CD172a), which contains three extracellular Ig domains) [44]. CD47–CD172a interaction recruits tyrosine phosphatases SHP-1 and SHP-2 with inhibition of phagocytosis and synthesis of the anti-inflammatory cytokine TGF-β (reviewed in [27]). Another NIRegs with a potential critical role in neuroprotection include the pair neuronal CX3CL1 and microglial CX3CR1 [45].

Interestingly, some membrane receptors contain immunoreceptor tyrosine-based activation motifs (ITAM). ITAM-containing signalling adaptor proteins are associated with receptor subunits. After ligand–receptor interaction, tyrosine residues of the ITAMs become phosphorylated by members of the Src kinase family and subsequently serve as docking sites for Src homology 2 (SH2) domains of Syk protein kinases which then mediate cellular activation via a number of downstream cascades (reviewed in [35]). Triggering receptor expressed on myeloid cells 2 (TREM-2) is an innate immune glycoprotein heterogeneously expressed in brain microglia [46]. It belongs to the immunoglobulin- and lectin-like superfamily with a short cytoplasmic tail [47]. Interaction of TREM-2 with its uncharacterized ligand induces its association with the ITAM-containing adaptor molecule DAP12 followed by the recruitment of ZAP70, SYK, PI3K, and phospholipase Cγ [48]. Activation of TREM-2/DAP12 leads to augmented phagocytosis with reduced expression of TNFα, IL-6, and iNOS (reviewed in [49]). Mutation of either TREM2 or DAP12 proteins leads to the rare Nasu–Hakola disease (NHD), also known as polycystic lipomembranous osteodysplasia with sclerosing leukoencephalopathy (PLOSL), an autosomal recessive inherited disease [50]. While NHD patients carrying TREM-2 mutation present an early-onset presenile dementia followed by delayed bone symptoms [51], patients with mutations in DAP12 display an early-onset combination of presenile dementia and systemic bone cysts [52]. Interestingly, Piccio et al. [53, 54] showed that a soluble form of TREM-2 was present in the cerebrospinal fluid (CSF) of patients with multiple sclerosis and that blockade of TREM-2 with a blocking antibody exacerbated experimental autoimmune encephalitis in rodents. These observations suggest a crucial role of TREM-2 in preventing neurodegenerative processes. So far the nature of the TREM-2 ligand remains unknown. It has been suggested that TREM-2 binds anionic ligands of the surface of Gram-positive and Gram-negative bacteria, including *Neisseria gonorrhoeae*, as well as an unidentified ligand in the astrocytoma cell line HTB12 (reviewed in [49]). Importantly, Hsieh et al. [55], using a TREM2/Fc chimera, showed that neurons undergoing apoptosis increased the expression of a TREM-2 ligand (TREM-2L), inducing their phagocytosis by BV2 cells. This effect could be reversed with the use of a blocking antibody against microglial TREM-2 [55].

To add to the complexities of neuron–microglia interactions, it has been recently described the existence of receptors containing inhibitory ITAM domains (ITAMi), and conversely some activating ITIMs have been described [56].

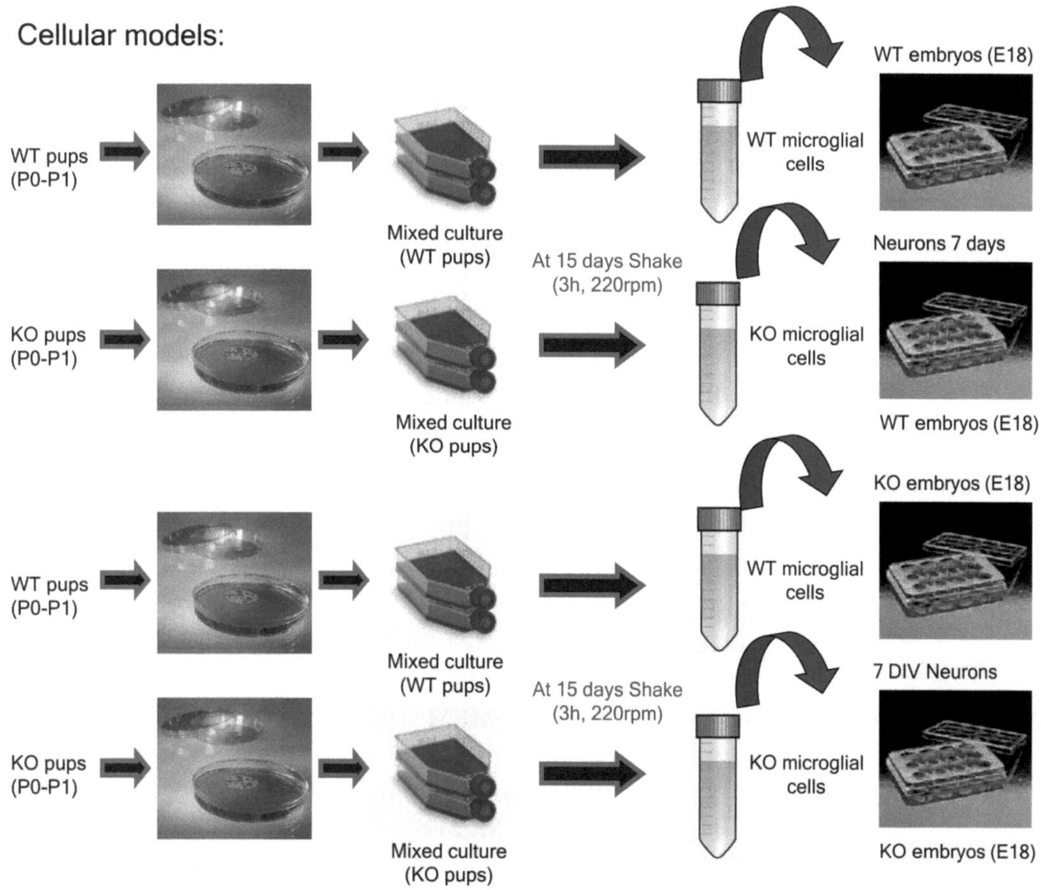

Fig. 3 Scheme showing the different arrays of neuron–microglia cocultures (*wt* microglia/*wt* neurons, KO microglia/*wt* neurons, *wt* microglia/KO neurons, KO microglia/KO neurons) that allow exploring different situations

Many different tools have been developed for the study of neuron–microglia interaction, ranging from transgenic mice to in vitro coculture models. One advantage of the cocultures is that it allows combining neuronal cells from wild-type or KO mice for the molecule of interest with microglial cells from *wild-type* or KO mice in a much simpler and less time and money-consuming way than developing double KO mice. In this sense, by combining the different possibilities (*wild-type* neurons/*wild-type* microglia, *wild-type* neurons/KO microglia, KO neurons/*wild-type* microglia, KO neurons/KO microglia), distinct scenarios can be explored. An example is summarized in Fig. 3. Moreover, since neurons are cultured separately from microglia prior the coculture is performed, these cells can be subjected to different treatments before being put together. This setting allows the dissection of drug effects on the neuron–microglia interaction by targeting one kind of cell only.

Here we describe a coculture model for studying the cell-to-cell contact between neurons and microglial cells [40].

2 Materials

The protocols described here are based on the ones by McCarthy and de Vellis [57] with minor modifications [58] and by Rose et al. [59] with modifications [40, 60]. All experiments involving live animals must conform to national and institutional regulations.

2.1 Reagents

1. Mice or rats in postnatal days 0–2 (P0–P2).
2. Dulbecco's modified Eagle's medium (DMEM).
3. Neurobasal medium.
4. Phosphate-buffered saline (PBS).
5. Fetal bovine serum (FBS).
6. Horse serum (HS).
7. Penicillin/streptomycin.
8. Poly-D-lysine.
9. Trypsin/EDTA.
10. MilliQ H_2O.

2.2 Reagent Setup

1. *DMEM 10:10:1.* (DMEM, 10 % FBS, 10 % HS, 1 % penicillin/streptomycin).
2. *Trypsin/EDTA solution.* (0.05 % Trypsin + 0.02 % EDTA in HBSS w/o Ca2+/Mg2+).
3. *Poly-D-lysine coating solution.* Dilute 5 mg of the poly-D-lysine in 50 ml of sterile MilliQ H_2O, to make a 20× stock solution. Aliquots of 10 ml each can be stored at −20 °C and diluted in 190 ml of MilliQ H_2O to make the 1× working concentration solution (5 μg/ml).

2.3 Equipments

1. Laminar flow hood.
2. Dissecting magnifying glass.
3. Water bath at 37 °C.
4. Humidified tissue culture incubator (37 °C, 5 % CO_2).
5. Sterilized microdissecting instruments: large dissecting scissors, mouse-teeth forceps, curved forceps, and fine Dumont forceps.
6. Orbital shaker.
7. Tabletop centrifuge.
8. 50 ml plastic conical tubes.
9. T75 cm^2 tissue culture flasks with plug seal.
10. Tissue culture plates.
11. Medium-size Petri dishes.
12. 50 μm sterile nylon mesh.

2.4 Equipment Setup

1. *Poly-D-lysine-coated flasks.* Coat the culture flasks with 5 ml of 1× poly-D-lysine (for 1–2 h or overnight) in a 37 °C incubator. Remove the coating solution and wash once with 5 ml of sterile MilliQ H_2O. Keep in the laminar flow hood until use.
2. *Poly-D-lysine-coated plates.* Coat the culture plates with 1× poly-D-lysine and keep overnight in a 37 °C incubator. Remove the coating solution, wash once with 5 ml of sterile MilliQ H_2O, and air dry in the laminar flow hood until use for neuronal cultures. Microglia cell cultures generally do not require coating; however, glass cover slips must be poly-D-lysine coated to ensure a proper adherence of neuronal and microglial cells to the glass surface. In all cases, plates can be stored at 4 °C under sterile conditions until use.

3 Methods

3.1 Preparation of Mixed Glial Cell Culture (Timing: 15 Days)

1. Dissect the brains: decapitate P0–P2 neonatal rat or mouse pups using sterile large scissors and gently place the head into a Petri dish containing 70 % ethanol. *A whole litter (8–12 pups) should be used to ensure high yield rates.*
2. Transfer the head to a Petri dish containing cold DMEM.
3. Use mouse-teeth forceps to hold the rostral portion of the head, and follow the midline to cut the skin and the skull from the nose to the foramen magnum using curved forceps (*see* **Note 1**).
4. Expose the whole brain and remove it from the skull base and place it in a new Petri dish containing cold DMEM.
5. Under a dissecting magnifying glass and using thin Dumont forceps, remove the meningeal layer from the brain. Briefly, remove the meningeal layer from the inner midbrain, cut the meninges between the hemispheres, and eliminate them. Then, clamp the olfactory bulbs and remove the meningeal layer from both hemispheres. Carefully open the hemispheres, remove the choroid plexus covering of the inside, and remove brainstem and cerebellum (*see* **Note 2**).
6. Place all the forebrains in new Petri dishes containing cold DMEM.
7. Using the Dumont forceps, transfer the forebrains to a 50 ml Falcon tube and add ~0.5 ml of DMEM medium to the tube.
8. Gently triturate and dissociate the nervous tissue with a serum-coated Pasteur pipette, adding small amounts of DMEM as long as the content is being aspirated and discarded, until a homogenate can be found in the medium (*see* **Note 3**).
9. Centrifuge the tubes for 10 min to 168 ×*g*.

10. Discard the supernatant by aspiration and suspend the pellet in 50 ml of warm DMEM 10:10:1.
11. Seed poly-D-lysine-coated 75 cm^2 flasks with 10 ml of cell suspension/each.
12. Incubate at 37 °C in water saturated with 5 % CO_2 and 95 % air atmosphere for 15 days, without changing the culture medium along this time.

3.2 Isolation of Microglial Cells (Timing ~4–24 h)

1. After the incubation time, properly close the flasks and place them in an orbital shaker.
2. Shake at 230 rpm for 3 h and collect the supernatants in 50 ml Falcon tube (*see* **Note 4**).
3. Centrifuge the cell suspension for 10 min to 168 × *g*.
4. Discard the supernatant by aspiration and suspend the pellet in 1 ml of warm DMEM 10:10:1.
5. Determine the number of viable cells by gently mixing 10 μl of homogenate with 80 μl of PBS and 10 μl of trypan blue.
6. Dilute the cell suspension to the desired cell concentration with warm DMEM 10:10:1. Incubate for 2 h in a tissue culture incubator with 5 % CO_2 at 37 °C. To eliminate nonadherent cells, replace the medium with warm DMEM 10:10:1, and incubate them for 24 h in a tissue culture incubator with 5 % CO_2 at 37 °C.
7. Assess the purity of microglial cultures by examining the characteristic cell morphologies under phase-contrast microscopy, and confirm it by immunostaining with Mac-1 anti-CD11b antibody.
8. At this point, microglial cells can be used to perform in vitro experiments.

3.3 Preparation of Cortical Neuronal Cell Culture (Timing: 7 Days)

1. Remove the E17-E18 embryos from pregnant rat or mouse: euthanize a pregnant rat or mouse with CO_2 or by cervical dislocation. Spray the abdomen region with 70 % EtOH and open the abdominal cavity with clean scissors. Remove the uterus containing the embryos and place it in a Petri dish containing DMEM.
2. Dissect the brains: remove individual embryos from the uterine horns and decapitate them using sterile large scissors and gently place the head into a Petri dish containing DMEM.
3. Using forceps gently remove the brain from the skull and place them in a new Petri dish containing cold DMEM.
4. Under a dissecting magnifying glass and using thin Dumont forceps, remove the meningeal layer from the brain. Briefly, remove the meningeal layer from the inner midbrain, cut the

meninges between the hemispheres, and eliminate them. Carefully open the hemispheres, remove the choroid plexus covering of the inside, and remove brainstem and cerebellum.

5. Dissect the cortices from the rest of the brain and transfer them into a 15 ml Falcon with 1 ml of PBS containing 0.25 % trypsin and 1 mg/ml DNAse I at 37 °C for 15 min.
6. Gently triturate and dissociate the nervous tissue with a serum-coated Pasteur pipette, adding small amounts of DMEM as long as the content is being aspirated and discarded, until a single-cell suspension is achieved.
7. Centrifuge at 500 rpm for 10 min, discard the supernatant by aspiration and resuspend the pellet in 1 ml of DMEM supplemented with 10 % heat-inactivated horse serum.
8. Determine the number of viable cells by gently mixing 10 μl of homogenate with 80 μl of PBS and 10 μl of trypan blue.
9. Dilute the cell suspension to the desired cell concentration with warm DMEM 10:10:1. Incubate for 3 h in a tissue culture incubator with 5 % CO_2 at 37 °C. To eliminate nonadherent cells, replace the medium with warm neurobasal medium supplemented containing 1 % penicillin/streptomycin, 2 mM glutamax, and 1× of the antioxidant B-27 supplement.
10. After 3 days in vitro (DIV), halt nonneuronal cell division by exposure to 10 μM cytosine-D-arabinofuranoside (Ara C), and continue growing neuronal cells in a humidified environment containing 5 % CO_2 and held at a constant temperature of 37 °C for about 7–8 days.
11. Assess the purity of neuronal cell cultures by examining the characteristic cell morphologies under phase-contrast microscopy, and confirm it by immunostaining with using anti-MAP-2. The amount of cells other than neurons can be quantified using GFAP for astrocytes, Mac-1, anti-CD11b antibody for microglia, and PDFGRα for OPCs. Discard any neuronal cell culture containing more than 2 % of nonneuronal cells.
12. At this point, neuronal cells can be used to perform in vitro experiments.

3.4 Preparation of Neuron–Microglia Coculture

1. In order to prepare neuron–microglia cocultures, it is extremely important to coordinate the timing of the preparation of neuronal cultures and of mixed glial cell cultures followed by the isolation of microglia.
2. Start by preparing a mixed glial cell culture as described above, and after 7 DIV, proceed to perform a cortical neuronal cell culture as previously described. This will ensure that the isolation of microglia (15 DIV) will correspond to the 7–8 DIV of the neuronal culture.

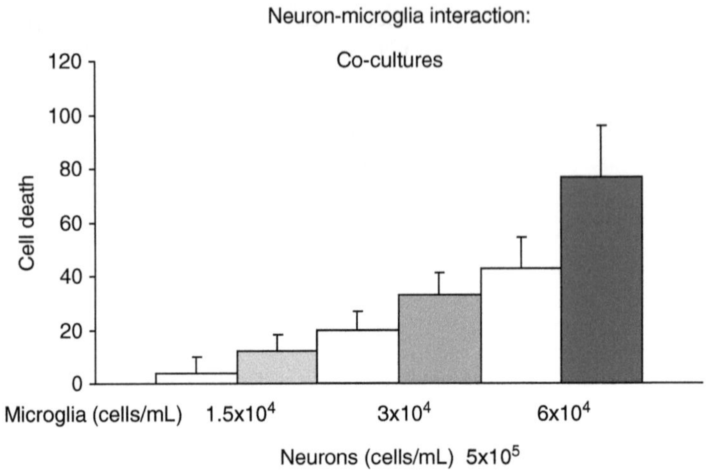

Fig. 4 The magnitude of neuronal cell death is dependent on the number of "resting" (*white bars*) or "activated" (*black bars*) microglial cells added to neuronal cultures (color figure online)

3. In order to establish the adequate coculture conditions, preliminary experiments varying the density of microglial cells and the neuron/microglia ratio should be performed.
4. In Hernangómez et al. [40], we described that the optimized coculture conditions for our study of CD200–CD200R1 interaction were achieved by adding 1.5×10^4 primary microglial onto 5×10^5 neuronal cultures that had been cultured for 7–8 days. Densities of 1.5×10^4, 3×10^4, or 6×10^4 cells per well in 12-well plates containing 5×10^5 neurons per well were tested (Fig. 4), obtaining the best results with the proportion of 1.5×10^4 microglial cells per 5×10^5 neurons (1:33 ratio).
5. However, the optimal ratio may vary depending on the experimental design and should be established accordingly.

4 Notes

Some tricks can be applied to improve the development of the culture:

1. In the dissection procedure, leave the olfactory bulbs intact so they can help you to remove quickly the meninges from the cerebral hemispheres.
2. If the meninges have not been completely removed, forebrain-cell suspension can be passed through a 150 μm nylon mesh after the mechanical dissociation, to prevent the contamination with fibroblasts.

3. If the mechanical dissociation of the nervous tissue is not enough to obtain a homogenate, a serum-coated fire-polished Pasteur pipette (with decreased diameter of the tip) can be used to ensure the triturating of the forebrains.
4. Avoid the use of plastic/glass pipettes by passing the cells directly from the flask to the Falcon or the Petri dish; if needed, they can be coated with horse serum or fetal serum simply taking some milliliters and releasing them, as this coating provides a smooth hydrophobic barrier so the adherence of cells to the glass/plastic surface is greatly reduced and minimizes physical damage to the cells.

Acknowledgments

The authors thank to REEM (Red Española de Esclerosis Múltiple) RD07/0060/0010, Instituto de Salud Carlos III, and Plan Nacional, SAF 2010/17501, Ministerio de Economía y Competitividad for financial support.

References

1. Yang I, Han SJ, Kaur G, Crane C, Parsa AT (2010) The role of microglia in central nervous system immunity and glioma immunology. J Clin Neurosci 17:6–10
2. del Río-Hortega P (1919) El tercer elemento de los centros nerviosos. I La microglia en estado normal. II Intervención de la microglia en los procesos patológicos. III Naturaleza probable de la microglia. Bol de la Soc esp de biol 9:69–120
3. Graeber MB (2010) Changing face of microglia. Science 330:783–788
4. Ling EA, Wong WC (1993) The origin and nature of ramified and amoeboid microglia: a historical review and current concepts. Glia 7:9–18
5. Saijo K, Glass CK (2011) Microglial cell origin and phenotypes in health and disease. Nat Rev Immunol 11:775–787
6. Barnham KJ, Cappai R, Beyreuther K, Masters CL, Hill AF (2006) Delineating common molecular mechanisms in Alzheimer's and prion diseases. Trends Biochem Sci 31: 465–472
7. Lawson LJ, Perry VH, Dri P, Gordon S (1990) Heterogeneity in the distribution and morphology of microglia in the normal adult mouse brain. Neuroscience 39:151–170
8. Mittelbronn M, Dietz K, Schluesener HJ, Meyermann R (2001) Local distribution of microglia in the normal adult human central nervous system differs by up to one order of magnitude. Acta Neuropathol 101:249–255
9. Olah M, Biber K, Vinet J, Boddeke HW (2011) Microglia phenotype diversity. CNS Neurol Disord Drug Targets 10:108–118
10. Hanisch UK, Kettenmann H (2007) Microglia: active sensor and versatile effector cells in the normal and pathologic brain. Nat Neurosci 10:1387–1394
11. Kraft AD, McPherson CA, Harry GJ (2009) Heterogeneity of microglia and TNF signaling as determinants for neuronal death or survival. Neurotoxicology 30:785–793
12. Ransohoff RM, Perry VH (2009) Microglial physiology: unique stimuli, specialized responses. Annu Rev Immunol 27:119–145
13. Davalos D, Grutzendler J, Yang G, Kim JV, Zuo Y, Jung S, Littman DR, Dustin ML, Gan WB (2005) ATP mediates rapid microglial response to local brain injury in vivo. Nat Neurosci 8:752–758
14. Nimmerjahn A, Kirchhoff F, Helmchen F (2005) Resting microglial cells are highly dynamic surveillants of brain parenchyma in vivo. Science 308:1314–1318

15. Gasque P, Dean YD, McGreal EP, VanBeek J, Morgan BP (2000) Complement components of the innate immune system in health and disease in the CNS. Immunopharmacology 49:171–186
16. Barnum SR (2002) Complement in central nervous system inflammation. Immunol Res 26:7–13
17. Wyss-Coray T, Mucke L (2002) Inflammation in neurodegenerative disease–a double-edged sword. Neuron 35:419–432
18. Medzhitov R, Janeway CA Jr (2002) Decoding the patterns of self and nonself by the innate immune system. Science 296:298–300
19. Elward K, Gasque P (2003) "Eat me" and "don't eat me" signals govern the innate immune response and tissue repair in the CNS: emphasis on the critical role of the complement system. Mol Immunol 40:85–94
20. Matzinger P (2007) Friendly and dangerous signals: is the tissue in control? Nat Immunol 8:11–13
21. Franc NC, White K, Ezekowitz RA (1999) Phagocytosis and development: back to the future. Curr Opin Immunol 11:47–52
22. Gregory CD (2000) CD14-dependent clearance of apoptotic cells: relevance to the immune system. Curr Opin Immunol 12:27–34
23. Imler JL, Hoffmann JA (2002) Toll receptors in Drosophila: a family of molecules regulating development and immunity. Curr Top Microbiol Immunol 270:63–79
24. Fadok VA, Bratton DL, Rose DM, Pearson A, Ezekewitz RA, Henson PM (2000) A receptor for phosphatidylserine-specific clearance of apoptotic cells. Nature 405:85–90
25. Amor S, Puentes F, Baker D, van der Valk P (2010) Inflammation in neurodegenerative diseases. Immunology 129(2):154–169
26. Parnaik R, Raff MC, Scholes J (2000) Differences between the clearance of apoptotic cells by professional and non-professional phagocytes. Curr Biol 10:857–860
27. Griffiths M, Neal JW, Gasque P (2007) Innate immunity and protective neuroinflammation: new emphasis on the role of neuroimmune regulatory proteins. Int Rev Neurobiol 82:29–55
28. Neumann H, Boucraut J, Hahnel C, Misgeld T, Wekerle H (1996) Neuronal control of MHC class II inducibility in rat astrocytes and microglia. Eur J Neurosci 8(12):2582–2590
29. McCluskey LP, Lampson LA (2000) Local neurochemicals and site-specific immune regulation in the CNS. J Neuropathol Exp Neurol 59:177–187
30. Neumann H, Wekerle H (1998) Neuronal control of the immune response in the central nervous system: linking brain immunity to neurodegeneration. J Neuropathol Exp Neurol 57:1–9
31. Grimsley C, Ravichandran KS (2003) Cues for apoptotic cell engulfment: eat-me, don't eat-me and come-get-me signals. Trends Cell Biol 13:648–656
32. Meri S, Pangburn MK (1990) Discrimination between activators and nonactivators of the alternative pathway of complement: regulation via a sialic acid/polyanion binding site on factor H. Proc Natl Acad Sci USA 87: 3982–3986
33. Crocker PR (2005) Siglecs in innate immunity. Curr Opin Pharmacol 5:431–437
34. Angata T, Kerr SC, Greaves DR, Varki NM, Crocker PR, Varki A (2002) Cloning and characterization of human Siglec-11. A recently evolved signaling molecule that can interact with SHP-1 and SHP-2 and is expressed by tissue macrophages, including brain microglia. J Biol Chem 277:24466–24474
35. Linnartz B, Wang Y, Neumann H (2010) Microglial immunoreceptor tyrosine-based activation and inhibition motif signaling in neuroinflammation. Int J Alzheimers Dis 2010
36. Wang Y, Neumann H (2010) Alleviation of neurotoxicity by microglial human Siglec-11. J Neurosci 30:3482–3488
37. Barclay AN, Wright GJ, Brooke G, Brown MH (2002) CD200 and membrane protein interactions in the control of myeloid cells. Trends Immunol 23:285–290
38. Wright GJ, Jones M, Puklavec MJ, Brown MH, Barclay AN (2001) The unusual distribution of the neuronal/lymphoid cell surface CD200 (OX2) glycoprotein is conserved in humans. Immunology 102:173–179
39. Koning N, Swaab DF, Hoek RM, Huitinga I (2009) Distribution of the immune inhibitory molecules CD200 and CD200R in the normal central nervous system and multiple sclerosis lesions suggests neuron-glia and glia-glia interactions. J Neuropathol Exp Neurol 68:159–167
40. Hernangómez M, Mestre L, Correa FG, Loría F, Mecha M, Íñigo PM, Docagne F, Williams RO, Borrell J, Guaza C (2012) CD200-CD200R1 interaction contributes to neuroprotective effects of anandamide on experimentally induced inflammation. Glia 60:1437–1450
41. Hoek RM, Ruuls SR, Murphy CA, Wright GJ, Goddard R, Zurawski SM, Blom B, Homola ME, Streit WJ, Brown MH, Barclay AN,

Sedgwick JD (2000) Down-regulation of the macrophage lineage through interaction with OX2 (CD200). Science 290:1768–1771
42. Gorczynski RM, Chen Z, Yu K, Hu J (2001) CD200 immunoadhesin suppresses collagen-induced arthritis in mice. Clin Immunol 101:328–334
43. de Vries HE, Hendriks JJ, Honing H, De Lavalette CR, van der Pol SM, Hooijberg E, Dijkstra CD, van den Berg TK (2002) Signal-regulatory protein alpha-CD47 interactions are required for the transmigration of monocytes across cerebral endothelium. J Immunol 168:5832–5839
44. Brown EJ, Frazier WA (2001) Integrin-associated protein (CD47) and its ligands. Trends Cell Biol 11:130–135
45. Lyons A, Lynch AM, Downer EJ, Hanley R, O'Sullivan JB, Smith A, Lynch MA (2009) Fractalkine-induced activation of the phosphatidylinositol-3 kinase pathway attenuates microglial activation in vivo and in vitro. J Neurochem 110(5):1547–1556
46. Schmid CD, Sautkulis LN, Danielson PE, Cooper J, Hasel KW, Hilbush BS, Sutcliffe JG, Carson MJ. Heterogeneous expression of the triggering receptor expressed on myeloid cells-2 on adult murine microglia. J Neurochem 83(6):1309–1320.
47. Bouchon A, Hernandez-Munain C, Cella M, Colonna M (2001) A DAP12-mediated pathway regulates expression of CC chemokine receptor 7 and maturation of human dendritic cells. J Exp Med 194:1111–1122
48. Lanier LL, Corliss BC, Wu J, Leong C, Phillips JH (1998) Immunoreceptor DAP12 bearing a tyrosine-based activation motif is involved in activating NK cells. Nature 391:703–707
49. Ford JW, McVicar DW (2009) TREM and TREM-like receptors in inflammation and disease. Curr Opin Immunol 21:38–46
50. Colonna M (2003) DAP12 signaling: from immune cells to bone modeling and brain myelination. J Clin Invest 111:313–314
51. Chouery E, Delague V, Bergougnoux A, Koussa S, Serre JL, Megarbane A (2008) Mutations in TREM2 lead to pure early-onset dementia without bone cysts. Hum Mutat 29:E194–E204
52. Paloneva J, Kestila M, Wu J, Salminen A, Bohling T, Ruotsalainen V, Hakola P, Bakker AB, Phillips JH, Pekkarinen P, Lanier LL, Timonen T, Peltonen L (2000) Loss-of-function mutations in TYROBP (DAP12) result in a presenile dementia with bone cysts. Nat Genet 25:357–361
53. Piccio L, Buonsanti C, Mariani M, Cella M, Gilfillan S, Cross AH, Colonna M, Panina-Bordignon P (2007) Blockade of TREM-2 exacerbates experimental autoimmune encephalomyelitis. Eur J Immunol 37:1290–1301
54. Piccio L, Buonsanti C, Cella M, Tassi I, Schmidt RE, Fenoglio C, Rinker J, Naismith RT, Panina-Bordignon P, Passini N, Galimberti D, Scarpini E, Colonna M, Cross AH (2008) Identification of soluble TREM-2 in the cerebrospinal fluid and its association with multiple sclerosis and CNS inflammation. Brain 131: 3081–3091
55. Hsieh CL, Koike M, Spusta SC, Niemi EC, Yenari M, Nakamura MC, Seaman WE (2009) A role for TREM2 ligands in the phagocytosis of apoptotic neuronal cells by microglia. J Neurochem 109:1144–1156
56. Barrow AD, Trowsdale J (2006) You say ITAM and I say ITIM, let's call the whole thing off: the ambiguity of immunoreceptor signalling. Eur J Immunol 36:1646–1653
57. McCarthy KD, de Vellis J (1980) Preparation of separate astroglial and microglial cell cultures from rat cerebral tissue. J Cell Biol 85:890–902
58. Mecha M, Íñigo PM, Mestre L, Hernangómez M, Borrell J, Guaza C (2011) An easy and fast way to obtain a high number of glial cells from rat cerebral tissue: a beginners approach. Protocol Exchange. doi:10.1038/protex.2011.218
59. Rose K, Goldberg MP, Choi DW (1993) Cytotoxicity in murine neocortical cell culture. In: Tyson CA, Frazier JM (eds) Methods in neurotoxicology. Academic, San Diego, pp 46–60
60. Loría F, Petrosino S, Hernangómez M, Mestre L, Spagnolo A, Correa F, Di Marzo V, Docagne F, Guaza C (2010) An endocannabinoid tone limits excitotoxicity in vitro and in a model of multiple sclerosis. Neurobiol Dis 37(1):166–176

Chapter 21

Preparation of Rodent Primary Cultures for Neuron–Glia, Mixed Glia, Enriched Microglia, and Reconstituted Cultures with Microglia

Shih-Heng Chen, Esteban A. Oyarzabal, and Jau-Shyong Hong

Abstract

Microglia, neurons, and macroglia (astrocytes and oligodendrocytes) are the major cell types in the central nervous system. In the past decades, primary microglia-enriched cultures have been widely used to study the biological functions of microglia in vitro. In order to study the interactions between microglia and other brain cells, neuron–glia, neuron–microglia, and mixed glia cultures were developed. The aim of this chapter is to provide basic and adaptable protocols for the preparation of these microglia-containing primary cultures from rodent. Meanwhile, we also want to provide a collection of tips from our collective experiences doing primary brain cell cultures.

Key words Microglia, Neuron–glia culture, Mixed glia culture, Microglia-enriched culture, Reconstituted neuron–microglia culture

1 Introduction

Although the in vitro study of microglia goes as far back as 1930 [1, 2], it was not until 1986, with the development of a new protocol to selectively isolate and culture microglia from mammalian brain [3], that the use of microglia in cell culture systems of the brain became popular. Since then, and with the advent of several different microglia-containing cell culture protocols and microglia-specific cell markers, our understanding of the physiology of microglia continues to vastly improve.

Several cell culture systems exist to study microglia derived from rodent brains, including mesencephalic neuron–glia, mixed glia, enriched microglia, and reconstituted neuron–microglia cultures. Each culture system can serve different purposes to study the molecular mechanisms and the interactions between microglia and neurons, microglia and astrocytes, or microglia themselves under normal physiological or pathological conditions.

Bertrand Joseph and José Luis Venero (eds.), *Microglia: Methods and Protocols*, Methods in Molecular Biology, vol. 1041, DOI 10.1007/978-1-62703-520-0_21, © Springer Science+Business Media New York 2013

Mesencephalic neuron–glia cultures serve as an excellent in vitro model of the midbrain—both in the diversity and ratio of different cell types when compared to the normal physiology of the brain. This culture is predominantly composed of astrocytes (~50 % GFAP immunopositive cells), neurons (~40 % NeuN immunopositive cells), and microglia (~10 % OX-42 immunopositive cells) [4] and is commonly used as a platform to study neuron–glia interaction, especially in culture models that focus on degeneration or neuroprotection of dopaminergic neurons (which account for 1–2 % of the neurons in this culture)—such as in toxin-induced models of Parkinson's disease [4–7].

Primary mixed glia cultures are composed of astrocytes (~80 % GFAP immunopositive cells) and microglia (~20 % OX-42 immunopositive cells) [8] and are commonly used to study the interaction between microglia and astrocytes. In mixed glia cultures, microglia grow loosely attached to the confluent monolayer of astrocytes; therefore, microglia can be shaken off to produce a microglia-enriched culture that is composed of more than 95 % microglia, as characterized by immunocytochemical staining with OX-42 [9]. Primary rodent microglia-enriched cultures are widely used to study microglial biology in vitro. Lastly, enriched microglia can be seeded on top of neuron-enriched cultures to generate reconstituted neuron–microglia cocultures which are a useful tool to investigate direct neuron and microglia interaction in vitro.

Over the past decade these protocols have successfully allowed our laboratory to investigate the physiology of microglia. Our hope is to share our experience to help other researchers establish their own microglia-containing primary cultures.

2 Materials

2.1 Equipment Component

1. A guillotine or an acrylic box connected to CO_2 tank for euthanizing animals (*see* **Note 1**).
2. One pair operating scissors (5½″).
3. One pair dissecting scissors (4¼″)
4. One pair microdissecting scissors (4½″, 17 mm blades).
5. One pair curved dissecting forceps (5″).
6. One pair microdissecting tweezers (no. 5).
7. One pair curved microdissecting tweezers (no. 5).
8. Dissection microscope.
9. Sterile petri dishes.
10. Sterile 50 mL conical vials.
11. Sterile 10 and 25 mL serological pipettes.
12. Sterile 200 μL and 1 mL pipette tips.

13. Sterile 70 μm cell strainer (BD Bioscience, San Jose, CA, USA).
14. Sterile tissue culture-grade plates (Corning Inc., Corning, NY, USA) or flasks (BD Bioscience, San Jose, CA, USA).
15. Hemocytometer, 0.4 % Trypan Blue Stain solution (Invitrogen, Carlsbad, CA, USA) and bright-field microscope.
16. Humidified cell culture incubator (37 °C, 5 % CO_2).

2.2 Reagents and Solutions Component

1. Neuron–Glia Dissection Buffer: Minimum Essential Medium (MEM) (cat. No. 11090, Invitrogen, Carlsbad, CA, USA).
2. Mixed Glia Dissection Buffer: Dulbecco's modified Eagle's medium/Nutrient Mixture F-12 (DMEM/F12) (cat. No. 11330, Invitrogen, Carlsbad, CA, USA).
3. Poly-D-lysine working solution: dilute 100 μg/mL poly-D-lysine (Sigma-Aldrich, St. Louis, MO, USA) in ddH_2O to a final concentration of 20 μg/mL. Store at 4 °C.
4. Cytosine β-D-arabinofuranoside (ara-C) (Sigma-Aldrich, St. Louis, MO, USA) solution: dilute 10 mM of stock solution in Neuron–Glia Maintenance Medium to the desired final concentration.
5. Neuron–Glia Maintenance Medium (for 500 mL): combine 380 mL of MEM, 50 mL of heat-inactivated fetal bovine serum (FBS), 50 mL of heat-inactivated horse serum (HS), 5 mL of 100× nonessential amino acids (Invitrogen, Carlsbad, CA, USA), 5 mL of 100 mM sodium pyruvate (Sigma-Aldrich, St. Louis, MO, USA), 5 mL of 200 mM L-glutamine, 5 mL of 5,000 U/mL penicillin and 5,000 μg/mL streptomycin, and 0.5 g of D-glucose. Stir gently to dissolve, filter sterilize, and store in the dark at 4 °C.
6. Mixed Glia Maintenance Medium (for 500 mL): combine 430 mL of DMEM/F12 (Invitrogen, Carlsbad, CA, USA), 5 mL of 100× nonessential amino acids, 5 mL of 100 mM sodium pyruvate, 5 mL of 200 mM L-glutamine, and 5 mL of 5,000 U/mL penicillin and 5,000 μg/mL streptomycin. Stir gently to dissolve, adjust pH to 7.2, add 50 mL of heat-inactivated FBS, filter sterilize, and store in the dark at 4 °C.

3 Methods

3.1 Preparing for Primary Culture

1. All cell culture plates and flasks are pre-coated with poly-D-lysine before seeding with single cell suspension. Coat 24- and 6-well plates with 0.5 and 1 mL of the poly-D-lysine working solution, respectively, and 25 mL to 175 cm^2 culture flasks. Place in 37 °C incubator for at least 1 h.
2. After coating, rinse the culture plates or flasks three times: twice with sterile ddH_2O and once with sterile PBS. The volume

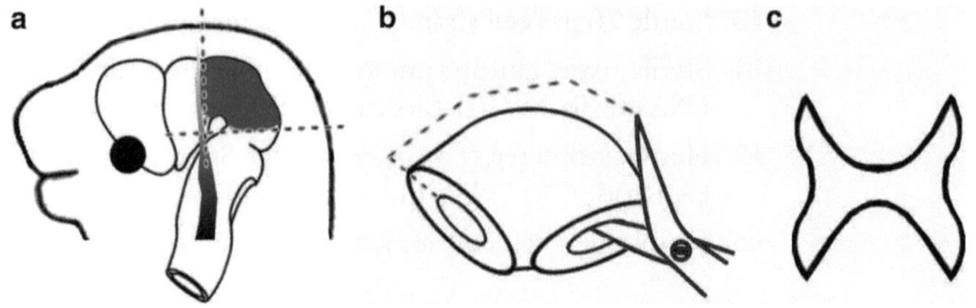

Fig. 1 Illustration of the steps to isolate the mesencephalic region. (**a**) Separate the rostral forebrain (use the traverse sinus as hallmark) and the caudal hindbrain from the mesencephalic region as indicated the *dotted lines*. (**b**) Butterfly the tissue by inserting the microdissection scissors through the inside of the mesencephalic region of the neural tubule and slicing open the tissue along the dorsal midline. (**c**) Remove the meninges from the butterflied mesencephalic tissue

of ddH_2O and PBS in different culture plates and flask is 1 mL in 24-well plate, 2 mL in 6-well plate, and 25 mL in 175 cm^2 flask. Store culture plates and flasks in PBS at 37 °C until immediately before cell seeding.

3. Preheat aliquoted culture medium in water bath at 37 °C, and sterilize dissection instruments and work surfaces with 70 % ethanol beforehand.

3.2 Mesencephalic Neuron–Glia Culture Protocol

1. Euthanize two to four rat dams at gestation day 14–15 and mouse dams at gestation day 13–14 (*see* **Notes 1** and **2**).
2. Lay the rodents on their back and clean the surface of abdominal area with 70 % ethanol. Lift the abdominal skin with the curved dissecting forceps, and using operating scissors make a vertical incision to expose the abdominal muscle layer. Using the dissection scissors make a vertical incision through the muscle layer to expose the abdominal cavity. Carefully remove the uterine horns and place them in a petri dish filled with ice-cold Neuron–Glia Dissection Buffer.
3. Carefully remove the embryos from the amniotic membrane and transfer them into a petri dish with ice-cold Neuron–Glia Dissection Buffer and place on ice. Rinse the embryos twice with ice-cold Neuron–Glia Dissection Buffer and transfer them into a new petri dish with ice-cold Neuron–Glia Dissection Buffer. Place this dish on top of a petri dish filled with ice and set on the floor of a dissection microscope.
4. Under the microscope, hold down the body of the embryo with a pair of microdissecting tweezers, and make two perpendicular incisions with the microdissecting scissor to cut out the mesencephalic region of the brain (*see* Fig. 1a) [10]. Insert one blade of the scissors into the tubular structure, and butterfly the tissue by making an incision down the dorsal midline (*see* Fig. 1b, c).

Carefully separate the tissue from the meninges, and transfer the ventral midbrain tissue into a 50 mL conical tube filled with 10 mL of ice-cold dissection Neuron–Glia Dissection Buffer.

5. In a biological safety cabinet, triturate the midbrain tissues into a single cell suspension by slowly passing the tissue through a 10 mL serological pipette until the tissues are smaller than the opening of the 1 mL pipette tip. Fit a sterile 1 mL pipette tip to the end of the 10 mL serological pipette, and continue to triturate the tissue until it is smaller than the opening of a 200 μL pipette tip. Remove the 1 mL pipette tip and fit a 200 μL pipette tip to the end of the 10 mL serological pipette and triturate the tissues three to five times and strain the cell suspension through a 70 μm cell strainer into a fresh 50 mL conical tube (*see* **Note 3**).
6. Centrifuge the cell suspension at 430 × *g* for 6 min at 4 °C.
7. Decant the supernatant and resuspend the pellet in 10 mL of warm Neuron–Glia Maintenance Medium.
8. Using Trypan Blue Stain solution and a hemocytometer, determine the cell count and viability of the cell suspension. Adjust the density of the viable cells with warm Neuron–Glia Maintenance Medium to 1×10^6 cells/mL. Cap and invert the conical tube to ensure a thorough mixture of the cell suspension. Add 0.5 mL per well of the cell suspension to a poly-D-lysine-coated 24-well plates (*see* **Note 4**). Gently agitate the plate in several directions to ensure even seeding, and place it in a humidified 37 °C, 5 % CO_2 incubator. Avoid stacking plates because it may affect the cellular air exchange of the plates.
9. Gently supplement each well with 0.5 mL of warm Neuron–Glia Maintenance Medium on the third day after seeding. Be sure to add medium along the side of the wells to avoid disturbing the cellular monolayer.
10. Cells are ready for treatment 7 days after their initial seeding. The cellular composition of the neuron–glia culture can be determined by immunocytochemistry staining (*see* **Note 5**).

3.3 Mixed Glia Culture Protocol

1. Wipe the head of 0–1-day-old rodent pups with 70 % ethanol and euthanize them by decapitation.
2. Remove the whole brain by securing the rostral end of the head (dorsal side upward) with a pair of tweezers and making an incision through the foramen magnum towards the ear and curving up (near the eye sockets) towards the coronal suture. Peel back the skin and skull to expose the brain with a pair of tweezers and gently scoop up the entire brain, and immerse it in a petri dish filled with Mixed Glia Dissection Buffer. Place this dish on top of a petri dish filled with ice and set on the floor of a dissection microscope.

Table 1
Seeding concentration for primary murine mixed glia cultures

	Mouse	Rat
6-well plate	1.5×10^6 cells/well	1×10^6 cells/well
24-well plate	1.5×10^5 cells/well	1×10^5 cells/well
175 cm^2 flask	5 brains/flask	2.5 brains/flask

3. Remove the olfactory bulbs, cerebellum, and brain stem using microdissecting tweezers and curved microdissecting tweezers. Dissociate the remaining brain tissue into cerebral hemispheres and midbrain. Carefully remove all the meninges and blood vessels from each tissue. Pool the brain tissue into a 50 mL conical tube with 15 mL of ice-cold Mixed Glia Dissection Buffer.
4. In a biological safety cabinet, triturate the pooled cerebral hemisphere and midbrain tissues into a single cell suspension by slowly passing the tissue through a 10 mL serological pipette until the tissues are smaller than the opening of the 1 mL pipette tip. Fit a sterile 1 mL pipette tip to the end of the 10 mL serological pipette, and continue to triturate the tissue until it is smaller than the opening of a 200 μL pipette tip. Remove the 1 mL pipette tip and fit a 200 μL pipette tip to the end of the 10 mL serological pipette and triturate the tissues three to five times and strain the cell suspension through a 70 μm cell strainer into a fresh 50 mL conical tube (*see* **Note 3**).
5. Centrifuge the cell suspension at 430 × *g* for 6 min at 4 °C.
6. Decant the supernatant and resuspend the pellet in 10 mL of warm mixed glia culture maintenance medium.
7. Using Trypan Blue Stain and a hemocytometer, determine the cell count and viability of the cell suspension. Adjust the density of the viable cells with warm Mixed Glia Maintenance Medium to acquire the proper seeding concentration (*see* Table 1).
8. Refresh the medium every three days by completely removing and replacing the spent medium. Add 25 mL for flasks, 1 mL/well for 24-well plates, and 3 mL/well for 6-well plates (*see* **Note 6**).
9. Cells are ready for treatment 7–14 days after their initial seeding, depending on size of the culture container. Cultures seeded on 6- or 24-well plates are ready for treatment 7 days after seeding (*see* **Note 7**).

3.4 Enriched Microglia Culture Protocol

1. Grow mixed glial cultures (*see* Subheading 3.3) in 175 cm^2 culture flasks for about 14–16 days after seeding, tightly cap the flasks, and seal them with parafilm (*see* **Note 8**).

2. Stack the flasks on a flat platform shaker and shake the flasks at 180 rpm for 30 min to 1 h at 37 °C (*see* **Note 9**). After shaking, collect the medium into 50 mL conical tubes, and centrifuge the cell suspension for 6 min at 430 × *g* at 4 °C.
3. Resuspend the cell pellet in an appropriate volume of Mixed Glia Maintenance Medium.
4. Using Trypan Blue Stain and a hemocytometer, determine the cell count and viability of the cell suspension. Adjust the density of the viable cells with warm Mixed Glia Maintenance Medium to 1×10^6 cells/mL. Seed the cell suspension into either a 96-, 24-, or 6-well plate with 0.1, 0.5, and 2 mL of the cell suspension per well, respectively. Store cells overnight in a humidified incubator (37 °C, 5 % CO_2).
5. Cells are ready for treatment the next day.

3.5 Reconstituted Neuron–Microglia Coculture Protocol

1. Grow neuron–glia cultures (*see* Subheading 3.2) in 24-well plates for 48 h, and gently supplement each well with 0.5 mL of warm Neuron–Glia Maintenance Medium containing ara-C with a final concentration of 5–15 μM (*see* **Notes 10** and **11**).
2. Refresh the medium at day 6 after seeding by completely removing and replacing the spent ara-C-containing medium with warm neuron–glia medium containing a cell suspension of enriched microglia, seeding at 5×10^4 cells/well. Store cells overnight in a humidified incubator (37 °C, 5 % CO_2).
3. Cells are ready for treatment the next day.

4 Notes

1. Rat and mouse dams are euthanized without anesthesia by decapitation and cervical dislocation, respectively, because anesthetic agents have been shown to affect brain chemistry [11]. Rapid asphyxiation with CO_2 gas is a suitable alternative method that appears to have a minimal impact on embryonic brain cells.
2. An accurate estimate of embryonic age is extremely important when culturing dopaminergic neurons—this is because cells must be transplanted during a critical 2-day window during which dopaminergic neurons are undergoing differentiation. Transplants outside this critical window result in poor survival. The embryonic age can be confirmed by comparing the crown-to-rump length (CRL) (*see* Fig. 2) of your embryos to the average lengths (*see* Table 2) [12, 13]. Differences in CRLs may also be attributed to differences among strains or animals with partial or complete genetic ablations. Thus it is important

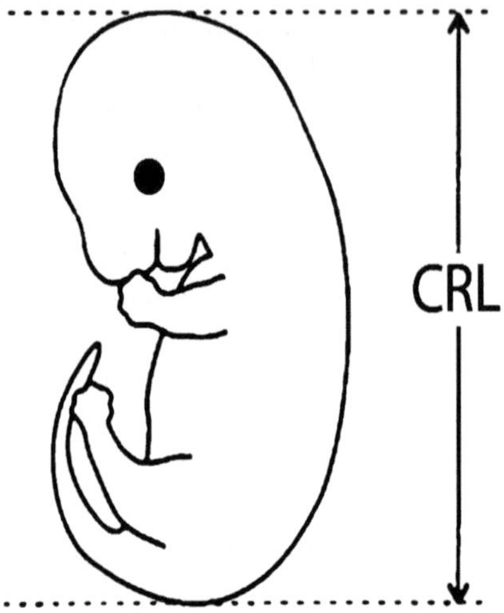

Fig. 2 The crown-to-rump length (CRL) of the embryo is the greatest distance from the top of the skull to the buttock

Table 2
Average CRL measurements from E13 to E15 rodents as derived from data compiled by Butler and Juurlink [12]

Gestation days	Mouse	Rat
E13	7–8 mm	7–9 mm
E14	9–10 mm	10–11 mm
E15	10.5–11.5 mm	12.5–13 mm

to conduct pilot studies to determine the most suitable critical time window for your cultures.

3. The frequency of tissue trituration varies among cell culture users. The number of passages may depend on the pipette-aid and the size and quantity of brain tissue. A gradual decrease in the size of the tip in respect to the decrease in the size of the tissue reduces the sheering pressure exerted on tissue resulting in less cellular rupturing (optimally less than 10 % of the total cells).
4. Based on our laboratory's experience, optimal growth of neuron–glia cultures has been observed in 24-well plate from Corning but not BD Falcon.
5. Immunocytochemical staining using cell type-specific markers (*see* Table 3) is routinely used to determine the cellular heterogeneity of the neuron–glia culture—which should be composed of ~50 % astrocytes, ~40 % neurons, and ~10 % microglia. A variation in the ratio of microglia, either above or below 10 % of

Table 3
Markers used to identify brain cells

Cell type	Marker
Astrocytes	GFAP
Microglia	OX-42 Iba-1
Neurons	NeuN MAP-2
Dopaminergic neurons	TH
Oligodendrocytes	MBP

the total cells, may result in highly variable experimental results. Thus, it is important to monitor the cellular heterogeneity of the neuron–glia culture to ensure cell culture stability from batch to batch.

6. The ability of the cells to adhere to the culture plates may be damaged if cultures are perturbed during the first three days. Gently agitate the plates or flasks before refreshing the medium to dislodge any cellular debris—reducing potential endogenous ligands that could activate microglia via damage-associated molecular pattern receptors.
7. The growth of microglia may be delayed in different strains or transgenic mice—pilot studies may be required to optimize the microglia composition in the culture.
8. Covering the filter cap of flasks with parafilm prevents the fluctuation of pH in medium and reduces the risk for contamination.
9. The speed and duration of shaking is relatively slower and shorter than other protocols, resulting in fewer but more pure cell suspensions of microglia. The time of shaking should be optimized by individual to reduce Type II astrocyte and oligodendrocyte contamination.
10. The consistency of the neuron–glia culture is important during the proper preparation of neuron-enriched cultures. Once the culture is consistent, optimize the treatment concentration of ara-C to ensure the highest glial toxicity and lowest neurotoxicity.
11. Ara-C and leucine methyl ester (LME) are routinely used to ablate glial cell and microglia, respectively, from culture systems. It is important to note that as little as 1 % microglial contamination can result in a detectable proinflammatory response.

Acknowledgements

This research was supported [in part] by the Intramural Research Program of the NIH, National Institute of Environmental Health Sciences. We would like to acknowledge Dr. Bin Liu for his contribution in developing these protocols.

References

1. Costero J (1930) Estudie del compotamento de la microlgia cultivade on vetro. Datos concernientes a su histogenesis. Mem R Soc cep Hist nat 14:125–182
2. Kettenmann H, Hanisch UK, Noda M et al (2011) Physiology of microglia. Physiol Rev 91(2):461–553
3. Giulian D, Baker TJ (1986) Characterization of ameboid microglia isolated from developing mammalian brain. J Neurosci 6(8):2163–2178
4. Gao HM, Hong JS, Zhang W et al (2002) Distinct role for microglia in rotenone-induced degeneration of dopaminergic neurons. J Neurosci 22(3):782–790
5. Gao HM, Jiang J, Wilson B et al (2002) Microglial activation-mediated delayed and progressive degeneration of rat nigral dopaminergic neurons: relevance to Parkinson's disease. J Neurochem 81(6):1285–1297
6. Qin L, Liu Y, Cooper C et al (2002) Microglia enhance beta-amyloid peptide-induced toxicity in cortical and mesencephalic neurons by producing reactive oxygen species. J Neurochem 83(4):973–983
7. Gao HM, Hong JS, Zhang W et al (2003) Synergistic dopaminergic neurotoxicity of the pesticide rotenone and inflammogen lipopolysaccharide: relevance to the etiology of Parkinson's disease. J Neurosci 23(4): 1228–1236
8. Chang RC, Chen W, Hudson P et al (2001) Neurons reduce glial responses to lipopolysaccharide (LPS) and prevent injury of microglial cells from over-activation by LPS. J Neurochem 76(4):1042–1049
9. Liu B, Du L, Hong JS (2000) Naloxone protects rat dopaminergic neurons against inflammatory damage through inhibition of microglia activation and superoxide generation. J Pharmacol Exp Ther 293(2):607–617
10. Paxinos G, Tork I, Tecott LH et al (1991) Atlas of the developing rat brain. Academic, San Diego
11. Hong JS, Wood PL, Gillin JC et al (1980) Changes of hippocampal Met-enkephalin content after recurrent motor seizures. Nature 285(5762):231–232
12. Butler H, Juurlink BHJ (1987) An atlas for staging mammalian and chick embryos, 1st edn. CRC, Boca Raton
13. Torres EM, Weyrauch UM, Sutcliffe R et al (2008) A rat embryo staging scale for the generation of donor tissue for neural transplantation. Cell Transplant 17(5):535–542

Part VIII

Analysis of Microglia Functions In Vivo

Chapter 22

Microglia Detection by Enzymatic Histochemistry

Beatriz Almolda, Berta González, and Bernardo Castellano

Abstract

Visualization of microglia by means of histochemistry has been for years a reliable method to demonstrate this population of cells in the central nervous system (CNS). Wide range of data on microglia has been published using lectin and enzymatic histochemistry. While at present, in most laboratories, the use of specific antibodies is the first choice, histochemical detection of microglia remains a powerful method as it has certain advantages upon immunohistochemical methods because it is faster, cheaper, and can be used in different species including human. In this chapter we want to present the detailed methodology for microglial staining using the histoenzymatic demonstration of the enzyme nucleoside-diphosphatase (NDPase), a phosphatase found in the plasma membrane of microglia that is absent in the plasma membrane of other glial cells and neurons. With this technique it is possible to visualize amoeboid microglia during development, ramified microglia in the adult brain, and also reactive microglia. As the technique also stains the blood vessels, it allows the analysis of the relationship between microglia and vasculature. This method can be performed in either histological sections or cell cultures for light microscopy analysis. Furthermore, we described how to combine this histochemical method with conventional immunohistochemistry for double labelling using other markers, and finally we give details to perform the procedure not only for optical microscopic studies but also for transmission electron microscopy (TEM).

Key words Microglia, Histochemistry, NDPase, Immunohistochemistry, Transmission electron microscopy, Cell culture, Double labelling

1 Introduction

It was during the first part of the twentieth century that Pio del Rio Hortega through the development of a specific technique such as the silver carbonate method [1] was able to describe the microglial cells and distinguish them from other cells in nervous tissue [2]. For many years silver carbonate has been the only tool that neuroscientist has had to visualize this population of cells. However, between 1981 and 1982, Murabe and Sano reported that using the method of Novikoff and Goldfisher [3], intense thiamine pyrophosphatase (TPPase) and nucleoside-diphosphatase (NDPase) enzymatic activities were selectively found at the plasma membrane of glial cells [4, 5]. Authors identified these glial cells as microglia

Bertrand Joseph and José Luis Venero (eds.), *Microglia: Methods and Protocols*, Methods in Molecular Biology, vol. 1041,
DOI 10.1007/978-1-62703-520-0_22,

because they were morphologically similar to microglial cells in the sections stained with Rio Hortega's silver impregnations. Although it is unclear whether a single enzyme performs both enzymatic activities or if there are different isoenzymes [6, 7], when performed in the same conditions, the reaction product was always found in the same location, without any difference, and therefore histochemical demonstration of either TPPase or NDPase has been used indiscriminately as a selective marker of microglia. In our hands, although we have used both thiamine pyrophosphate and nucleoside diphosphate as substrate, and all worked well, in most studies we have used nucleoside diphosphates because stains showed up cleaner and stronger, also requiring less incubation time.

For years, NDPase histochemistry has been used for the selective labelling of microglial cells in the central nervous system of different species including fishes and even humans [8–14]. NDPase histochemistry stains not only amoeboid and ramified microglia in the normal intact nervous system [15, 16] but also activated-reactive microglia in the experimentally lesioned brain [11, 17] and in neurological diseases such as Alzheimer's disease [18] and other disorders [12, 19]. In front to immunohistochemical methods, NDPase histochemistry is a faster and cheaper technique that can be developed not only on histological sections but also in cell cultures [20] allowing the fast identification of microglial cells, which are usual contaminants of "pure" neuronal or astrocyte cultures.

The method is based on the detection of the enzyme that hydrolyzes the nucleoside diphosphates such as 5′-inosine-diphosphate (IDP) to 5′-inosine-monophosphate (IMP). To accomplish that, histological sections obtained with a vibratome are incubated in a buffered medium containing the substrate (IDP), Mn^{2+} ions as cofactor of the enzymatic reaction, and a lead salt that act as donor of Pb^{2+} ions. When the enzyme NDPase acts on the substrate, it yields inorganic phosphate that reacts with the Pb^{2+} ions and precipitate "in status nascendi" as lead phosphate which is insoluble in aqueous solution. As the lead phosphate is white and cannot be distinguished at the light microscope, it is necessary to develop the histochemical reaction by using a sulfide salt which will yield lead sulfide that is brown and therefore visible at the light microscope. An ulterior treatment with a silver salt will yield silver sulfide, a more stable reaction product (Fig. 1). Another advantage of the NDPase histochemistry is that as the first reaction product of the enzymatic activity is electron dense, the sections can be analyzed at the ultrastructural level using an electron microscope.

Finally, although not obvious, this technique can be combined with other markers allowing the double labelling of microglia and other cells [21]. For instance, double labelling of microglia and astrocytes can be obtained by combining NDPase with

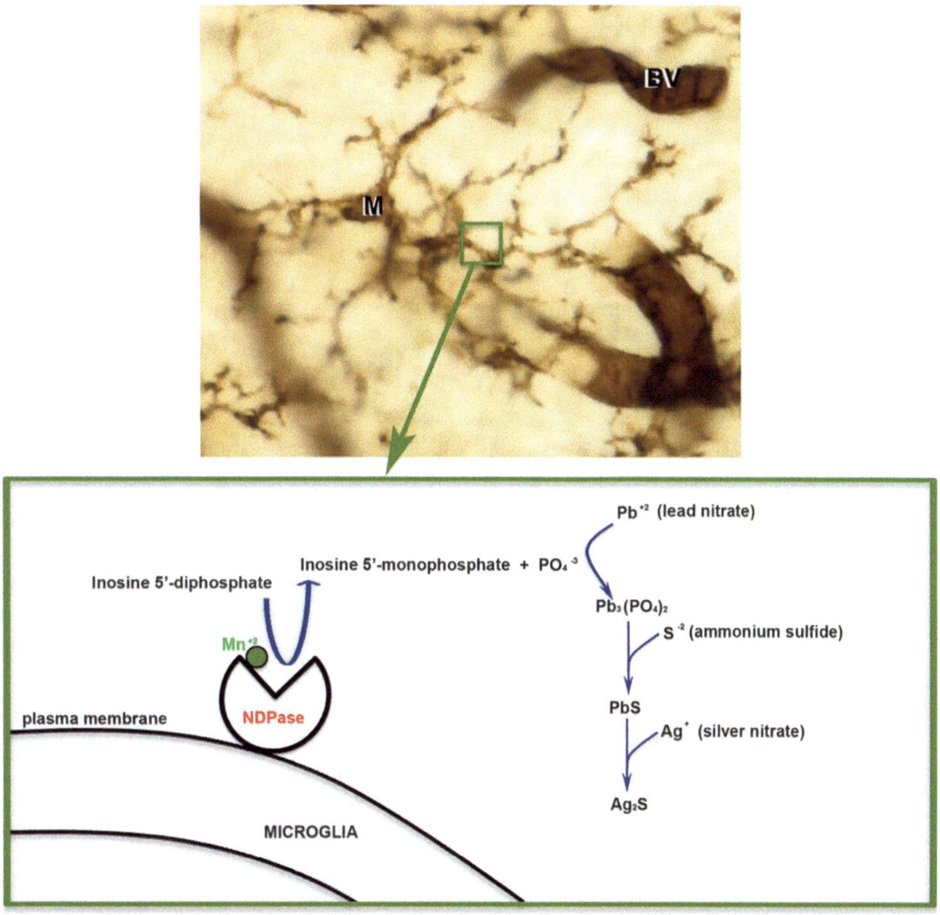

Fig. 1 Histochemical demonstration of NDPase. In the CNS, NDPase enzyme is located on the plasma membrane of both microglial cells (M) and endothelial cells of the blood vessels (BV). The enzyme NDPase, which requires manganese ions (Mn^{+2}) as cofactor, catalyzes the transformation of the substrate inosine 5′-diphosphate into inosine 5′-monophosphate and inorganic phosphate (PO_4^{-3}). If lead ions (Pb^{+2}) are present in the incubation medium, then the inorganic phosphate precipitates "in status nascendi" as lead phosphate ($Pb_3(PO_4)_2$). This salt is white and therefore not observable at the light microscope. Then to visualize it, it is necessary to treat the samples with ammonium sulfide that reacts with the lead phosphate producing lead sulfide (PbS), an insoluble salt with brown color. Finally, treatment with silver nitrate yields silver sulfide (Ag_2S), a more stable salt that allows indefinite storage of samples

immunohistochemical detection of GFAP (a recognized astrocyte marker). To accomplish that, after the histoenzymatic incubation and before the treatment with sulfide, the process is interrupted and followed with immunohistochemical staining. Only after the immunohistochemical staining is finished, the development of the histoenzymatic reaction is done. As the final product of the histochemical technique is brownish black, it is necessary to develop the immunohistochemical reaction in a completely different color (e.g., light blue) to distinguish both labels.

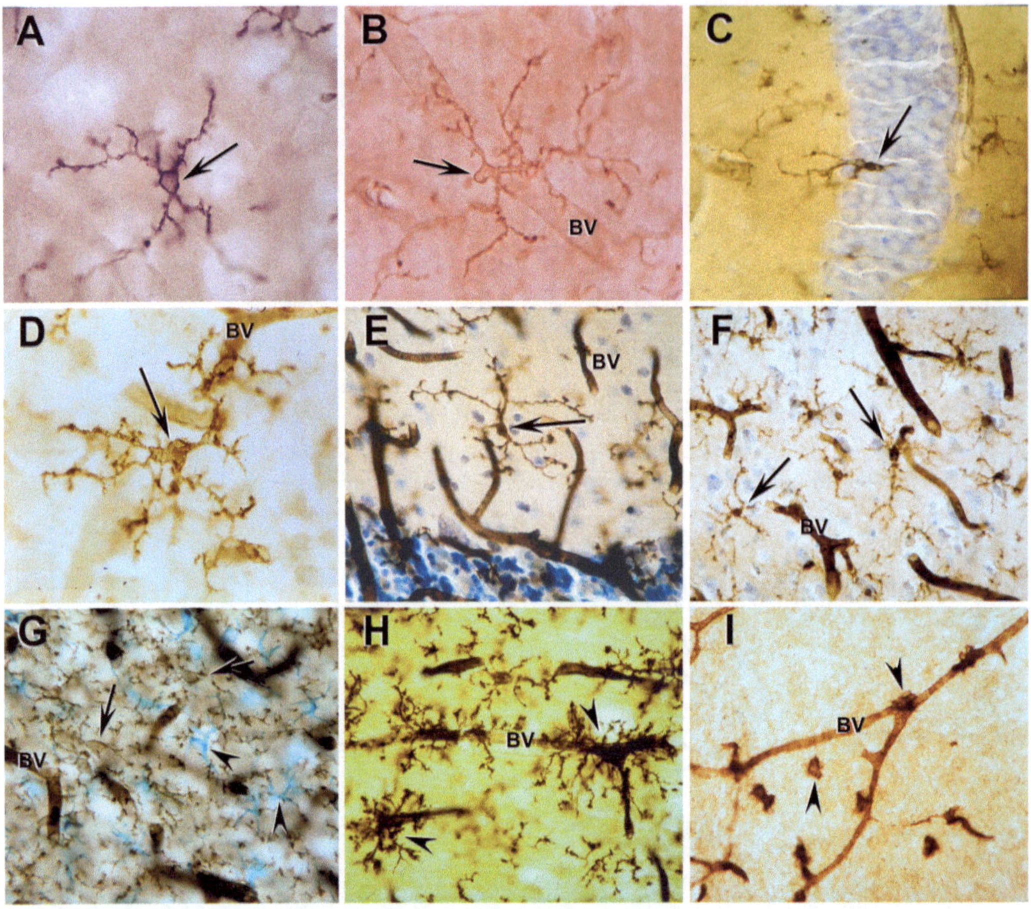

Fig. 2 Microglia labelling by NDPase histochemistry. Photographs showing the NDPase-labelled microglial cells [*arrows* in (**a**–**b**)] observed in the CNS of different species: frog (**a**), quail (**b**), lizard (**c**), rat (**d**, **g**–**i**), mouse (**e**), and human (**f**). In photographs (**c**, **e**, and **f**), toluidine blue was used as counterstain. (**g**) Double labelling combining NDPase histochemistry and GFAP immunohistochemistry. NDPase-labelled microglia (*arrows*) are seen in *brown*, whereas GFAP-stained astrocytes (*arrowheads*) are visualized in *blue*. In addition to microglia, NDPase histochemistry also stains blood vessels (BV in **b**, **d**–**i**) allowing the study of microglial interaction with the vasculature. Activated microglia (*arrowheads*) are found in close relationship with blood vessels during aging in RLA rats (**h**). Microglial precursors and amoeboid microglia (*arrowheads*) in the developing rat brain (**i**)

In this chapter we describe how to perform the NDPase enzymatic histochemistry in vibratome sections and cultures for quick identification of microglia (Figs. 2 and 3); how to combine this histochemical technique with other immunohistochemical labelling, in particular GFAP immunohistochemistry that allows to the visualization of astrocytes (Figs. 2g and 3b); and finally how to process the NDPase histochemically reacted sections for electron microscopy analysis (Fig. 4).

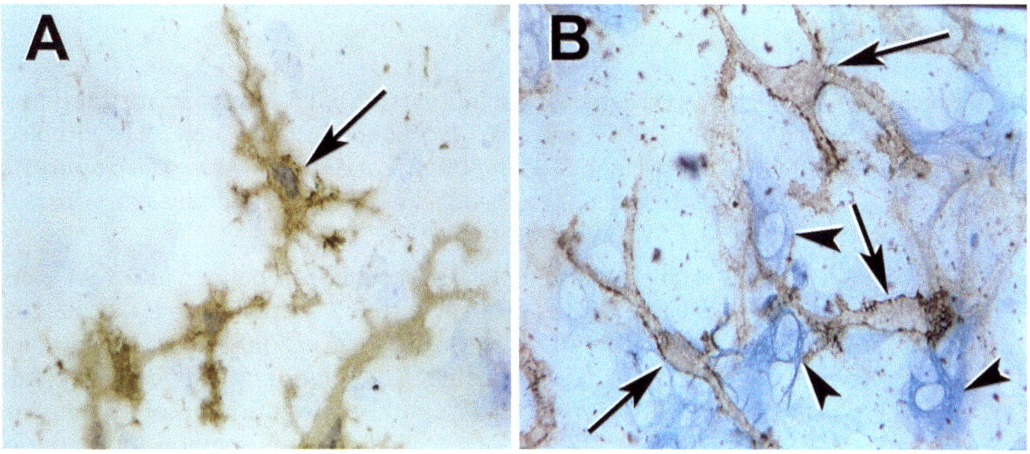

Fig. 3 NDPase labelling of microglia in cell cultures. Primary astroglial cell cultures from brains of newborn mice showing the presence of contaminating microglia (*arrows*) visualized by NDPase histochemistry. In (**a**) the cell body and microglial processes are seen in *brown* and the nucleus in *blue* because the culture has been counterstained with toluidine blue. In (**b**) NDPase histochemistry has been combined with GFAP immunohistochemistry: microglia are visualized in *brown,* whereas astrocytes are stained in *blue*

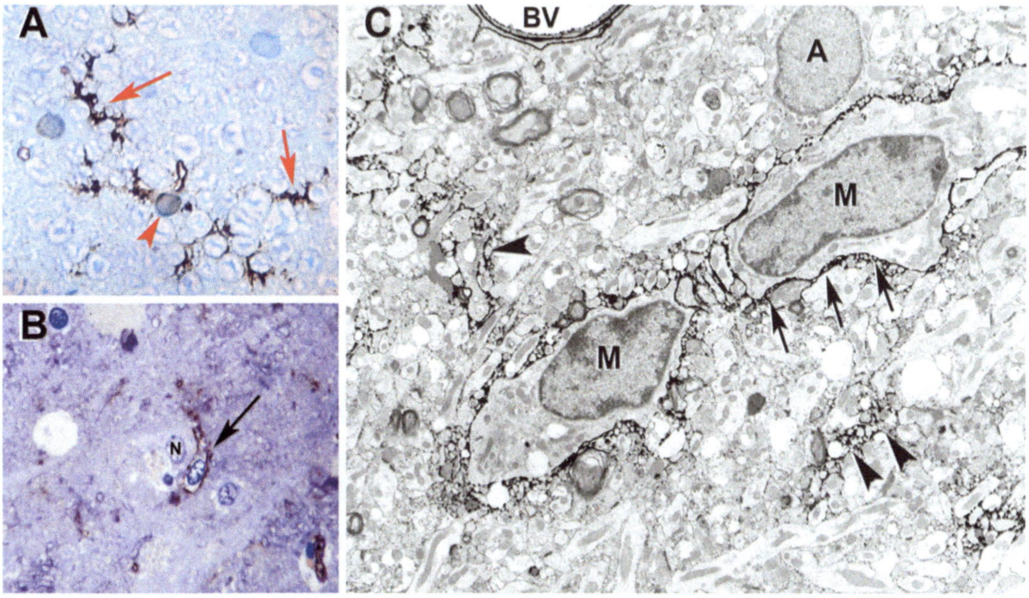

Fig. 4 NDPase histochemistry for TEM studies. (**a**) Semithin section showing NDPase-labelled microglia in the white matter of mice spinal cord. The section has been counterstained with toluidine blue. Note the product of the NDPase enzymatic activity in both the prolongations (*arrows*) and cell body (*arrowheads*) of microglial cells. (**b**) Semithin section showing NDPase-labelled microglia (*arrow*) in a sample from human cerebral cortex. Note that other cells like neurons (N) are not stained with this histochemical method. The section has been counterstained with toluidine blue. (**c**) Ultrathin section from mice spinal cord showing the reaction product of NDPase activity, which precipitates at the plasma membrane (*arrows*) of microglia (M). Note that the histochemical reaction allows the detection of microglial processes (*arrowheads*) in the neuropilum. NDPase labelling was also observed in the wall of blood vessels (BV). Note that other glial cells (A) do not show NDPase stain

2 Materials

2.1 Solutions for Fixation and Washing

1. Cacodylate stock solution (CSS): 0.5 M cacodylate solution at pH 7.4. Dissolve 107 g of sodium cacodylate × 3 H_2O (molecular weight 214) in 800 mL of distilled water. Make up to 1 L with distilled water and adjust the pH by adding some drops of 0.2 M NaOH. Store at 4 °C.
2. Fixative solution (FS): 4 % paraformaldehyde in 0.1 M cacodylate buffer at pH 7.4, with 5 % sucrose. For 1 L of FS, dissolve 40 g of paraformaldehyde in 700 mL of hot distilled water (70 °C) by stirring. The solution will appear milky. Add 0.2 M NaOH dropwise until the solution becomes transparent. Add 200 mL of CSS and filter the solution. Add 50 g of sucrose to the solution and make up to 1 L with distilled water. Adjust the pH to 7.4 and store at 4 °C.
3. Fixative solution with glutaraldehyde (FSG): 2 % paraformaldehyde in 0.1 M cacodylate buffer and 0.5 % glutaraldehyde at pH 7.4, with 5 % sucrose. For 1 L of FSG, dissolve 20 g of paraformaldehyde in 700 mL of hot distilled water (70 °C) by stirring. The solution will appear milky. Add 0.2 M NaOH dropwise until the solution becomes transparent. Add 200 mL of CSS and filter the solution. Add 50 mL of 25 % glutaraldehyde. Finally, add 50 g of sucrose to the solution and make up to 1 L with distilled water. Adjust the pH to 7.4 and store at 4 °C until use.
4. Cacodylate washing buffer with sucrose (CWBS): 0.1 M cacodylate buffer at pH 7.4, with 7.5 % sucrose. For 1 L, dissolve 75 g of sucrose in 700 mL of distilled water. Add 200 mL of CSS, make up to 1 L with distilled water, and adjust the pH by adding some drops of 0.2 M NaOH. Store at 4 °C.

2.2 Solutions for the Nucleoside-Diphosphatase Histochemical Reaction

1. Substrate: 5′-inosine-diphosphate sodium salt (Sigma-Aldrich, I4375; molecular weight 428.19) stored at −20 °C.
2. Trizma maleate buffer (TMB) stock solution: 0.2 M Trizma maleate, pH 7.4. Dissolve 3.083 g of Trizma maleate (Sigma-Aldrich, T3128: molecular weight 237.21) in 65 mL of distilled water and adjust the pH to 7.4. Store at room temperature (RT) but do not use after 4 weeks.
3. Manganese (II) chloride solution: Dissolve 0.25 g of $MnCl_2$ in 50 mL of distilled water to obtain a 0.5 % solution. Store at RT.
4. Lead (II) nitrate solution: Dissolve 0.4 g $Pb(NO_3)_2$ in 40 mL of distilled water to obtain a 1 % solution. Store at RT.
5. Ammonium sulfide solution: Add 5 mL of 20 % $(NH_4)_2S$ to 50 mL of distilled water to obtain a 2 % ammonium sulfide solution. Prepare fresh just before use.
6. Silver nitrate solution: Dissolve 0.5 g $Ag(NO_3)$ in 50 mL of distilled water to obtain a 1 % solution. Prepare fresh just before use.

2.3 Solutions for Toluidine Blue Counterstaining

1. Acetic acid solution (AAS): Add 2.75 mL of glacial acetic acid (≥99.7 %) to 240 mL of distilled water to obtain a 0.2 M solution.
2. Sodium acetate solution (SAS): Dissolve 2.63 g of sodium acetate (molecular weight 82.03) in 160 mL of distilled water to obtain a 0.2 M solution.
3. Walpole buffer (WB): 0.2 M Walpole buffer solution, pH 4.5. Mix three parts of AAS with two parts of SAS and adjust the pH by adding some drops of the 0.2 M AAS.
4. Toluidine blue solution: Dissolve 0.1 g of toluidine blue in 100 mL of WB in order to obtain a 0.1 % solution. Filter and store at RT.

2.4 Solutions for Resin-Embedding Procedure for Electron Microscopy

1. Cacodylate buffer (CB): 0.2 M cacodylate buffer, pH 7.4. Mix 40 mL CSS with 60 mL of distilled water. Adjust the pH and store at 4 °C.
2. Osmium tetroxide solution (OTS): 1 % osmium tetroxide solution in 0.1 M cacodylate buffer. Mix 5 mL of 2 % OsO_4 aqueous solution (Electron Microscopy Sciences, 19172) with 5 mL CB. Store at 4 °C in the dark.
3. Uranyl acetate solution (UAS): 2 % uranyl acetate solution in 70 % ethanol. Mix 2 g uranyl acetate powder with 100 mL 70 % ethanol. Filter and store at 4 °C.
4. Araldite I: Mix 10 g of component AM epoxy resin (Sigma-Aldrich ACM Fluka, 44611) with 10 g of component B (Sigma-Aldrich ACM Fluka, 44612) and 0.3 g of component D (Sigma-Aldrich ACM Fluka, 44614) in a glass container. Mix well using a glass rod and maintain in the oven at 55 °C. Before use, be sure that there are no air bubbles in the mixture. Araldite I should be freshly prepared 1–2 h before starting the embedding process (*see* **Note 1**).
5. Araldite II: Mix the same components used to prepare Araldite I and add 0.3 g of component C (Sigma-Aldrich ACM Fluka, 44613). Mix well with a glass rod and stand in the oven at 55 °C. Prepare fresh just before use. Do not use until all air bubbles are gone (*see* **Note 1**).

2.5 Solutions for Double Labelling: NDPase Histochemistry with GFAP Immuno-histochemistry

1. Tris-buffered solution (TB): 0.05 M tris-buffered solution, pH 7.4. Add 12.11 g of Trizma base (Sigma-Aldrich, T1503; molecular weight 121.14) and 7.2 mL of 37 % chlorhydric acid solution (HCl) to 2 L of distilled water. Adjust the pH to 7.4 and store at RT.
2. Tris-buffered saline solution (TBS): 0.05 M tris-buffered saline solution, pH 7.4. Add 12.11 g of Trizma base (Sigma-Aldrich, T1503; molecular weight 121.14), 17.53 g of NaCl

(Sigma-Aldrich, 71379; molecular weight: 58.44), and 7.2 mL of 37 % HCl to 2 L of distilled water. Mix well, adjust the pH to 7.4 and store at RT.

3. TBS with Triton X-100 (TBST): 0.05 M tris-buffered saline solution with 1 % Triton X-100, pH 7.4. Add 10 mL of Triton X-100 to 900 mL of TBS and mix well until the Triton X-100 is completely dissolved. Make up to 1 L with TBS. Store at RT. Discard when the solution is not transparent.
4. Buffer blocking solution (BBS): 0.05 M TBST, pH 7.4 with 10 % fetal bovine serum (FBS), and 3 % bovine serum albumin (BSA). Add 10 mL of FBS (Sigma-Aldrich, F7524) to 90 mL of TBST pH 7.4 and dissolve 300 mg of BSA (Sigma-Aldrich, A9647) in the previous solution. Mix well until the BSA is completely dissolved. Prepare fresh just before use.
5. Primary antibody: Polyclonal rabbit anti-GFAP antibody (DAKOPatts, Z0334).
6. Secondary antibody: Biotinylated anti-rabbit Ig (affinity purified) antibody from goat (Vector Laboratories, BA-1000).
7. Horseradish peroxidase streptavidin (Vector Laboratories, SA-5004).
8. 1-Naphthol working solution: 1-Naphthol solution with 0.033 % hydrogen peroxide. Dissolve 50 mg of 1-naphthol (Sigma-Aldrich, N1000; molecular weight 144.17) in 0.5 mL ethanol. Mix this solution with 10 mL of 1 % ammonium carbonate and 89.5 mL of TBS. After filtering the solution, add 0.01 mL 30 % hydrogen peroxide. Prepare fresh just before use.
9. Azur A working solution: 0.05 % Azur A in 0.05 M TBS pH 8. Dissolve 0.05 g of Azur A chloride (Sigma-Aldrich, 861049) in 100 mL of TBS. Adjust the pH to 8.00 with some drops of 0.2 M NaOH.

3 Methods

3.1 Intracardiac Perfusion, Postfixation, and Vibratome Sections

1. Anesthetized animals (*see* **Note 2**) are intracardially perfused with the FS (FSG for TEM analysis) at 4 °C for 10 min (*see* **Note 3**). In case of human samples from biopsies or autopsies, the samples should be immediately submerged in FS (or FSG) at 4 °C. The sample size should be not thicker than 1 cm for best results.
2. The areas of interest (telencephalon, cerebellum, spinal cord, and others) should be dissected out as quickly as possible and immersed in glass vials containing the same fixative solution (4 °C) to complete a total time of 4 h of fixation for light microscopy (1 h for TEM analysis). Take into account that

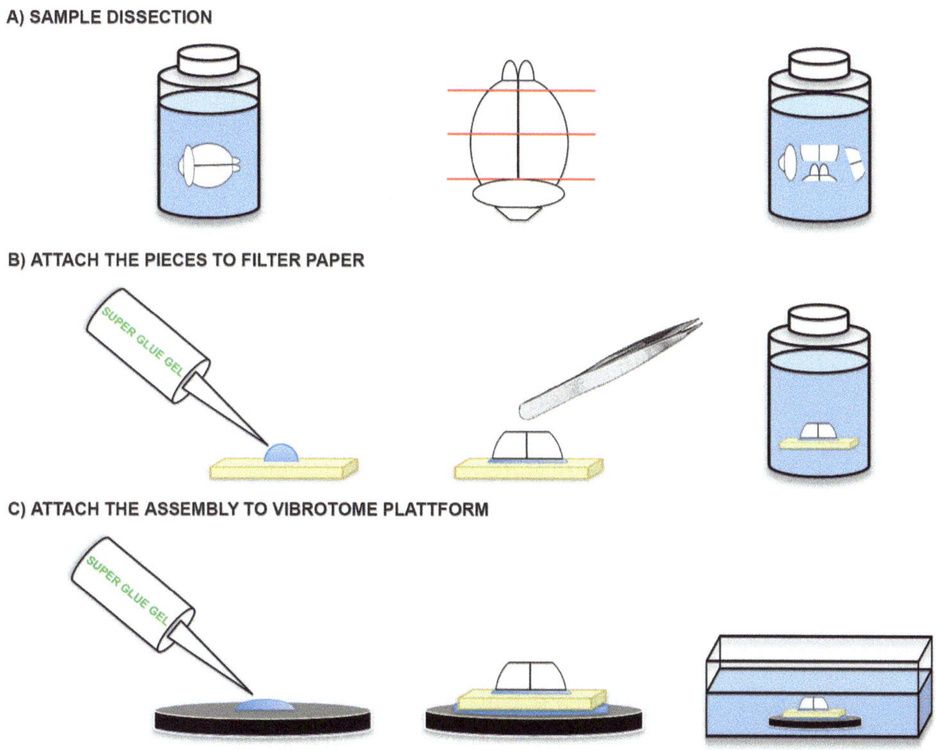

Fig. 5 Diagrams showing the sample handling previous to vibratome sectioning. (**a**) Sample dissection during postfixation time facilitates the penetration of the fixative solution and prepares the samples for ulterior vibratome sectioning. (**b**) Tissue samples are attached to squares of filter paper using super glue gel. (**c**)The assembly is attached to the vibratome platform and submerged in the vibratome container previously filled with the buffer

volume of FS (or FSG) should be at least ten times the volume of the tissue pieces. Remove the meninges when possible. Special care should be taken in order to prevent any desiccation during all the process including the extraction of the brain (*see* **Note 4**), and the subsequent dissection because otherwise the NDPase enzymatic activity in the plasma membrane of microglia may be completely destroyed.

3. Dissect out the samples to be cut with the vibratome (Fig. 5a). This process can be done during the time of fixation (*see* **Note 5**).
4. Once completed the fixation, wash the samples in CWBS (3×10 min, 4 °C) and store them in this buffer (4 °C) until cutting (*see* **Note 6**).
5. Attach the piece to be cut to a small square of filter paper using a super glue gel (*see* **Note** 7). To achieve that, place some drops of glue depending on the size of the tissue sample, on the surface of a square piece of filter paper to produce a little dome. Immediately, with the aid of tweezers, take the piece of tissue

to be attached to the filter paper, and before placing on it, quickly absorb the liquid from its basis by gently touching it during 2 s with a strip of filter paper. This is important to assure an appropriated attachment. Once the tissue piece is placed on the center of the glue dome, press gently down with a flat object (as the handle of the tweezers) to be sure that the basis is attached as flat as possible to the filter paper. Wait for two additional seconds and quickly take the square and the attached tissue sample and immediately submerge the assembly in cold (4 °C) CWBS until to be cut (*see* **Notes 8** and **9**) (Fig. 5b).

6. With the aid of some drops of super glue gel, attach the assembly (square of the filter paper plus the tissue sample) with glue to the vibratome platform and quickly submerge the platform with the assembly in the specific container of the vibratome that should be previously filled with cold (4 °C) CWBS (*see* **Note 10**) (Fig. 5c). Obtain sections 30–50 μm thick for optical microscope analysis and 60–100 μm thick for resin-embedding and ulterior transmission electron microscopy (TEM) studies. Free-floating sections obtained should be progressively taken from the vibratome container with the aid of a soft brush and introduced in 5 mL glass vials filled with cold CWBS (4 °C) until the histochemical reaction will be performed.

3.2 Histochemical Reaction for Single NDPase Demonstration and Ulterior Counterstaining with Toluidine Blue for Light Microscopy

1. Mix 50 mL of 0.2 M TMB stock solution with 50 mL of distilled water in order to obtain 0.1 M TMB solution.
2. Remove the CWBS from the glass vials with the aid of a Pasteur pipette and immediately add enough 0.1 M TMB solution, usually 2 mL per vial. The TMB solution should always cover the free-floating sections within the glass vial.
3. Bring the glass vials into an oven at 37 °C for 15 min (meanwhile prepare the fresh incubation medium).
4. Prepare the incubation medium taking into account that you will need around 2 mL of solution per vial containing around 20 vibratome sections. To prepare the medium (for around ten vials) mix always the components in this order:
   (a) Dissolve 25 mg of substrate in 7 mL of distilled water.
   (b) Add 10 mL of 0.2 M TMB stock solution.
   (c) Add 5 mL manganese (II) chloride solution.
   (d) Add 3 mL lead (II) nitrate solution (*see* **Note 11**).
5. Filter the incubation medium. This should become mainly transparent then.
6. Take into account that as negative control of the histochemical reaction, some sections should be incubated in a medium containing the same components but lacking the substrate.

7. Remove the TMB solution from the glass vials with a Pasteur pipette and add the incubation medium.
8. Incubate at 37 °C in the oven for 20–25 min. Shake the vials every 5 min to ensure a homogeneous histochemical reaction in all the sections.
9. Stop the histochemical reaction by removing the incubation medium and adding cold (4 °C) 0.1 M TMB for 1 min.
10. Wash 3 × 10 min with 0.1 M CWBS at RT (*see* **Note 12**).
11. Visualize the histochemical reaction by removing the CWBS and adding ammonium sulfide solution at RT for 1 min. Immediately the sections become brown, which is an indicative that the histochemical reaction has worked properly.
12. Wash 3 × 3 min in distilled water.
13. Remove the distilled water and stabilize the histochemical reaction product by adding to the vials the silver nitrate solution for 1 min at RT.
14. Wash 3 × 3 min in distilled water.
15. Mount the sections on gelatin-coated slides and allow to dry for at least 24 h before applying the toluidine blue counterstaining (*see* **Note 13**).
16. Submerge the slides in distilled water for 3 min.
17. Counterstain with toluidine blue for 1 min (*see* **Note 14**).
18. Wash with distilled water for 1 min to remove the excess of colorant.
19. Dehydrate by submerging the slides in graded ethanol: 50, 70, 90, and 3× 100 %. No more than 30 s in each alcohol.
20. Differentiate with 1-butanol for 1 min.
21. Submerge in xylol (2× 2 min).
22. Finally, coverslip the slides using DPX as permanent mounting medium.
23. Wait at least 24 h in order to allow complete polymerization of the DPX before proceeding with the analysis at the light microscope.

3.3 NDPase Histochemistry for Transmission Electron Microscopy

This technique can also be used for visualization of microglia at the ultrastructural level. To accomplish that, go ahead with the steps detailed in the previous part until **step 10**. As the reaction product of the histochemical reaction, that is to say, the lead phosphate, is electrodense, this product will be already seen at the electron microscope; however, it is not possible to see it in semithin sections. If microglia need to be analyzed in semithin sections, then it is necessary to visualize the reaction product by treating the sections with ammonium sulfide solution (**step 11**) in order to obtain lead sulfide that can be visualized with the light microscope.

Subsequently in both cases, follow the steps detailed below:

1. Mix 50 mL of 0.2 M of CB with 50 mL distilled water in order to obtain 0.1 M CB.
2. Wash the free-floating sections by adding to the vials containing the free-floating sections, the 0.1 M CB (3 × 10 min).
3. Postfix the sections in OTS for 2 h at RT in the dark.
4. Wash sections 3 × 10 min in 0.1 M CB at 4 °C.
5. Remove the 0.1 M CB from the glass vials and start the dehydration process with graded ethanol as indicated:
   (a) 30 % ethanol for 15 min at 4 °C.
   (b) 50% ethanol for 15 min at 4 °C.
   (c) 70 % ethanol for 15 min at 4 °C.
6. En bloc stain with UAS for 1.5 h at 4 °C.
7. Wash in 70 % ethanol for 15 min at 4 °C.
8. Continue the dehydration as follows:
   (a) 96 % ethanol for 15 min at 4 °C.
   (b) 100 % ethanol 3 × 10 min at 4 °C.
9. Remove the last 100 % ethanol and add propylene oxide 3 × 10 min at 4 °C.
10. Start the embedding process by removing the propylene oxide and adding to the glass vials a mixture of one part of Araldite I and three parts of propylene oxide for 2 h at RT (*see* **Note 15**). Prepare the mixture just before use.
11. Remove the previous mixture and add a new mixture of one part of Araldite I and one part of propylene oxide for 2 h at RT. Prepare the mixture just before use.
12. Again remove the previous mixture and add a new mixture of three parts of Araldite I and one part of propylene oxide for 2 h at RT. Prepare the mixture just before use.
13. After removing the last mixture, add Araldite I to the glass vials and maintain them in the oven at 55 °C overnight.
14. Change the Araldite I and maintain at 55 °C for 3 h. Meanwhile prepare the Araldite II (*see* Subheading 2).
15. Remove the Araldite I and add Araldite II for 2 h at 55 °C.
16. Take the samples from the glass vials with the aid of a wood spatula and carefully transfer the sections to silicone molds previously filled with Araldite II. Make sure that all sections are completely flat and correctly oriented.
17. Maintain the silicone molds with the sections in the oven at 65 °C for 48 h to ensure the correct polymerization of the resin.

3.4 Double Labelling Technique for Light Microscopy Combining NDPase Histochemistry and GFAP Immunohistochemistry

Simultaneous demonstration of microglial and astroglial cells can be achieved easily by combining NDPase histochemistry with GFAP immunohistochemistry through three sequential steps: (1) incubation for histochemical demonstration of NDPase activity, (2) incubation for immunohistochemical detection of GFAP, and (3) visualization of the histochemical and the immunohistochemical reaction products for light microscopy.

1. Incubation for histochemical demonstration of NDPase activity: Follow the **steps 1–10** previously specified in Subheading 3.2 (*see* **Note 16**).
2. Incubation for the immunohistochemical detection of GFAP:
   (a) Remove the CWBS from the glass vials and wash the sections 3 × 10 min with TBS at RT.
   (b) Wash 3 × 10 min with TBST at RT.
   (c) Block the unspecific binding with BBS for 1 h at RT.
   (d) Incubate with the primary GFAP antibody diluted in BBS (1:1,800) overnight at 4 °C (*see* **Note 17**).
   (e) Wash the sections 3 × 10 min in TBST at RT.
   (f) Incubate sections in the secondary biotinylated anti-rabbit Ig antibody diluted in BBS (1:500) for 1 h at RT.
   (g) Wash the sections 3 × 10 min in TBST at RT.
   (h) Incubate in the horseradish peroxidase streptavidin diluted in BBS (1:500) for 1 h at RT.
   (i) Wash the sections 3 × 10 min in TBS at RT.
3. Visualization of the histochemical and the immunohistochemical reaction products for light microscopy. To achieve a successful visualization of the double staining, the final steps should be executed in this specific sequence:
   (a) Remove TBS from glass vials and incubate 15 min in 1-naphthol working solution at RT (*see* **Note 18**).
   (b) Wash the sections 3 × 10 min in TBS at RT.
   (c) Visualize the NDPase histochemistry by incubating the sections with 2 % ammonium sulfide solution for 1 min at RT.
   (d) Wash the sections 3 × 10 min in TBS at RT.
   (e) Remove the TBS and add the Azur A working solution for 30 min at RT.
   (f) Wash the sections 4 × 10 min with TBS at RT with continuous stirring.
   (g) Wash 2 × 5 min in distilled water.
   (h) Stabilize the histochemical reaction product with 1 % silver nitrate solution for 1 min at RT.

(i) Wash 2 × 5 min in distilled water.

(j) Mount the sections on gelatin-coated slides and allow dry for 24 h at RT.

(k) Dehydrate by submerging the slides in graded ethanol: 50, 70, 90, and 3× 100 %. No more than 30 s in each alcohol.

(l) Submerge in xylol (2× 2 min).

(m) Finally, coverslip the slides using DPX as permanent mounting medium.

(n) Wait for at least 24 h in order to allow complete polymerization of the DPX before proceeding with the analysis at the light microscope.

3.5 Cytochemical Demonstration of NDPase in Cell Cultures for Microglia Visualization

Visualization of microglia in cell cultures can also be performed by demonstration of NDPase enzymatic activity as follows:

1. Fixation should be done by removing the cell culture medium from culture plates or flasks and adding enough FS at 4 °C to cover the cells in the culture.
2. Fix the cultures for 2 h at 4 °C and then wash in CWBS (3× 10 min, 4 °C).
3. Follow the **steps** indicated in Subheading 3.2 from **1** to **14** taking into account that you are not using sections mounted on slides but cell plates; therefore, all changes should be done accordingly (*see* **Note 19**).
4. If you want to counterstain with toluidine blue, add toluidine blue and stain for 1 min, then wash with distilled water for 1 min.
5. Follow the **steps** indicated in Subheading 3.2 from **19** to **23** (*see* **Note 19**).
6. In the case you want to demonstrate the presence of microglia and astroglia combining the NDPase visualization with the GFAP immunostaining, follow the steps indicated in Subheading 3.4. Again, take into account that you are not using sections mounted on slides but cell plates (*see* **Note 19**).

4 Notes

1. Take into account how much volume you will need in function of the number of changes to do, samples, volume of glass vials, and volume of the silicon molds to be used.
2. Animals can be anesthetized using any anesthetic as isofluorane, pentobarbital, ketamine/xylazine, and others.

3. Ideally the perfusion in rats and mice should be done by inserting the canule through the left ventricle until reaching the aorta and tie up the blood vessel with a forceps. Then make a cut in the right auricula and open the flux of the fixative. To assure that the cold (4 °C) fixative does not cause any undesirable vasoconstriction, it is necessary to use an appropriated device that provides a constant pressure and enough flux of the FS (or FSG) to push out the blood during the first 10 s of perfusion.
4. After perfusion, during bone removal and brain extraction, it is important to maintain the exposed brain surfaces always wet by adding some drops of FS (or FSG) with a Pasteur pipette.
5. To facilitate the dissection of the samples, it is recommendable to do it after some time (30 min) of fixation. Dissections can be done in a Petri dish using a razor blade and, if necessary, under the binocular lens. Tissue pieces should be covered all the time by the FS (or FSG) at 4 °C, in order to prevent desiccation. The size of the pieces should be enough to be appropriately cut with the vibratome. Coronal sections 5–6 mm thick of a whole rat brain may be obtained and cut without problems. Preferably the basis of the pieces should be as flat as possible to assure a good and stable attachment to the vibratome. Once the samples are dissected out, they should be transferred back from the Petri dish to the glass vials to complete the total time of fixation (4 h for light microscopic study and 1 h for TEM) established in Subheading 3.
6. Best results are obtained when samples are cut on the same day and immediately stained. However, enzymatic activity can still remain during at least the next 24 h. If it is not possible to do the enzymatic reaction on the same day of perfusion, it is recommendable to keep the dissected samples in CWBS (4 °C) and to obtain the vibratome sections in the following day.
7. Although any super glue can be used, the gel type is the most recommendable because it does not spread when placed in the filter paper and it holds better the tissue piece, especially if the piece is not completely flat in the base.
8. If there are many pieces, they can be attached to different filter paper squares and maintain them submerged in cold CWBS until they are cut.
9. All the process should be as fast as possible to minimize the possible desiccation of the samples. Do your best to ensure that the time spent in attaching the samples to the filter paper is only some few seconds.
10. If the vibratome device you use does not have a container where temperature can be regulated, then change frequently

the CWBS, or if the vibratome device allows it, add some ice around the container to maintain cold the CWBS.

11. When lead (II) nitrate is added to the incubation medium, it becomes whitish.
12. If necessary, sections can be stored in 0.1 M CWBS at 4 °C for some hours before treatment with ammonium sulfide.
13. To check if the histochemical NDPase reaction has been successful, add a drop of glycerine to one of the slides, coverslip, and have a look in the microscope. Profiles of microglia and blood vessels should be distinguished.
14. Toluidine blue solution for counterstaining can be reused, but it is important to filter always the solution before use. Note that the time of counterstaining should be adjusted in every reuse; therefore, try first with one slice to determine the appropriated time.
15. From now, remove the lids of the glass vials in order to facilitate the evaporation of propylene oxide.
16. Visualization and stabilization of the reaction with ammonium sulfide and silver nitrate are omitted now, as these reagents may interfere with antigenicity. However, to check the histochemical reaction, some sections can be immediately treated with ammonium sulfide to visualize the reaction product.
17. As negative control of the immunohistochemical reaction, some sections should be processed in parallel without the primary antibody.
18. 1-Naphthol basic dye method of Mauro et al. [22] has been chosen for demonstration of the horseradish peroxidase in the immunohistochemical staining for GFAP, because this gives a blue HRP reaction product that was easily distinguished from the brownish-black NDPase staining.
19. The addition and removal of the different reagents to the cell plates should be done carefully using a Pasteur pipette. It has to pay special care not to touch the cell culture with the pipette tip.

References

1. Rio HP (1918) Noticia de un nuevo y fácil método para la coloración de la neuroglia y del tejido conjuntivo. Trab Lab Invest Biol 15:367–378
2. Rio HP (1920) El tercer elemento de los centros nerviosos. I. La microglía en estado normal. Bol Soc Esp Biol 8:68–82
3. Novikoff AB, Goldfisher S (1961) Nucleosidediphosphatase activity in the Golgi apparatus and its usefulness for cytological studies. Proc Natl Acad Sci 47:802–810
4. Murabe Y, Sano Y (1981) Thiaminepyrophosphatase activity in the plasma membrane of microglia. Histochemistry 71:45–52
5. Murabe Y, Sano Y (1982) Morphological studies on neuroglia V. Microglial cells in the cerebral cortex of the rat, with special reference to their possible involvement in synaptic function. Cell Tissue Res 223:493–506
6. Yamazaki M, Hayaishi O (1968) Allosteric properties of nucleoside diphosphatase and its

identity with thiamine pyrophosphatase. J Biol Chem 243–2934
7. Sano S, Matsuda Y, Miyamoto S et al (1984) Thiamine pyrophosphatase and nucleoside diphosphatase in rat brain. Biochem Biophys Res Commun 118:292–298
8. Salvador-Silva M, Vidal-Sanz M, Villegas-Pérez MP (2000) Microglial cells in the retina of Carassius auratus: effects of optic nerve crush. J Comp Neurol 417:431–447
9. Castellano B, González B, Dalmau I et al (1991) Identification and distribution of microglial cells in the cerebral cortex of the lizard: a histochemical study. J Comp Neurol 311:434–444
10. Lopez-Garcia C, Nacher J, Castellano B et al (1994) Transitory disappearance of microglia during the regeneration of the lizard medial cortex. Glia 12:52–61
11. Jensen MB, González B, Castellano B et al (1994) Microglial and astroglial reactions to anterograde axonal degeneration: A histochemical and inmunocytochemical study of the adult rat fascia dentata after entorhinal perforant path lesions. Exp Brain Res 98:245–260
12. Almolda B, Costa M, Montoya M et al (2009) CD4 microglial expression correlates with spontaneous clinical improvement in the acute Lewis rat EAE model. J Neuroimmunol 209:65–80
13. Vela JM, Dalmau I, González B et al (1995) Morphology and distribution of microglial cells in the young and adult mouse cerebellum. J Comp Neurol 361:602–616
14. Schnitzer J (1989) Enzyme-histochemical demonstration of microglial cells in the adult and postnatal rabbit retina. J Comp Neurol 282:249–263
15. Dalmau I, Finsen B, Zimmer J et al (1998) Development of microglia in the postnatal rat hippocampus. Hippocampus 8:458–474
16. Dalmau I, Vela JM, González B et al (2003) Dynamics of microglia in the developing rat brain. J Comp Neurol 458:144–157
17. Jørgensen MB, Finsen BR, Jensen MB et al (1993) Microglial and astroglial reactions in ischemic and kainic acid induced lesions in the adult rat hippocampus. Exp Neurol 120:70–88
18. Vostrikov VM (1985) Electron-cytochemical study of microglia in Alzheimer's disease and senile dementia. ZhNevropatol PsikhiatrIm S SKorsakova 85:974–976
19. Vela JM, Dalmau I, González B et al (1998) Glial abnormalities in genetically determined disorders of myelin. In: Castellano B, González B, Nieto M (eds) Understanding glial cells, Chapter 18. Kluwer Academics Publisher, Boston, pp 363–384
20. Dalmau I, Castellano B, Pedersen EB et al (1996) Reduction of the microglial cell number in rat primary glial cell cultures by exogenous addition of dibutyryl cyclic adenosine monophosphate. J Neuroimmunol 70:123–129
21. Castellano B, González B, Jensen MB et al (1991) A double staining technique for simultaneous demonstration of astrocytes and microglia in brain sections and astroglial cell cultures. J Histochem Cytochem 39:561–568
22. Mauro A, Germano I, Giaccone G et al (1985) 1-Naphthol basic dye (1-NBD). An alternative to diaminobenzidine (DAB) in immuno peroxidase techniques. Histochem 83:97

Chapter 23

Tomato Lectin Histochemistry for Microglial Visualization

Nàdia Villacampa, Beatriz Almolda, Berta González, and Bernardo Castellano

Abstract

The use of different lectins for the study of microglial cells in the central nervous system (CNS) is a valuable tool that has been extensively used in the last years for the selective staining of this glial cell population, not only in normal physiological conditions, but also in a wide range of pathological situations where the normal homeostasis of the parenchyma is disturbed. In this chapter we accurately describe the methodology for the selective labelling of microglial cells by using the tomato lectin (TL), a protein lectin obtained from *Lycopersicum esculentum* with specific affinity for poly-*N*-acetyl lactosamine sugar residues which are found on the plasma membrane and in the cytoplasm of microglia. Here we describe how to perform this technique on vibratome, frozen, and paraffin sections for optical microscopy, as well as for transmission electron microscopy (TEM) studies. Using this methodology it is possible to visualize amoeboid microglia in the developing brain, ramified microglia in the adult, and activated/reactive microglia in the experimentally damaged brain. In addition, as TL also recognized sugar residues in endothelial cells, this technique is very useful for the study of the relationship established between microglia and the CNS vasculature.

Key words Microglia, Histochemistry, Tomato lectin, Transmission electron microscopy, Double labelling, Fluorescence

Abbreviations

AAS	Acetic acid solution
BBS	Buffer blocking solution
CNS	Central nervous system
DS	Developing solution
DSN	Developing solution with nickel
DSNC	Developing solution with nickel-cobalt-enhanced DAB
EPBS	Endogenous peroxidase blocking solution
FS	Fixative solution
FSG	Fixative solution with glutaraldehyde
GFAP	Glial fibrillary acidic protein

In memoriam of Laia Acarin Pérez-Simó (1970–2011)

Bertrand Joseph and José Luis Venero (eds.), *Microglia: Methods and Protocols*, Methods in Molecular Biology, vol. 1041, DOI 10.1007/978-1-62703-520-0_23, © Springer Science+Business Media New York 2013

GSA-IB4	Griffonia simplicifolia B4 isolectin
MHC	Major histocompatibility complex
ML-1	Mistletoe lectin
OS	Olmos solution
OTS	Osmium tetroxide solution
PBS	Phosphate buffer solution
PWB	Phosphate washing buffer
RCA-1	Ricinus communis agglutinin-1
SAS	Sodium acetate solution
SS	Sucrose solution
TB	Tris-buffered solution
TBS	Tris-buffered saline solution
TBST	TBS with Triton X-100
TEM	Transmission electron microscopy
TL	Tomato lectin
UAS	Uranyl acetate solution
WB	Walpole buffer
WGA	Wheat germ agglutinin

1 Introduction

Lectins are sugar-binding proteins that have been of great interest for their use in the characterization of the membranes composition and the identification of different cell types [1]. In the central nervous system (CNS), several lectins have been used as selective microglial markers. Some of them are wheat germ agglutinin (WGA), mistletoe lectin (ML-1), Ricinus communis agglutinin-1 (RCA-1), Griffonia simplicifolia B4 isolectin (GSA-IB4), tomato lectin (TL) [2], and more recently galectins [3]. The TL specifically recognizes the poly-*N*-acetyl lactosamine residues, a sugar found in the plasma membrane and cytoplasm of microglia and endothelial cells [2] (Fig. 1).

Since its first description by Acarin et al. in 1994, the use of TL as microglial marker has been a useful tool for the characterization of this glial cell population in the rodent adult normal brain during development [4, 5] and aging [6] but also after different experimental injuries, such as excitotoxicity [7–10], aspiration lesions [11], experimental autoimmune encephalomyelitis [12], as well as for microglial studies in genetically modified animals such as myelin mutants [13, 14] and transgenic mice [15]. In addition to microglia, TL histochemistry also visualized the blood vessels allowing the study of the relationship between microglial cells and vasculature. As activated microglia in addition to changes in morphology also increase the expression of *N*-acetyl lactosamine residues, TL histochemistry easily demonstrates the reactive changes in this glial cell population.

Methodologically, the use of TL as microglial marker has some advantages in front of other techniques such as immunodetection of specific antigens and those based on the detection of specific

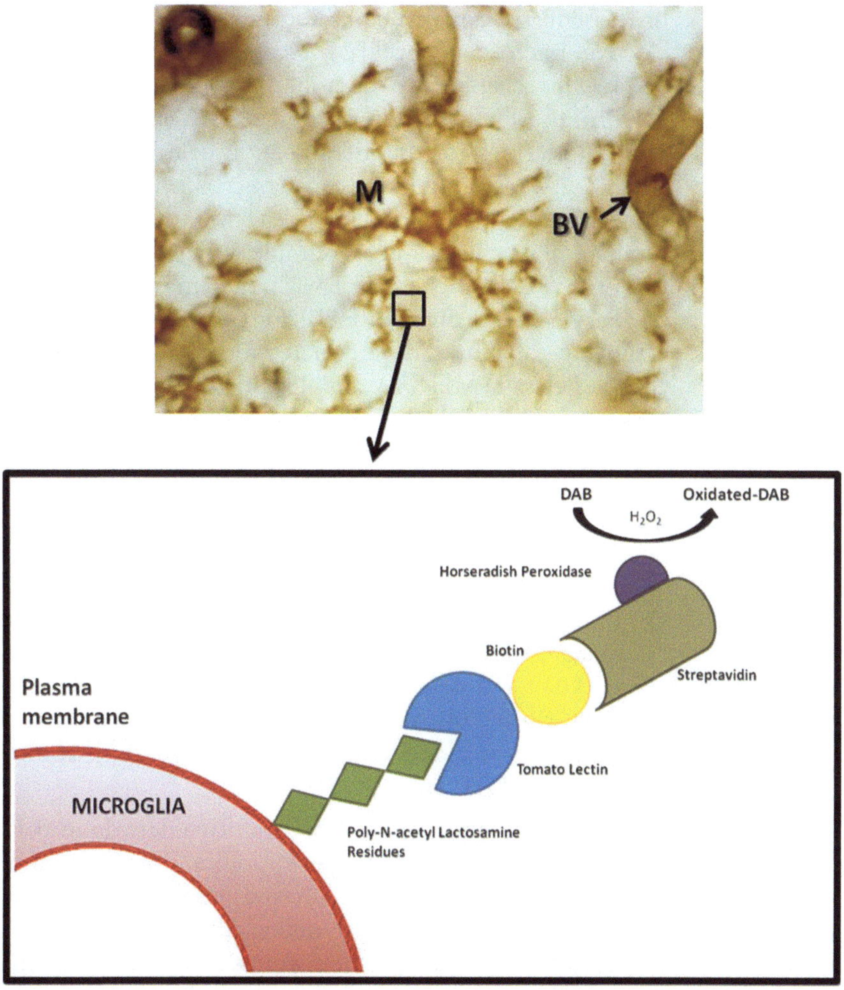

Fig. 1 Histochemical demonstration of tomato lectin. TL is a carbohydrate-binding protein that recognizes specifically poly-*N*-acetyl lactosamine residues located on the plasma membrane and cytoplasm of both microglial cells (M) and endothelial cells of the blood vessels (BV). To visualize these poly-*N*-acetyl lactosamine residues, biotinylated TL is used. Afterwards, biotin is recognized by a streptavidin-horseradish complex added to the samples, which in turn carries conjugated the peroxidase enzyme. Finally, visualization of the reaction product is performed by incubating the samples with 3,3′-diaminobenzidine (DAB). In presence of peroxidase enzyme and hydrogen peroxide (H_2O_2), DAB will be oxidated and quickly precipitated in the place where peroxidase enzyme is located, showing the characteristic *brown staining* observed in the picture

phosphatases such as PNPase [16] and NDPase enzymatic histochemistry (*see* Chapter 22), because sugar residues are very stable and therefore can be easily detected after a wide range of fixation and histological processing. Therefore TL technique can be performed on both frozen sections and in sections obtained from paraffin-embedded tissue. Moreover, TL histochemistry not only allows the specific staining of microglia in histological sections but also has been used for the characterization of microglia in organotypic cultures [17, 18]. Finally, we want to point out that TL histochemistry can also be used for the analysis of microglia at the electron microscope level [2].

In this chapter we described the detailed methodology to perform the TL histochemistry method in vibratome, paraffin, and cryostat sections for the identification of microglial cells for both light (Fig. 2a–d) and fluorescence microscopy (Fig. 3a). Moreover, we give the details to combine this technique with any immunohistochemical labelling, and as example we have detailed how to combine the TL histochemistry with the immunodetection of glial fibrillary acidic protein (GFAP) for specific visualization of astrocytes (Fig. 2e, f) and immunodetection of major histocompatibility complex class I (MHC-I), a molecule involved in antigen presentation (Fig. 3b). Finally we describe how to process vibratome TL-stained samples for electron microscopy studies (Fig. 4).

2 Materials

2.1 Solutions for Fixation and Washing

1. Phosphate buffer stock solution (PBS): 0.2 M phosphate buffer at pH 7.4. Mix 6.6 g of $NaH_2PO_4{\cdot}2H_2O$ (molecular weight: 156.01) and 43.58 g of $Na_2HPO_4{\cdot}7H_2O$ (molecular weight: 268.07) in 700 mL of distilled water. Make up to 1 L with distilled water and adjust the pH by adding some drops of 0.2 M NaOH. Store at 4 °C.
2. Fixative solution (FS): 4 % paraformaldehyde in 0.1 M phosphate buffer at pH 7.4. For 1 L of FS, dissolve 40 g of paraformaldehyde in 500 mL of hot distilled water (70 °C) by stirring. The solution will appear milky. Add 0.2 M NaOH dropwise until the solution becomes transparent. Add 500 mL of PBS and filter the solution. Adjust the pH to 7.4 and store at 4 °C until use during the following hours.
3. Phosphate washing buffer (PWB): phosphate buffer 0.1 M at pH 7.4. Mix 500 mL of PBS with 500 mL distilled water. Store at 4 °C.
4. Sucrose solution (SS): sucrose 30 % in phosphate buffer 0.1 M at pH 7.4. Dissolve 90 g of sucrose in 200 mL of PWB and wait until sucrose is completely dissolved. Make up to 300 mL with PWB. Store at 4 °C.
5. Fixative solution with glutaraldehyde (FSG): 2 % paraformaldehyde in 0.1 M phosphate buffer and 0.5 % glutaraldehyde at pH 7.4. For 1 L of FSG, dissolve 20 g of paraformaldehyde in 700 mL of hot distilled water (70 °C) by stirring. The solution will appear milky. Add 0.2 M NaOH dropwise until the solution becomes transparent. Add 200 mL of PBS and filter the solution. Add 50 mL of 25 % glutaraldehyde. Finally, make up to 1 L with distilled water and adjust the pH to 7.4. Store at 4 °C until use during the following hours.

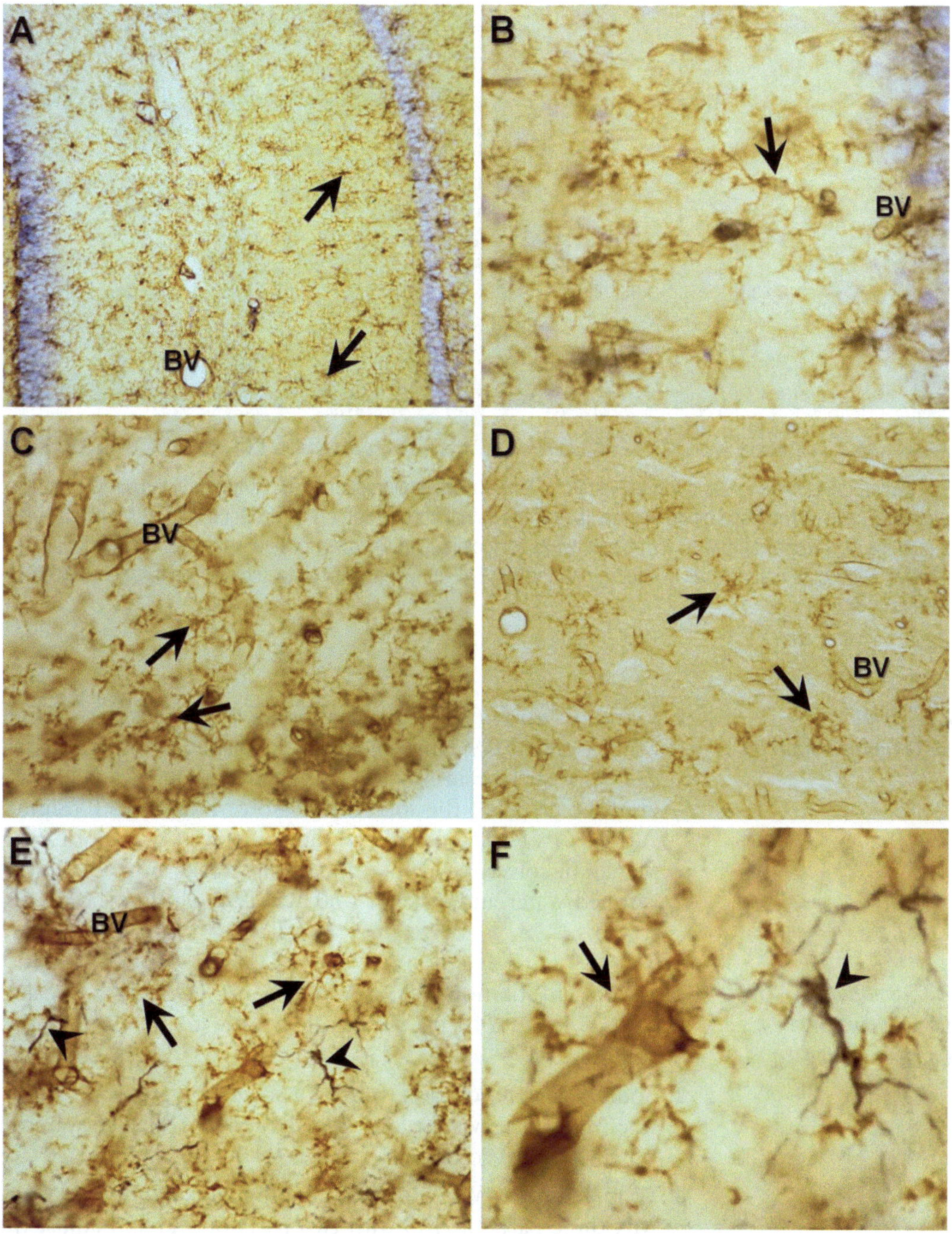

Fig. 2 Microglial labelling using tomato lectin histochemistry for light microscopy. Photographs showing TL-labelled microglial cells (*arrows*) and blood vessels (BV) in vibratome sections (**a**, **b**), cryostat sections (**c**), and paraffin sections (**d**). Note that vibratome sections (**a**, **b**) have been counterstained with toluidine blue. (**e**) Double labelling combining TL histochemistry and GFAP immunohistochemistry allows visualization of microglial cells in *brown* (*arrows*) and astrocytes in *bluish black* (*arrowheads*). (**f**) Magnification of the same area shown in (**e**) where microglial cells in brown (*arrow*) and astrocytes in bluish black (*arrowhead*) can be seen with more detail

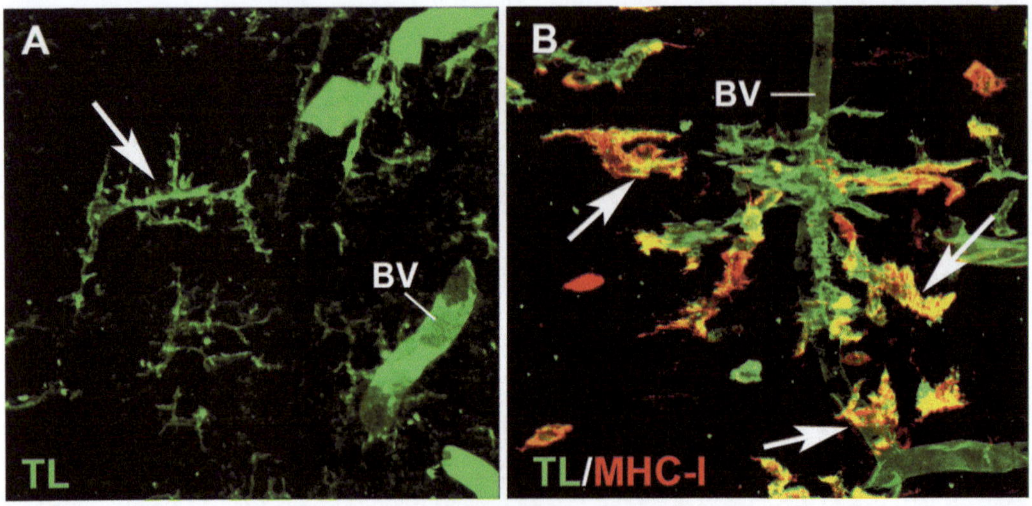

Fig. 3 Microglial labelling using tomato lectin for fluorescence microscopy. (**a**) Photograph showing TL-labelled microglial cells (*arrow*) and blood vessels (BV) in the spinal cord of adult rats. (**b**) Double fluorescence labelling combining TL (*green*) and MHC-I (*red*) showing co-localization (*arrows*) of both markers in some activated/reactive microglial cells found in the spinal cord of EAE-induced rats. Note that blood vessels (BV) do not expressed MHC-I

2.2 Solutions for Cryopreserving

1. Olmos solution (OS): 20 % sucrose, 1 % polyvinylpyrrolidone (PVP), and 30 % ethylene glycol in distilled water. Dissolve 200 g of sucrose in 400 mL of distilled water. Add 300 mL ethylene glycol to the solution. Dissolve 10 g of PVP (Sigma-Aldrich, PVP40, molecular weight: 40) and wait until completely dissolved (*see* **Note 1**). Finally, make up to 1 L with distilled water and store at −20 °C. The solution should not freeze.

2.3 Solutions for Toluidine Blue Counterstaining

1. Acetic acid solution (AAS): Add 2.75 mL of glacial acetic acid (≥99.7 %) to 240 mL of distilled water to obtain a 0.2 M solution.
2. Sodium acetate solution (SAS): Dissolve 2.63 g of sodium acetate (molecular weight 82.03) in 160 mL of distilled water to obtain a 0.2 M solution.
3. Walpole buffer (WB): 0.2 M Walpole buffer solution, pH 4.5. Mix three parts of AAS with two parts of SAS and adjust the pH by adding some drops of the 0.2 M AAS.
4. Toluidine blue solution: Dissolve 0.1 g of toluidine blue in 100 mL of WB in order to obtain a 0.1 % solution. Filter and store at RT.

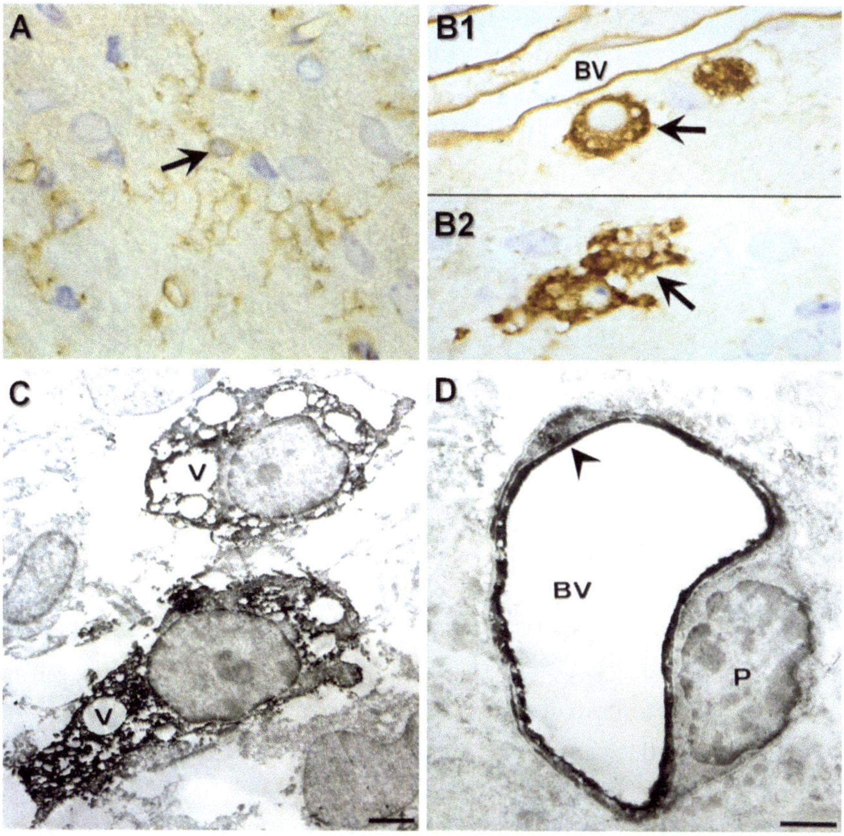

Fig. 4 Tomato lectin histochemistry for TEM studies. Semithin sections reacted with TL and counterstained with toluidine blue of (**a**) adult rat brain showing resting microglial cells (*arrows*) and (**b1**, **b2**) 5-day-old rat brain showing strongly stained amoeboid microglial cells (*arrows*) and blood vessels (BV). (**c**) Electron microphotograph showing amoeboid microglial cells from 5-day-old rat. TL binding shows an intracellular localization, whereas intracytoplasmic vacuoles (V) remain unstained (Reproduced from ref. 2). (**d**) Electron micrograph showing a blood vessel with intense TL staining (*arrowhead*), whereas pericytes (P) remain TL-negative (Reproduced from ref. 2)

2.4 Solutions for Single Labelling with TL Histochemistry and Double Labelling Combining TL Histochemistry with Either GFAP or MHC-I Immunohistochemistry

1. Tris-buffered solution (TB): 0.05 M tris-buffered solution, pH 7.4. Add 12.11 g of Trizma base (Sigma-Aldrich, T1503; molecular weight 121.14) and 7.2 mL of 37 % chlorhydric acid solution (HCl) to 1.8 L of distilled water. Mix well and make up to 2 L. Adjust the pH at 7.4 and store at RT.
2. Tris-buffered saline solution (TBS): 0.05 M tris-buffered saline solution, pH 7.4. Add 12.11 g of Trizma base (Sigma-Aldrich, T1503; molecular weigh 121.14), 17.53 g of NaCl (molecular weight 58.44), and 7.2 mL of 37 % HCl to 1.8 L of distilled water. Mix well, make up to 2 L, adjust the pH at 7.4, and store at RT.
3. TBS with Triton X-100 (TBST): 0.05 M tris-buffered saline solution with 1 % Triton X-100, pH 7.4. Add 10 mL of Triton X-100 to 900 mL of TBS and mix well until the Triton X-100

is completely dissolved. Make up to 1 L with TBS. Store at RT. Discard when the solution is not transparent.

4. Endogenous peroxidase blocking solution (EPBS): 70 % methanol solution with 0.66 % hydrogen peroxide in distilled water. Mix 70 mL methanol with 28 mL distilled water. Finally add 2 mL 30 % hydrogen peroxide. Prepare fresh just before use.
5. Buffer blocking solution (BBS): 0.05 M TBST, pH 7.4 with 10 % fetal bovine serum (FBS) and 3 % bovine serum albumin (BSA). Add 10 mL of FBS (Sigma-Aldrich) to 90 mL of TBST pH 7.4. Add to this solution 300 mg of BSA (Sigma-Aldrich). Mix well until the BSA is completely dissolved. Prepare fresh just before use.
6. Biotinylated TL from Lycopersicum esculentum (Sigma-Aldrich, L-9389).
7. Primary polyclonal rabbit anti-GFAP antibody (DAKOPatts, Z0334).
8. Primary monoclonal mouse anti-MHC-I antibody (AbD Serotec, MCA51G).
9. Secondary biotinylated anti-rabbit Ig (affinity purified) antibody from goat (Vector Laboratories, BA-1000).
10. Secondary Alexa Fluor® 555 donkey anti-mouse IgG (Molecular Probes, A-31570).
11. Horseradish peroxidase streptavidin (Vector Laboratories, SA-5004).
12. Streptavidin-Alexa Fluor® 488 conjugate (Molecular Probes, S-32354).
13. Developing solution (DS): 0.5 mg/mL 3,3′-diaminobenzidine solution with 0.011 % hydrogen peroxide in TBS. Dissolve 50 mg 3,3′-diaminobenzidine (Sigma-Aldrich, ≥96 %, molecular weight 360.11) in 100 mL TBS. Mix well and add 33 μL 30 % hydrogen peroxide. Prepare fresh just before use.
14. Developing solution with nickel (DSN): 0.5 mg/mL 3,3′-diaminobenzidine solution with 0.011 % hydrogen peroxide and 0.026 g/mL nickel ammonium sulfate in TBS. Add 2.6 g nickel ammonium sulfate (Sigma-Aldrich, 03750, molecular weight 62.07) to 100 mL of DS.
15. Developing solution with nickel-cobalt-enhanced DAB (DSNC): 0.5 mg/mL 3,3′-diaminobenzidine solution with 0.011 % hydrogen peroxide, 0.02 % nickel ammonium sulfate, and 0.025 % cobalt chloride. Add 4 mL 1 % of nickel ammonium sulfate (Sigma-Aldrich, molecular weight 62.07) and 5 mL 1 % of cobalt chloride (Sigma-Aldrich) to 200 mL of DS.

2.5 Solutions for Resin-Embedding Procedure for Electron Microscopy

1. Osmium tetroxide solution (OTS): 1 % osmium tetroxide solution in distilled water. Mix 5 mL of 2 % OsO_4 aqueous solution (Electron Microscopy Sciences, 19172) with 5 mL distilled water. Store at 4 °C in the dark.
2. Uranyl acetate solution (UAS): 2 % uranyl acetate solution in 70 % ethanol. Mix 2 g uranyl acetate powder with 100 mL 70 % ethanol. Filter and store at 4 °C.
3. Araldite I: Mix 10 g of component AM epoxy resin (Sigma-Aldrich ACM Fluka, 44611) with 10 g of component B (Sigma-Aldrich ACM Fluka, 44612) and 0.3 g of component D (Sigma-Aldrich ACM Fluka, 44614) in a glass container. Mix well using a glass rod and maintain in the oven at 55 °C. Before use, be sure that there are not air bubbles in the mixture. Araldite I should be freshly prepared 1–2 h before starting the embedding process (*see* **Note 2**).
4. Araldite II: Mix the same components used to prepare Araldite I and add 0.3 g of component C (Sigma-Aldrich ACM Fluka, 44613). Mix well with a glass rod and stand in the oven at 55 °C. Prepare fresh just before use. Do not use until all air bubbles are gone (*see* **Note 2**).

3 Methods

3.1 Intracardiac Perfusion and Postfixation

1. Anesthetized animals (*see* **Note 3**) are intracardially perfused with the FS (FSG for TEM analysis) at 4 °C for 10 min (*see* **Note 4**).
2. The areas of interest (telencephalon, cerebellum, spinal cord, and others) should be dissected out as quickly as possible and immersed in glass vials containing the same fixative solution (4 °C) to complete a total time of 4 h of fixation for light and fluorescence microscopy (1 h for TEM analysis). Take into account that volume of FS (or FSG) should be at least ten times the volume of the tissue pieces. Remove the meninges when possible. Special care should be taken in order to prevent any desiccation during all the process including the extraction of the brain (*see* **Note 5**).
3. Dissect out the samples to be cut. This process can be done during the time of fixation (*see* **Note 6**).

3.2 Obtaining Vibratome Sections

1. Once completed the fixation, wash the samples in PWB (3×10 min, 4 °C) and store them in this buffer (4 °C) until cutting (*see* **Note** 7).
2. Attach the piece to be cut to a small square of filter paper using a super glue gel (*see* **Note 8**). To achieve that, place some drops of glue depending on the size of the tissue sample on the surface of a square piece of filter paper to produce a

little dome. Immediately, with the aid of tweezers, take the piece of tissue to be attached to the filter paper and, before placing on it, quickly absorb the liquid from its basis by gently touching it during 2 s with a strip of filter paper. This is important to assure an appropriated attachment. Once the tissue piece is placed on the center of the glue dome, press gently down with a flat object (as the handle of the tweezers) to be sure that the basis is attached as flat as possible to the filter paper. Wait for two additional seconds and quickly take the square and the attached tissue sample and immediately submerge the assembly in cold (4 °C) PWB until to be cut (*see* **Notes 9** and **10**).

3. With the aid of some drops of super glue gel, attach the assembly (square of the filter paper plus the tissue sample) with glue to the vibratome platform and quickly submerge the platform with the assembly in the specific container of the vibratome, which should be previously filled with cold (4 °C) PWB (*see* **Note 11**). Obtain sections 30–50 μm thick for optical microscope analysis and 50–80 μm thick for resin-embedding and ulterior transmission electron microscopy (TEM) studies. Free-floating sections obtained should be progressively taken from the vibratome container with the aid of a soft brush and introduced in 5 mL vials filled either with cold PWB (4 °C) if the histochemical reaction will be performed immediately after cutting or with OS to store the sections at −20 °C in case of posterior use.

3.3 Obtaining Cryostat Sections

1. Once completed the fixation, wash the samples in cold PWB (4 °C) (2 × 5 min).
2. Remove the PWB and submerge samples in SS for 48 h at 4 °C (*see* **Note 12**).
3. Place the tissue in the correct orientation and glue on a filter paper strip with Tissue-Tek® OCT™ Compound (Sakura, 4583) (*see* **Note 13**).
4. Carefully drop the assembly of the sample and the filter paper into cold 2-methylbutane solution (Sigma-Aldrich, 320404, molecular weight 72.15) (*see* **Note 14**). Freeze for 30–45 s before inserting the sample into precooled storage tubes and store at −80 °C.
5. With the aid of a cryostat, obtain sections 25–30 μm thick. Free-floating sections obtained should be progressively taken from the cryostat device with the aid of a soft brush and introduced in 2 mL plastic vials filled with cold OS (−20 °C) (*see* **Note 15**). Store at −20 °C until their posterior use.

3.4 Obtaining Paraffin Sections

1. After the 4 h of fixation, wash the samples in cold PWB (4 °C) (2 × 5 min).

2. Submerge the samples in graded ethanols as follows:
   (a) 50 % ethanol for 1 h at RT.
   (b) 70 % ethanol for 1 h at RT.
   (c) 95 % ethanol for 1 h at RT.
   (d) 100 % ethanol 3 × 1 h at RT.
3. After the last ethanol, submerge samples in xylene, 2 × 1 h at RT.
4. Remove the xylene and with the aid of the tweezers put each samples in a metal mold previously filled with hot paraffin (58 °C). Maintain samples in paraffin for 2 h, changing the paraffin after the first hour.
5. Embed samples in paraffin blocks (Fig. 5):
   (a) Choose a new mold with an appropriate size for the tissue (*see* **Note 16**).
   (b) Dispense a small amount of molten paraffin in the mold.
   (c) Transfer tissue into the mold with the aid of warm forceps, taking into account that the side to be cut must face the bottom of the mold (Fig. 5a).
   (d) To hold the tissue in the correct position, transfer the mold with the tissue to a cold plate, and gently press tissue flat. Paraffin will solidify in a thin layer holding the tissue.
   (e) Once the tissue is in the right position, place a labelled tissue cassette on the top of the mold. Press firmly (Fig. 5b).
   (f) Gently add hot paraffin to the mold using the paraffin dispenser. The face of the plastic cassette must be completely covered with paraffin (Fig. 5c).
   (g) After 30 min, paraffin should be solidified. When the wax is cold and hard, the paraffin block can be easily removed from the mold and should not stick (*see* **Note 17**) (Fig. 5d). In case of cracked wax or tissue not well aligned, melt them again and start over. Paraffin blocks can be stored at room temperature for years.
6. Obtain sections 10 μm thick with the aid of a microtome and collect them on gelatine-coated slides. Keep the slides stored at RT until their use for the staining procedure.
7. Dewax sections just before starting the TL staining procedure following the next steps:
   (a) Heat the slides in dry oven at 55–60 °C for 20 min.
   (b) Immerse slides in xylene (3 × 10 min).
   (c) Immerse slides in 100 % ethanol for 2 min.
   (d) Immerse slides in 95 % ethanol for 1 min.
   (e) Immerse slides in 70 % ethanol for 1 min.
   (f) Immerse slides in 50 % ethanol for 1 min.

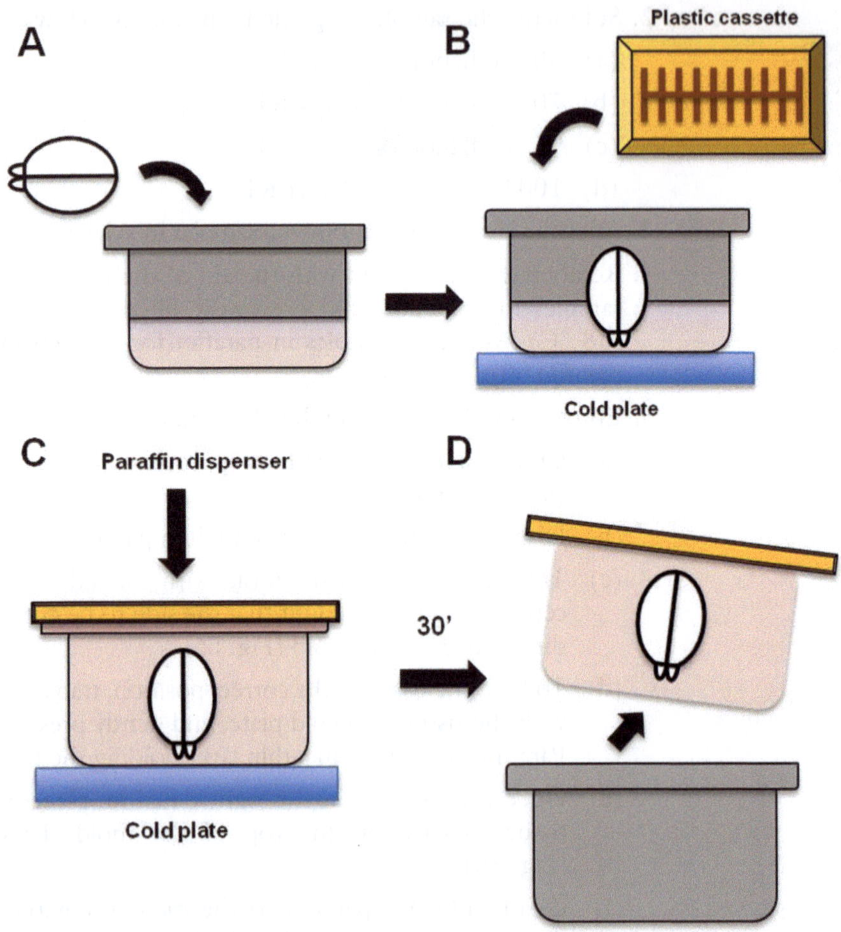

Fig. 5 Schematic representation of the different steps for the embedding of samples in paraffin blocks. (**a**) Place tissue in a metal mold with a small amount of molten paraffin. Be sure the tissue is in the correct orientation. (**b**) Transfer mold to a cold plate to allow hardening of paraffin and place the cassette on top of the mold. (**c**) Move the mold to the cold plate and refill completely with molten paraffin using the paraffin dispenser. Be sure paraffin covers all the cassette holes. (**d**) After 30 min of cooling, remove the cassette with the paraffin block from the metal mold

(g) Immerse slides in 30 % ethanol for 1 min.

(h) Immerse slides in 1× PBS (Gibco, 14190) for 2 min.

(i) Proceed with the staining procedure.

3.5 Histochemical Reaction Using Single Tomato Lectin Labelling for Light Microscopy and Fluorescence

1. Wash the sections or the gelatine-coated slides with TBS at RT (3×10 min).
2. Incubate the sections or slides in EPBS at RT for 10 min (*see* **Notes 18** and **19**).
3. Wash 3×10 min with TBST at RT.
4. Incubate sections with the TL diluted in TBST (1:150) overnight at 4 °C and afterwards 1 h at RT (*see* **Note 20**).

5. Wash the sections 3 × 10 min in TBST at RT.
6. Incubate sections with the horseradish peroxidase streptavidin diluted in TBST (1:500) for 1 h at RT. In case of fluorescence microscopy, incubate with the streptavidin-Alexa Fluor® 488 conjugate diluted in TBST (1:1,000) for 1 h at RT (*see* **Note 21**).
7. Wash the sections 3 × 10 min in TBS at RT.
8. Wash the sections 3 × 10 min in TB at RT.
9. For light microscopy, visualize the peroxidase reaction using the DS. Incubate sections in the DS solution for exactly 3 min. Be accurate over time in this step. For fluorescence microscopy, skip this step. If desired in these fluorescence-processed sections, a nuclei counterstaining can be performed by incubating the sections in 4′,6-diamidino-2-phenylindole dihydrochloride (DAPI; Sigma-Aldrich, D9542) in TB (1:10,000) for 3 min.
10. In case of vibratome and cryostat free-floating samples, mount the sections on gelatine-coated slides. Allow dry for at least 24 h. In case of light microscopy, proceed with the toluidine blue counterstaining following **steps 11–18**. In case of fluorescence, skip **steps** from **11–13** and proceed with the dehydration (**step 14**) and coverslip the slides as specified below (**steps 17** and **18**).
11. Submerge the slides in distilled water for 3 min.
12. Counterstain with toluidine blue for 1 min (*see* **Note 22**).
13. Wash with distilled water for 1 min to remove the excess of colorant.
14. Dehydrate by submerging the slides in graded ethanol: 50, 70, 90, and 3 × 100 %. No more than 30 s in each alcohol.
15. Differentiate with 1-butanol for 1 min.
16. Submerge in xylene (2 × 2 min).
17. Finally, coverslip the slides using DPX as permanent mounting medium (*see* **Note 23**).
18. Wait at least 24 h in order to allow complete polymerization of the DPX before proceeding with the analysis at the light microscope.

3.6 Double Labelling Technique Combining Tomato Lectin Histochemistry and Either GFAP or MHC-I Immunohistochemistry

Double labelling can be visualized by using either light or fluorescence microscopy. In case of light microscopy, follow the **steps 1**, **3**, and **4**; in case of fluorescence microscopy, follow the **steps 2–4**:

1. Incubation for the immunohistochemical detection of GFAP for light microscopy:
   (a) Remove the sections stored at −20 °C on Olmos solution or the previously dewax slides and wash 3 × 10 min with TBS at RT.

(b) After washing, block the endogenous peroxidase incubating for exactly 10 min with the EPBS at RT (*see* **Note 18**).

(c) Remove EPBS and wash 3 × 10 min with TBST at RT.

(d) Block the unspecific binding with BBS for 1 h at RT.

(e) Incubate with the primary GFAP antibody diluted in BBS (1:1,800) overnight at 4 °C and 1 h at RT (*see* **Note 24**).

(f) Wash the samples 3 × 10 min in TBST at RT.

(g) Incubate samples in the secondary biotinylated anti-rabbit Ig antibody diluted in BBS (1:500) for 1 h at RT.

(h) Wash the samples 3 × 10 min in TBST at RT.

(i) Incubate in the horseradish peroxidase streptavidin diluted in BBS (1:500) for 1 h at RT.

(j) Wash the sections 3 × 10 min in TBS at RT.

(k) Visualize the peroxidase reaction incubating the samples for 3 min in DSN at RT.

(l) Remove the DSN and wash sections 3 × 10 min in TB at RT.

(m) After washing, block the endogenous peroxidase incubating for exactly 10 min with the EPBS at RT.

2. Incubation for the immunohistochemical detection of MHC-I for fluorescence microscopy:

(a) Remove the sections stored at −20 °C on Olmos solutions or the previously dewax slides and wash 3 × 10 min with TBS at RT.

(b) Wash 3 × 10 min with TBST at RT.

(c) Block the unspecific binding with BBS for 1 h at RT.

(d) Incubate with the primary MHC-I antibody diluted in BBS (1:1,000) overnight at 4 °C (*see* **Note 24**).

(e) Wash the sections 3 × 10 min in TBST at RT.

(f) Incubate sections in the secondary Alexa Fluor® 555 donkey anti-mouse IgG antibody diluted in BBS (1:1,000) for 1 h at RT.

(g) Wash the sections 3 × 10 min in TBST at RT.

3. For both light and fluorescence microscopy, after the GFAP or MHC-I immunohistochemistry demonstration, incubate for the histochemical detection of TL. For this purpose, follow the **steps 3–9**, specified previously in Subheading 3.5 taking into account which type of microscopy will be used to visualize the staining.

4. Finally, in case of cryostat or vibratome samples, mount the sections on gelatine-coated slides and allow dry for at least 24 h. Skip this step in case of paraffin-processed samples.
5. Dehydrate by submerging the slides in graded ethanol: 50, 70, 90, and 3×100 %. No more than 30 s in each alcohol.
6. Submerge in xylene (2×2 min). Skip this step in case of fluorescence microscopy, as the treatment with xylene can eliminate the fluorescence signal.
7. Finally, coverslip the slides using DPX as permanent mounting medium.
8. Wait at least 24 h in order to allow complete polymerization of the DPX before proceeding with the analysis at the microscope.

3.7 Tomato Lectin Histochemistry for Transmission Electron Microscopy

This technique can also be used for visualization of microglia at the ultrastructural level as follows:

1. Fixation must be done following the steps specified in Subheading 3.1 but using FSG instead of FS. To obtain the sections to be processed, follow the steps detailed above in Subheading 3.2.
2. Once 50–80 μm thick sections will be obtained, process the samples for the histochemical demonstration of TL following the **steps** indicated previously in Subheading 3.5 from **1** to **8** using the solutions needed for light microscopy.
3. For better visualization of the peroxidase reaction product with the electron microscope, perform the **step 9** specified in Subheading 3.5 but using DSNC instead of DS.
4. Transfer the sections to glass vials and wash by adding 0.1 M PWB (3×10 min) to the vials.
5. Postfix the sections in OTS for 2 h at RT in the dark.
6. Wash sections 3×10 min in 0.1 M PWB at 4 °C.
7. Remove the 0.1 M PWB from the glass vials and start the dehydration process with graded ethanol as indicated:
   (a) 30% ethanol for 15 min at 4 °C.
   (b) 50% ethanol for 15 min at 4 °C.
   (c) 70% ethanol for 15 min at 4 °C.
8. En bloc stain with UAS for 1.5 h at 4 °C.
9. Wash in 70 % ethanol for 15 min at 4 °C.
10. Continue the dehydration as follows:
    (a) 96% ethanol for 15 min at 4 °C.
    (b) 100% ethanol 3×10 min at 4 °C.

11. Remove the last 100 % ethanol and add propylene oxide 3×10 min at 4 °C.
12. Start the embedding process by removing the propylene oxide and adding to the glass vials a mixture of one part of Araldite I and three parts of propylene oxide for 2 h at RT (*see* **Note 25**). Prepare the mixture just before use.
13. Remove the previous mixture and add a new mixture of one part of Araldite I and one part of propylene oxide for 2 h at RT. Prepare the mixture just before use.
14. Again remove the previous mixture and add a new mixture of three parts of Araldite I and one part of propylene oxide for 2 h at RT. Prepare the mixture just before use.
15. After removing the last mixture, add Araldite I to the glass vials and maintain them in the oven at 55 °C overnight.
16. Change the Araldite I and maintain at 55 °C for 3 h. Meanwhile prepare the Araldite II (*see* Subheading 2).
17. Remove the Araldite I and add Araldite II for 2 h at 55 °C.
18. Take the samples from the glass vials with the aid of a wood spatula and carefully transfer the sections to silicone molds previously filled with Araldite II. Make sure that all sections are completely flat and correctly oriented.
19. Maintain the silicone molds with the sections in the oven at 65 °C for 48 h to ensure the correct polymerization of the resin.
20. Using an ultramicrotome, obtain either semithin (1–2 μm) or ultrathin (300–500 Å) sections. Semithin sections can be visualized with a conventional light microscope whereas for the study of ultrathin sections a transmission electron microscope should be used.

4 Notes

1. Take into account that polyvinylpyrrolidone is a polymer and is difficult to dissolve. To make it easy, add gradually in 1 g portions with constant stirring and wait until the first gram is completely dissolved to continue adding. Make sure there are no lumps in the solution.
2. Take into account how much volume you will need in function of the number of changes to do, samples, volume of glass vials, and volume of the silicon molds to be used.
3. Animals can be anesthetized using any anesthetic as isoflurane, pentobarbital, ketamine/xylazine, and others.
4. Ideally the perfusion in rats and mice should be done inserting the cannula through the left ventricle until reaching the aorta

and tie up the blood vessel with a forceps. Then make a cut in the right auricle and open the flux of the fixative. To assure that the cold (4 °C) fixative does not cause any undesirable vasoconstriction, it is necessary to use an appropriated device that provides a constant pressure and enough flux of the FS (or FSG) to push out the blood during the first 10 s of perfusion.

5. After perfusion, during bone removal and brain extraction, it is important to maintain the exposed brain surfaces always wet by adding some drops of FS (or FSG) with a Pasteur pipette.
6. To facilitate the dissection of the samples, it is recommendable to do it after some time (30 min) of postfixation. Dissections can be done in a Petri dish using a razor blade and, if necessary, under the binocular lens. Tissue pieces should be covered all the time by the FS (or FSG) at 4 °C, in order to prevent desiccation. Preferably the basis of the pieces should be as flat as possible to assure a good and stable attachment to the cutting device. Once the samples are dissected out, they should be transferred back from the Petri dish to the glass vials to complete the total time of fixation (4 h for light microscopic study and 1 h for TEM) established in the method section. In the case of vibratome sections, the size of the pieces should be enough to be appropriately cut. In this sense, coronal sections 5–6 mm thick of a whole rat brain may be obtained and cut without problems.
7. Best results are obtained when samples are cut on the same day. If not possible to do the histochemistry the same day of perfusion, it is recommendable to keep the dissected samples in PWB (4 °C) and to obtain the vibratome sections in the following day.
8. Although any super glue can be used, the gel type is the most recommendable because it does not spread when placed in the filter paper and it holds better the tissue piece, especially if the piece is not completely flat in the base.
9. If there are many pieces, they can be attached to different filter paper squares and maintained submerged in cold PWB until they are cut.
10. All the process should be as fast as possible to minimize the possible desiccation of the samples. Do your best to ensure that the time spend in attaching the samples to the filter paper is only some few seconds.
11. If the vibratome device you use does not have a container where temperature can be regulated, then change frequently the PWB, or if the vibratome device allows it, add some ice around the container to maintain the PWB cold.
12. SS is a cryopreservative solution and avoids the formation of water crystals that can break the tissue structure. It is important

to maintain the samples at list 48 h in this solution to ensure a good preservation. Samples sunk at the bottom of the glass vial indicate completely penetration of sucrose in the tissue.

13. Tissue-Tek® OCT™ Compound is special medium that freezes and provides a correct support of the sample during the cutting process.
14. The freezing process should be as quick as possible to ensure that all the tissue freezes at the same time. For that purpose, allow 2-methylbutane to reach cold enough temperature by putting the solution surrounded by dry ice at least 15 min before starting the freezing procedure. In order to be sure that 2-methylbutane is cold enough, drop a piece of tissue into it. Tissue must become white in 1–2 s indicating a rapid freezing.
15. Keep the plastic vials filled with OS inside the cryostat to maintain the temperature of OS at −20 °C during the cutting process.
16. An appropriate mold for a tissue is one that ensures that all the sides of the tissue will be surrounded by paraffin. This is important because it gives a best cutting support.
17. To increase the speed of paraffin block cooling, the mold can be submerged into cold water.
18. The incubation time should be the same for all the samples, especially if you want to perform a quantitative analysis of the staining. Skip this step in case of fluorescence microscopy.
19. In case of paraffin sections mounted on gelatine-coated slides, it may be useful to draw a hydrophobic mark around the tissue using a specific pencil (IHCworld, Super PAP Pen Liquid Blocker, SPM0928) in order to prevent the solutions from spreading. In addition, the different incubations should be performed in a humid chamber which will ensure that the solutions will not evaporate during the incubation time.
20. As a negative control of the histochemistry, some sections should be processed in parallel without the lectin.
21. From now on, keep the sections away from direct light, as fluorochrome has been used and fluorescence may fade easily.
22. Toluidine blue solution for counterstaining can be reused, but it is important to filter always the solution before use. Note that the time of counterstaining should be adjusted in every reused; therefore, try first with one slice to determine the appropriated time.
23. DPX could be used as well for fluorescence microscopy, but fluorescent mounting medium (Dako, S3023) is strongly recommended. In this case avoid **steps 15** and **18** and coverslip the slides just after mounting (**step 11**).

24. As negative control of the immunohistochemical reaction, some sections should be processed in parallel without the primary antibody.
25. From now, remove the lids of the glass vials in order to facilitate the evaporation of propylene oxide.

References

1. Roth J (2011) Lectins for histochemical demonstration of glycans. Histochem Cell Biol 136(2):117–130
2. Acarin L, Vela JM, Gonzalez B et al (1994) Demonstration of poly-N-acetyl lactosamine residues in ameboid and ramified microglial cells in rat brain by tomato lectin binding. J Histochem Cytochem 42(8):1033–1041
3. Pesheva P, Urschel S, Frei K et al (1998) Murine microglial cells express functionally active galectin-3 in vitro. J Neurosci Res 51(1): 49–57
4. Dalmau I, Vela JM, Gonzalez B et al (1997) Expression of LFA-1alpha and ICAM-1 in the developing rat brain: a potential mechanism for the recruitment of microglial cell precursors. Brain Res Dev Brain Res 103(2): 163–170
5. Dalmau I, Vela JM, Gonzalez B et al (2003) Dynamics of microglia in the developing rat brain. J Comp Neurol 458(2):144–157
6. Campuzano O, Castillo-Ruiz MM, Acarin L et al (2009) Increased levels of proinflammatory cytokines in the aged rat brain attenuate injury-induced cytokine response after excitotoxic damage. J Neurosci Res 87(11): 2484–2497
7. Acarin L, Gonzalez B, Castellano B et al (1996) Microglial response to N-methyl-D-aspartate-mediated excitotoxicity in the immature rat brain. J Comp Neurol 367(3): 361–374
8. Acarin L, Gonzalez B, Castellano B et al (1997) Quantitative analysis of microglial reaction to a cortical excitotoxic lesion in the early postnatal brain. Exp Neurol 147(2): 410–417
9. Acarin L, Gonzalez B, Castro AJ et al (1999) Primary cortical glial reaction versus secondary thalamic glial response in the excitotoxically injured young brain: microglial/macrophage response and major histocompatibility complex class I and II expression. Neuroscience 89(2):549–565
10. Acarin L, Gonzalez B, Castellano B (2000) Neuronal, astroglial and microglial cytokine expression after an excitotoxic lesion in the immature rat brain. Eur J Neurosci 12(10):3505–3520
11. Sanz O, Acarin L, Gonzalez B et al (2001) Expression of 27 kDa heat shock protein (Hsp27) in immature rat brain after a cortical aspiration lesion. Glia 36(3):259–270
12. Almolda B, Gonzalez B, Castellano B (2010) Activated microglial cells acquire an immature dendritic cell phenotype and may terminate the immune response in an acute model of EAE. J Neuroimmunol 223(1–2):39–54
13. Vela JM, Dalmau I, Gonzalez B et al (1996) The microglial reaction in spinal cords of jimpy mice is related to apoptotic oligodendrocytes. Brain Res 712(1):134–142
14. Vela JM, Hidalgo J, Gonzalez B et al (1997) Induction of metallothionein in astrocytes and microglia in the spinal cord from the myelin-deficient jimpy mouse. Brain Res 767(2): 345–355
15. Minten C, Terry R, Deffrasnes C et al (2012) IFN regulatory factor 8 is a key constitutive determinant of the morphological and molecular properties of microglia in the CNS. PLoS One 7(11):e49851
16. Castellano B, Gonzalez B, Finsen BR et al (1990) Histochemical demonstration of purine nucleoside phosphorylase (PNPase) in microglial and astroglial cells of adult rat brain. J Histochem Cytochem 38(11):1535–1539
17. Montero Dominguez M, Gonzalez B, Zimmer J (2009) Neuroprotective effects of the anti-inflammatory compound triflusal on ischemia-like neurodegeneration in mouse hippocampal slice cultures occur independent of microglia. Exp Neurol 218(1):11–23
18. Montero M, Gonzalez B, Zimmer J (2009) Immunotoxic depletion of microglia in mouse hippocampal slice cultures enhances ischemia-like neurodegeneration. Brain Res 1291: 140–152

Chapter 24

Immunohistochemical Detection of Microglia

Rocío M. de Pablos, Ana M. Espinosa-Oliva, and Antonio J. Herrera

Abstract

Immunohistochemistry (IHC) is a technique that allows the localization of antigens or proteins in tissue sections using the high specificity and affinity of antibodies to recognize molecules and join them. The commercial offer and the standardization of protocols make this technique a simple, fast, and powerful method. Microglia, the resident macrophage cells of the central nervous system, can exist in three different forms that can be identified using different antibodies. The aim of this chapter is to describe the methods to perform IHC using these different antibodies.

Key words Iba1, Immunohistochemistry, Microglia, OX-6, OX-42

1 Introduction

Immunohistochemistry (IHC) is a technique that allows the localization of antigens or proteins in tissue sections; it uses labelled antibodies as specific reagents through antigen–antibody interactions visualized by a variety of markers, such as enzymes, fluorescent dyes, or colloidal gold. Since the first experiments back in the 1930–1940s, methods of tissue fixation, protein conjugation, labels, and microscopy have improved greatly, making IHC a simple, powerful, and essential tool in diagnostic and research laboratories.

The technique is based on the high specificity and affinity of antibodies to recognize molecules and join them. It enables the visualization (using light or confocal microscopy) of the tissue distribution of specific antigens (epitopes). Specific monoclonal and polyclonal antibodies target proteins of interest in a process called *antibody incubation* (Fig. 1). The use of avidin–biotin immunohistochemistry on recent years takes advantage of the high affinity that avidin (or streptavidin) has for biotin, making possible the use of multiple binding sites on the target molecules, thus resulting in the amplification of tissue signal, which is very helpful to detect low amounts of antigens. Conjugation or combination of the antibodies with fluorescing substances enables the detection of minute amounts in the tissue sections.

Bertrand Joseph and José Luis Venero (eds.), *Microglia: Methods and Protocols*, Methods in Molecular Biology, vol. 1041, DOI 10.1007/978-1-62703-520-0_24, © Springer Science+Business Media New York 2013

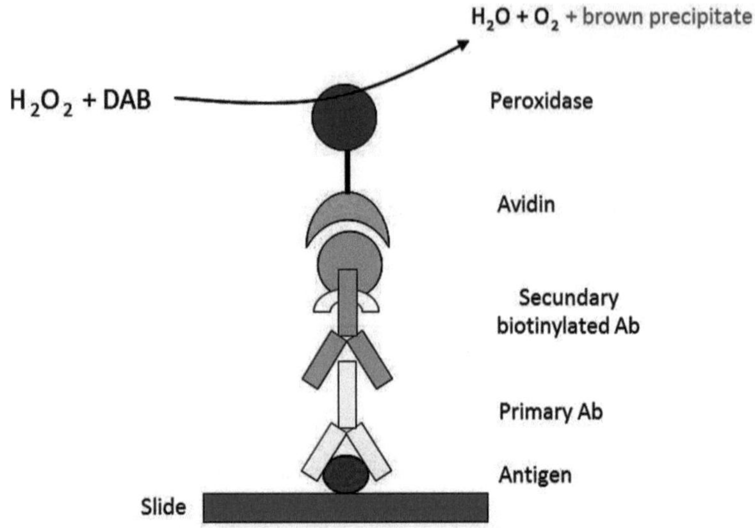

Fig. 1 Schematic representation of the immunohistochemical staining using antibodies. Primary antibody specifically binds the antigen through its variable region, whereas the biotinylated secondary antibody recognizes the constant region of the primary one. The avidin–peroxidase conjugate binds biotin attached to the secondary antibody. The peroxidase was visualized with a standard diaminobenzidine/hydrogen peroxide reaction

Microglia is the resident macrophage cell of the central nervous system, showing a highly specialized morphology and unusual antigenic phenotype. Microglia can exist in three different forms serving different functional roles: amoeboid, ramified, and reactive/activated microglia. It has generally been accepted that resting microglia—with small cell bodies and thin processes—increase in their soma size as well as thicken and shorten their processes in early stage of activation to become activated microglia, which are finally transformed into amoeboid microglia with phagocytic activity [1–4]. Microglia share many properties with peripheral monocytes and macrophages but retain unique morphological and molecular characteristics [5, 6]. Microglia can bind several lectins, including *Griffonia simplicifolia* agglutinin I B4, *Ricinus communis* agglutinin, and mistletoe agglutinin, as well as express complement receptor 3 (CR3), even in the resting state. In addition, activated microglia expresses high levels of major histocompatibility complex (MHC) class I and II [1, 2, 7, 8]. These different forms of microglial cells can be identified using different antibodies. Iba1 is a 17-kDa EF hand protein that is specifically expressed in macrophages/microglia and is upregulated during the activation of these cells. Iba1 is markedly expressed in microglia (Fig. 2), but is never expressed in neurons or astrocytes. These observations indicate that Iba1 is a macrophage-specific protein

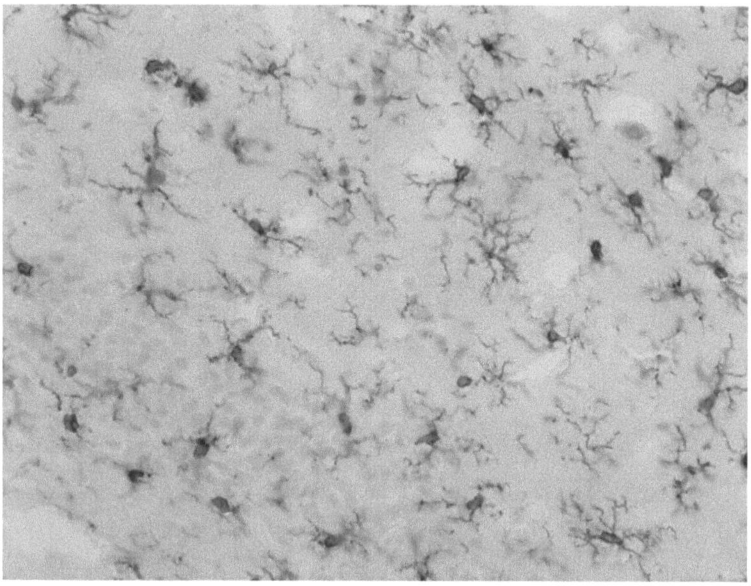

Fig. 2 Immunostaining of microglial cells using the Iba1 antibody

and that anti-Iba1 antibody is valuable for identification of microglia and macrophages. In addition, anti-Iba1 antibody has a great advantage over other marker antibodies since it is so far the only polyclonal antibody applicable to dual labelling studies of the macrophage/microglia lineage even in paraffin sections [9]. OX-42 is an antibody to CR3, a marker expressed by microglia (Fig. 3), whereas OX-6 is an antibody recognizing the MHC II antigen expressed by activated microglia (Fig. 4).

The aim of this chapter is to describe the methods to perform an IHC using the three antibodies mentioned above.

2 Materials

2.1 General

1. Section-staining tray (humid chamber for sections staining).
2. Coplin jars.
3. Hydrophobic pen (Dako, Glostrup, Denmark).
4. Coverslips.
5. Tissue-Tek® OCT™ compound (Sakura, Alphen aan den Rijn, the Netherlands).
6. Ethanol solutions (50–100 %).
7. Methanol.
8. Hydrogen peroxide (30 %).
9. Triton® X-100.

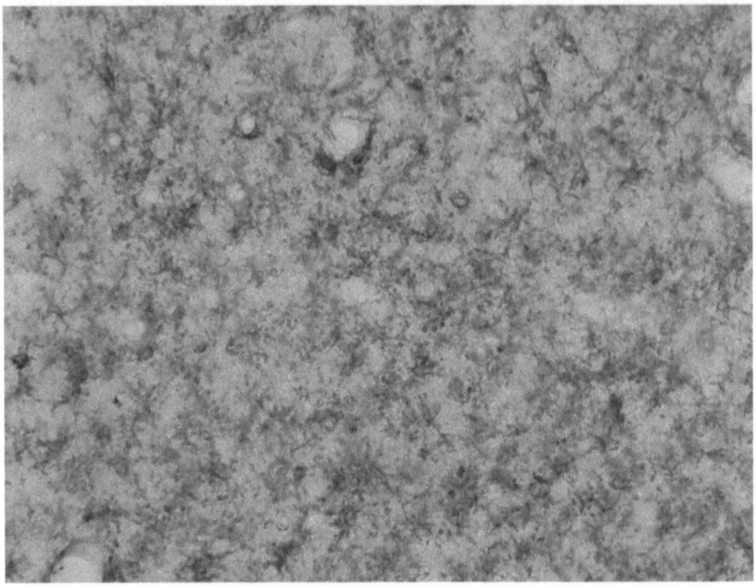

Fig. 3 Immunostaining of microglial cells using the OX-42 antibody. The picture shows resting microglial cells with small cell bodies and thin processes

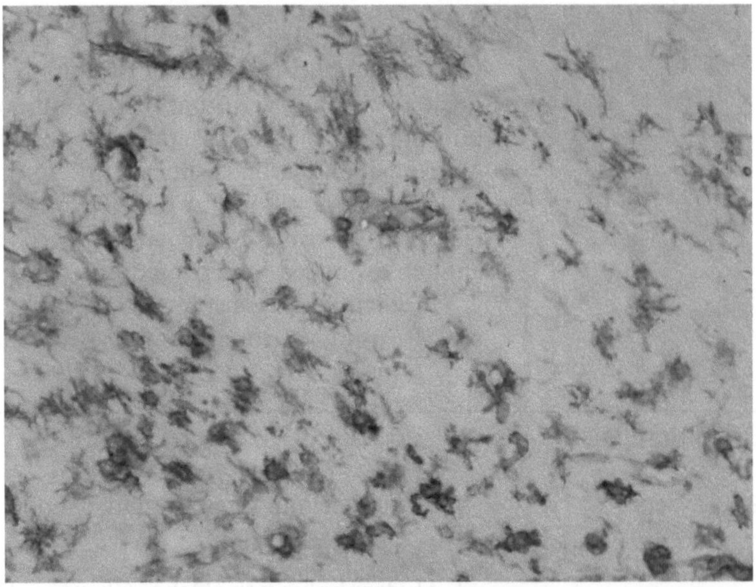

Fig. 4 Immunohistochemistry of activated microglial cells using the OX-6 antibody. The picture shows microglial cells in different states of activation

10. Histolemon (Carlo Erba, Milano, Italy).
11. DPX mounting medium (VWR International Ltd, Poole, England).

2.2 Reagents

1. ABC solution (Vectastain®; Vector Laboratories Inc.).
2. Proteinase K (Sigma-Aldrich).

3. 4 % paraformaldehyde in PBS, pH 7.4. Weigh 20 g paraformaldehyde and 7.1 g Na_2HPO_4 in 100–150 mL of distilled water. Heat to 55 °C with continuous stirring. Add a few pellets of NaOH. Add 25 g sucrose and 2 mL 25 % glutaraldehyde. Adjust pH, top up to 500 mL with distilled water, and filter. Store at 4 °C.
4. Tris-buffered saline (TBS), pH 7.4, 10×: 121 g Tris base, and 1,470 g NaCl in 7 L of distilled water. Adjust pH, top up to 10 L with distilled water, and filter. Store at room temperature.
5. Tris-EDTA (TE) buffer for proteinase K, pH adjusted to 8.0: 6.1 g Tris base, 0.37 g EDTA, and 5 mL Triton X-100 in 1 L of distilled water. Store at room temperature.
6. Phosphate-buffered saline (PBS), pH 7.4, 10×: 7.933 g $NaH_2PO_4{\cdot}H_2O$, 8.9 g $Na_2HPO_4{\cdot}2H_2O$, and 75.86 NaCl in 500 mL of distilled water. Adjust pH, top up to 1 L with distilled water, and filter. Store at room temperature.
7. Diaminobenzidine (DAB; Sigma-Aldrich): Dissolve 1 tablet (10 mg) in 100 mL TBS 1×. Filter and make aliquots of 2 mL, approximately. Store at −20 °C protected from the light. DAB–hydrogen peroxidase reaction solution should be prepared freshly, adding 2 μL of H_2O_2 to 2 mL of DAB solution just before use.
8. Proteinase K: 2 mg of proteinase K, 2.5 mL TE buffer, and 2.5 mL glycerol. Gently mix proteinase K and TE buffer and then add glycerol. Mix and make aliquots (75 μL) stored at −20 °C. In this condition, aliquots are viable for 2–3 years.

2.3 Antibodies and Serums

1. Primary antibodies: Mouse-derived OX-6 (Serotec, Oxford, UK; 1:200), mouse-derived OX-42 (Serotec; 1:100), and rabbit-derived anti-Iba1 (Wako, Osaka, Japan; 1:500).
2. Secondary antibodies: Biotinylated horse anti-mouse IgG (1:200) and biotinylated goat anti-rabbit IgG (1:200) from Vector Laboratories (Burlingame, CA).
3. Horse and goat serum (Vector Laboratories, 1:100).

3 Methods

3.1 Tissue Preparation

To obtain good results, fixation and processing of the tissue must be taken into consideration.

1. Perfuse through the heart (left ventricle) under deep anesthesia with approximately 100 mL of saline solution to remove brain blood.
2. Perfuse with 150–200 mL of 4 % paraformaldehyde in 0.1 M PBS, pH 7.4.

3. Remove the brain from the skull and then cryoprotect it serially in sucrose in PBS, pH 7.4 (*see* **Note 1**).
4. Freeze the brains in isopentane at −80 °C (*see* **Note 2**). If brains are not used immediately, store them at −40 °C (*see* **Note 3**).
5. Mount specimens with OCT; cut 20-μm sections in a cryostat (−20 °C) and mount them in gelatine-coated slides. Sections can be stored at −20 °C for years (*see* **Note 4**).

3.2 Antigen Retrieval

Most proteins and peptides retain their antigenicity with 4 % paraformaldehyde, although low molecular weight antigens (e.g., biogenic amines) may lose it. Reaction with many antigens can be significantly improved by the pretreatment with an antigen retrieval reagent that breaks the protein cross-links formed by formalin fixation, uncovering hidden antigenic sites. Different antigens and antibodies may require different antigen retrieval methods, as the heat-induced epitope retrieval (HIER) or the proteolytic-induced epitope retrieval (PIER), among others.

Proteinase K: Formalin and other aldehyde fixatives form protein cross-links that mask antigenic sites in tissue specimens, thereby producing weak or false-negative staining in the immunohistochemical detection process of some proteins. Proteinase K solutions (PIER) break these protein cross-links, unmasking antigens and epitopes in formalin-fixed and paraffin-embedded tissue sections, which enhances staining intensity of antibodies. This method is highly recommended for Iba1 immunostaining (to be done before immunostaining procedure), proceeding as follows:

1. Place a humid chamber in an incubator at 37 °C for 10 min.
2. Prepare the proteinase K solution by adding 1,425 μL of TE buffer to an aliquot of 75 μL proteinase K prepared as described below (dilution 1:20).
3. Incubate sections with this solution in the humid chamber for 10 min at 37 °C and then for 10 min at room temperature (*see* **Note 5**).
4. Rinse twice in TBS 1×, 10 min each (*see* **Note 6**).

3.3 Immunohistochemistry

1. Completely thaw the sections for at least 15–30 min (*see* **Note 7**).
2. Surround sections with the hydrophobic pen (*see* **Note 8**).
3. Rinse in TBS 1× for 10 min in a jar.
4. Block endogenous peroxidase with 0.3 % H_2O_2 in methanol in a jar.
5. Rinse twice in TBS 1× for 10 min each in a jar.
6. Incubate in a solution containing TBS 1× and 1 % normal serum (from the same animal species as the biotinylated antibody) for 60 min in a humid chamber. Add 50 μL approximately to each section (*see* **Note 9**).

7. Drain slides without washing.
8. Incubate overnight at 4 °C in the humid chamber with 50 μL per section of the primary antibody in TBS 1× containing 1 % serum and 0.25 % Triton X-100 (*see* **Note 10**).
9. Rinse three times in TBS 1× for 10 min each in a jar.
10. Incubate with the secondary biotinylated antibody (1:200 dilution) in TBS 1× containing 0.25 % Triton X-100 (50 μL per section) for 2 h at room temperature in the humid chamber.
11. Rinse three times in TBS 1× for 10 min each in a jar.
12. Allow to react with ABC solution (50 μL per section) for 1 h at room temperature (*see* **Note 11**).
13. Rinse three times in TBS 1× for 10 min each in a jar.
14. Allow reacting with DAB–hydrogen peroxide solution (50 μL per section) for 3–6 min.
15. Rinse twice in TBS 1× for 10 min each, changing the jar after the first wash (*see* **Note 12**).
16. Dehydrate sections with 50, 70, and 90 % ethanol (5 min each) and with 100 % ethanol for 10 min, then twice in Histolemon, 10 min each (*see* **Note 13**).
17. Mount sections in a permanent medium (DPX) with cover glasses and observe under microscope (*see* **Note 14**).

4 Notes

1. Sucrose displaces part of the water in the tissue, reducing freezing artifacts during sectioning. Put the brain in 10 % sucrose for 24 h and then in 20 and 30 % sucrose until it sunk (2–5 days).
2. To prevent deformation of the brain during the process, put it and remove it slowly several times before dropping it in the frozen isopentane.
3. Slow freezing can form ice crystal, altering the normal architecture of the tissue and producing holes, in a sort of "Swiss cheese" effect. Thus, freezing should be made quickly to avoid ice crystals keeping water in a vitreous form that does not expand when solidifies. Make sure freezing medium is cold enough. Brains can be frozen in isopentane at −80 °C to obtain a fast freezing and then kept at −40 °C. It might be thought that liquid nitrogen (nearly −200 °C, one of the coldest liquids routinely available; it does not mix with tissue) would be very appropriate for fast freezing. However, it boils in contact with objects over their own temperature (virtually all the objects in the lab, including specimens), creating a vapor barrier that causes freezing in a slower, unpredictable pattern.

Because of this, tissue in OCT often cracks due to the unpredictable freezing pattern. We do not recommend liquid nitrogen to freeze brain specimens.

4. Sometimes, sections can detach from slides during the long incubation process. This can be due to the handling of sections while cutting; if sections are not well stretched over the glass, they can be moved smoothly by using a fine brush and saline solution. Sections should not be frozen until the saline solution has dried; keep sections at room temperature for 15–30 min before storing them frozen.
5. This method can damage under-fixed tissues. An appropriate incubation time (5–20 min) and temperature (20–60 °C) are required for specific application; avoid over-digesting tissues, especially under-fixed ones.
6. Other types of antigen retrieval system (HIER) greatly enhance immunohistochemistry results after microwave irradiation of formalin-fixed, paraffin-embedded specimens in buffer. The energy provided by irradiation breaks some of the bonds formed during fixation, increasing the intensity of reactions and thus the number of positive cells available, although the exact mechanism involved is unclear.

   Introduce the slides in a coplin jar with citrate solution, pH 6 (0.65 g citric acid and 0.48 g sodium citrate in 2 L of distilled water. Store at room temperature), and put this jar within a bigger jar with water. Heat in the microwave for 4 min at approximately 500 W. Repeat replacing water and citrate solution. Place sections in a coplin jar with TBS for 10 min to cool.
7. As mentioned in **Note 4**, sections can detach from slides during the incubation process. To avoid this, sections must be absolutely dried before starting the protocol. Putting them in a hot plate during 15–30 min could be helpful.
8. Alternatively, you can drain and dry the slides around the sections with filter paper after every wash. Once you start the protocol, sections should not dry in any moment. Be especially careful in long incubations.
9. The humid chamber can be prepared by placing some filter paper at the bottom of a staining tray, wetting it with water or buffer.
10. Detergents (e.g., Triton X-100) should be used for antiserum raised against an intracellular antigen or domain. As a general rule, antibodies may be stored frozen but repeated thaw/frozen cycles should be avoid. Prepare small aliquots adjusted to the amounts normally used for the staining and store those frozen.

    OX-6 can recognize a monomorphic determinant of the rat I-A antigen present on B lymphocytes, dendritic cells, macrophages, and certain epithelial cells. The anti-Iba1 antibody also recognized a number of tissue-resident macrophages, including alveolar macrophages, Kupffer cells, and splenocytes. OX-42 recognizes most macrophages (including resident peritoneal and activated macrophages), Kupffer cells, and about 35 % alveolar macrophages [10]. This antibody also labels dendritic cells extensively, granulocytes, and cells with the morphology of microglia in the brain. Note that when using these antibodies, you can be staining not only microglial cells. To distinguish microglial cells from macrophages, other techniques, such as flow cytometry, are required.

11. ABC solution should be made 30 min before use by mixing A and B solutions (10 μL each in 1 mL of TBS 1×).
12. Remove DAB solution carefully with a pipette, discarding it in an appropriate vial for waste; stop the reaction by transferring to TBS 1×. Check the reaction (brown color) by observing under a light microscope.

    DAB is a carcinogenic compound that must be manipulated carefully. We recommend to develop the DAB reaction by placing the staining tray on aluminum foil and discard it when finished. The coplin jar used to the first wash after the DAB reaction must be cleaned thoroughly.
13. For proper conservation of the slides, ethanol and Histolemon should not be reused many times (they accumulate water) to ensure the dehydrating process.
14. When mounting the slides with mounting medium, remove bubbles by pressing the coverslip with Dumont forceps.

References

1. Streit WJ, Kreutzberg GW (1988) Response of endogenous glial cells to motor neuron degeneration induced by toxic ricin. J Comp Neurol 268:248–263
2. Mor G, Nilsen J, Horvath T et al (1999) Estrogen and microglia: a regulatory system that affects the brain. J Neurobiol 40: 484–496
3. Stence N, Waite M, Dailey ME (2001) Dynamics of microglial activation: a confocal time-lapse analysis in hippocampal slices. Glia 33:256–266
4. Cho BP, Song DY, Sugama S et al (2006) Pathological dynamics of activated microglia following medial forebrain bundle transection. Glia 53:92–102
5. Perry VH, Gordon S (1991) Macrophages and the nervous system. Int Rev Cytol 125:203–244
6. Sedgwick JD, Schwender S, Imrich H et al (1991) Isolation and direct characterization of resident microglial cells from the normal and inflamed central nervous system. Proc Natl Acad Sci USA 88:7438–7442
7. Kaur C, Ling EA (1992) Activation and re-expression of surface antigen in microglia following an epidural application of kainic acid in the rat brain. J Anat 180:333–342
8. Kreutzberg GW (1996) Microglia: a sensor for pathological events in the CNS. Trends Neurosci 19:312–318
9. Imai Y, Kohsaka S (2002) Intracellular signaling in M-CSF-induced microglia activation: role of Iba1. Glia 40:164–174
10. Robinson AP, White TM, Mason DW (1986) Macrophages heterogeneity in the rat as delineated by two monoclonal antibodies MRC OX-41 and MRC OX-42, the latter recognizing complement receptor type 3. Immunology 57:239–247

Chapter 25

Intrathecal Infusion of Microglia Cells

Takahiro Masuda, Makoto Tsuda, Hidetoshi Tozaki-Saitoh, and Kazuhide Inoue

Abstract

Spinal microglia have been implicated in the pathogenesis of neuropathic pain after peripheral nerve injury concomitant with diseases such as diabetes and cancer. To reveal the etiological roles of microglia in behavioral pain hypersensitivity or neuronal excitability, technical approaches have been used. Here, we describe a technique for intrathecal transfer of cultured microglial cells through a catheter surgically implanted into the spinal subarachnoid space.

Key words Microglia, Pain, Spinal cord

1 Introduction

Following nerve injury arising from bone compression in cancer, diabetes, infection, autoimmune disease, or physical injury, spinal microglia transform into a reactive state through a progressive series of cellular and molecular changes and contribute to the formation of debilitating chronic pain conditions (referred to as neuropathic pain) [1, 2]. Various approaches have been used to determine the microglial genes crucial for neuropathic pain, such as those encoding cell surface receptors and diffusible mediators (e.g., proinflammatory cytokines), and a number of microglial genes have been characterized as "pain-related molecules" [1, 2]. Here, we demonstrate the intrathecal infusion of primary cultured microglia as a tool for verifying the direct link between pain hypersensitivity and microglial genes, in combination with behavioral or electrophysiological approaches.

2 Materials

Prepare and store the following at room temperature (unless otherwise specified). We do not add sodium azide to the reagents.

Bertrand Joseph and José Luis Venero (eds.), *Microglia: Methods and Protocols*, Methods in Molecular Biology, vol. 1041, DOI 10.1007/978-1-62703-520-0_25, © Springer Science+Business Media New York 2013

1. 10–12-week-old male mice (25–30 g).
2. Isoflurane.
3. Intrathecal catheter: connect 2.9 cm of 32-gauge catheter (ReCathCo, Allison Park, PA) (indwelling end) and 5 cm of polyethylene catheter (PE-10) (BD) (external end) using surgical adhesive.
4. Primary microglial cells: prepare according to the method described previously [3, 4].
5. Culture medium for microglial cells: 10 % fetal bovine serum (FBS), penicillin (100 U/mL), streptomycin (100 μg/mL), and L-glutamine (2 mM) in Dulbecco's modified Eagle medium (DMEM). Store at 4 °C.
6. Lentiviral particles: prepare as previously described (ref. 5, or *see* the Chapter 8). Store at −80 °C.
7. 35-mm cell culture dish.
8. 24-well plate.
9. Cell lifter.
10. Cell counter.
11. Microsyringe.

3 Methods

3.1 Intrathecal Catheterization in Mice

1. Anesthetize mice with 2 % isoflurane.
2. Clean the skin over the nape of the neck using 75 % ethanol and incise for about 1 cm.
3. Expose the atlanto-occipital membrane by detaching the muscle on either side of the external occipital crest.
4. Incise the membrane using the tip of a needle (*see* **Note 1**).
5. Insert an indwelling 32-gauge catheter through the membrane into the intrathecal space in the midline, dorsal to the spinal cord (*see* **Note 2**).
6. Externalize the exit end of the catheter (PE-10) through a separate opening in the skin.
7. After more than 5 days for recovery, administer 1.5 μl of 2 % lidocaine intrathecally to confirm that the tip of catheter was placed appropriately for the intended target (*see* **Note 3**).

3.2 Intrathecal Infusion of Gene-Transduced Microglial Cells

1. Seed primary microglial cells in a cell culture dish (*see* **Note 4**).
2. Transduce the microglial cells by adding an appropriate volume of viral particles onto the cells (*see* **Note 5**).
3. After 72 h for transduction, aspirate culture medium.

4. Gently wash the dish twice with PBS to remove culture medium (*see* **Note 6**).
5. Harvest microglial cells using a cell lifter and transfer to a 1.5-mL tube.
6. Spin down the cells by centrifuging at 400 × *g* for 1 min.
7. Discard the supernatant.
8. Resuspend the cells by gently pipetting with 100 μl of PBS.
9. Measure the density of microglia using a cell counter, and adjust the cell density to 2.0×10^4 cells per 5 μl by adding an appropriate volume of PBS (*see* **Note** 7).
10. Infuse 5 μl of cell suspension using a microsyringe through an intrathecal catheter implanted into the intrathecal space, followed by injection of PBS (5–10 μl).

3.3 Intrathecal Infusion of Microglial Cells After Drug Treatment

1. Seed primary microglial cells in a culture dish (*see* **Note 4**).
2. After they have adhered to the bottom, aspirate culture medium.
3. Gently wash the dish twice with PBS to remove culture medium (*see* **Note 6**).
4. Harvest microglial cells by using a cell lifter and transfer to a 1.5-mL tube.
5. Spin down the cells by centrifuging at 400 × *g* for 1 min.
6. Discard the supernatant.
7. Resuspend the cells by gently pipetting with 100 μl of PBS.
8. Measure the density of microglia using a cell counter, and adjust the cell density according to the experimental condition (*see* **Note** 7).
9. Treat the cells with drug, such as a neutralizing antibody.
10. Infuse the cells plus supernatant as described above (*see* Subheading 3.2, **step 10**).

4 Notes

1. The escape of cerebrospinal fluid can be observed when the atlanto-occipital membrane is incised (for Subheading 3.1, **step 4**).
2. To place the tip of the catheter in proximity to the L3–L4 dorsal surface of the spinal cord, insert the catheter intrathecally about 2.9 cm (for Subheading 3.1, **step 5**).
3. Mice showing severe motor weakness under the normal condition should be excluded from the study (for Subheading 3.1, **step** 7).

4. We usually seed primary microglial cells at a density of 5×10^5 cells per 35-mm culture dish or 1.2×10^5 cells per well (24-well plate) (for Subheading 3.2, **step 1,** and Subheading 3.3, **step 1**).
5. Transduce the cells with viral particles according to our previous paper [5] (or see the Chapter 8) (for Subheading 3.2, **step 2**).
6. PBS can be used at room temperature. To avoid adding culture medium to the cell suspension, remove culture medium as carefully as possible (for Subheading 3.2, **step 4,** and Subheading 3.3, **step 3**).
7. Adjust the density of microglial cells according to experimental condition (for Subheading 3.2, **step 9**, and Subheading 3.3, **step 8**).

References

1. McMahon SB, Malcangio M (2009) Current challenges in glia-pain biology. Neuron 64:46–54
2. Inoue K, Tsuda M (2009) Microglia and neuropathic pain. Glia 57:1469–1479
3. Nakajima K, Shimojo M, Hamanoue M et al (1992) Identification of elastase as a secretory protease from cultured rat microglia. J Neurochem 58:1401–1408
4. Tsuda M, Masuda T, Kitano J et al (2009) IFN-gamma receptor signaling mediates spinal microglia activation driving neuropathic pain. Proc Natl Acad Sci USA 106:8032–8037
5. Masuda T, Tsuda M, Yoshinaga R et al (2012) IRF8 is a critical transcription factor for transforming microglia into a reactive phenotype. Cell Rep 1:334–340

Chapter 26

Intracranial Injection of LPS in Rat as Animal Model of Neuroinflammation

Ana M. Espinosa-Oliva, Rocío M. de Pablos, and Antonio J. Herrera

Abstract

Animal models of neuroinflammatory processes are needed to study the involvement of inflammation in neurodegenerative disorders such as Parkinson's and Alzheimer's diseases. One of the models used is based on lipopolysaccharide (LPS) as brain inflammation-inducing agent. This toxin is a potent inducer of inflammation and has different effects on cells of the immune system, as microglial cells. This chapter describes a protocol for the model of brain inflammation in rats based on the unilateral stereotaxic injection of LPS, which mimics the inflammatory milieu produced in some brain diseases.

Key words Animal models, Lipopolysaccharide, Neurodegeneration, Neuroinflammation, Stereotaxic surgery

1 Introduction

The possible role of neuroinflammatory processes in the development of neurodegenerative diseases (i.e., Parkinson's disease, PD) is receiving growing support. Early on 1988, McGeer's group showed the presence of activated microglia in the substantia nigra (SN) of PD patients, suggesting the implication of neuroinflammation as a component of the neurodegenerative progress of the disease [1]; for review, *see* refs. 2, 3. In fact, they showed a huge proliferation of reactive amoeboid macrophages and microglia in the SN of postmortem parkinsonian brains; this finding, along with the oxidative stress marks observed, was compatible with the existence of inflammation, as it had been previously described for Alzheimer's disease (AD) [4] and multiple sclerosis [5].

Animal models of neuroinflammation have become essential to study the possible implication of neuroinflammatory processes in the development of these disorders. Stereotaxic injection of lipopolysaccharide (LPS) is now commonly used to produce neuroinflammation within the brain. LPS, a component of the Gram-negative bacteria cell wall [6], is a potent inducer of inflammation and has

Bertrand Joseph and José Luis Venero (eds.), *Microglia: Methods and Protocols*, Methods in Molecular Biology, vol. 1041, DOI 10.1007/978-1-62703-520-0_26, © Springer Science+Business Media New York 2013

diverse effects on cells of the immune system [7]. Several works have shown that LPS activated glial cells in vivo after intracranial injection [8–12]; microglial cells release several neurotoxins [13–18], as inflammatory cytokines, *N*-methyl-D-aspartate (NMDA) receptor agonists, oxygen free radicals, and nitric oxide (NO), which could contribute to neurotoxicity. All these data suggest that the injection of LPS in different areas of the brain could provide a good experimental model to ascertain the importance of inflammation in this organ. Stereotaxic surgery allows the experimenter to reach any part of the brain with great accuracy and thus is widely used for delivering drugs and other substances within the brain. Two fundamental things are required to carry out stereotaxic injection of LPS: a stereotaxic apparatus and a stereotaxic brain atlas.

The stereotaxic apparatus allows the researcher to place an object (cannula, electrode, etc.) or substance at a specific location within the brain. It includes a support to keep the animal's skull in the appropriate position, as well as a holder for the syringe or the electrode. The holder is attached to a calibrated mechanism that moves the holder through the three spatial axes, the so-called stereotaxic coordinates: anterior-posterior (AP), lateral-medial (LM), and dorsal-ventral (DV).

The stereotaxic brain atlas is a set of photographs or drawings of brain sections, ordered along one of the three axes mentioned above (Fig. 1a). As it will be explained later on, injections are performed through a small drill hole made on the skull, trying to minimize the damage inflicted to the animal. Thus, brain coordinates have to be referred to an external reference point. Two of these points are used: bregma (the anatomical point on the skull at which the coronal suture is intersected perpendicularly by the sagittal suture) and lambda (a dense, fibrous connective tissue joint on the posterior part of the skull that connects the parietal and temporal bones with the occipital bone; Fig. 1b). For a discussion on bregma and lambda points, see Paxinos and Watson [19].

Pages of a stereotaxic atlas typically correspond to coronal sections taken at various distances anterior and posterior to bregma and lambda. Each plate includes the LM and the DV distances to these reference points. Sagittal and horizontal sections are also provided. Brain structures can be easily found in an index (list of structures) included in the stereotaxic atlas. Thus, brain structures (that we cannot see in the animal alive) are easily located following the three coordinates (with respect to bregma or lambda points on the skull) provided by the atlas.

Stereotaxic surgery can be made either unilateral or bilateral. In the first case, the intact side serves as absolute control. Bilateral surgery is performed when sham and true surgery are required in the same animal.

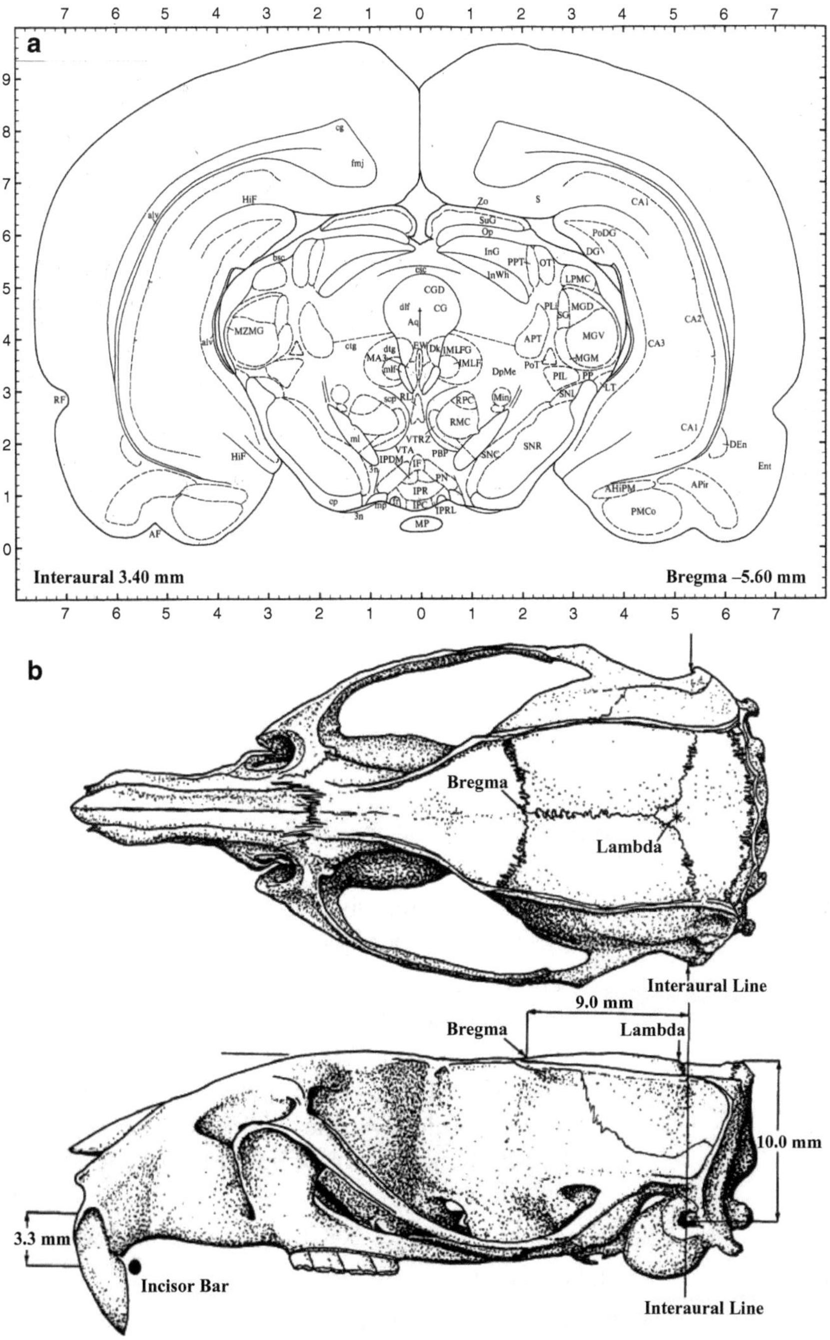

Fig. 1 (**a**) Diagram showing the three stereotaxic coordinates and the structures at 5.6 mm posterior to bregma. (**b**) Skull diagram showing bregma and lambda points as well as the position of the tooth bar to get a flat position of the skull (Reproduced with permission from The Rat Brain in Stereotaxic Coordinates, Paxinos G. and Watson C.; 1986. Academic Press)

2 Materials

2.1 Surgical Apparatus and Tools

1. Stereotaxic instrument for small animals supplemented with ear bars.
2. Ten-microliter Hamilton syringe (Hamilton Bonaduz AG, Bonaduz, Switzerland).
3. Electric shaver.
4. Scissors.
5. Scalpel.
6. Metal clips.
7. Sharp forceps.
8. Needles.
9. Suture thread.
10. Forceps.
11. Disinfectant.
12. Syringe.
13. Sterile cotton pads.
14. Cloth.
15. Small drill.

2.2 Reactives

1. Anesthetics: Several anesthetic agents can be used as described in the literature:
   - Chloral hydrate: Dissolve 6 g in 100 ml of distilled water; inject 400 mg per kg of body weight intraperitoneal (i.p.).
   - Fentanyl citrate salt, 50 μg/ml + medetomidine hydrochloride (DormitorVet®, Pfizer Animal Health) 50 μg/ml, 20:1 mixture; inject 6.3 ml per kg of body weight (i.p.).
   - Ketamine 80 mg/kg + xylazine 12 mg/kg (ketamine hydrochloride/xylazine hydrochloride solution, Sigma-Aldrich); inject 1 ml of solution per kg of body weight intramuscular.
2. Saline solution (0.9 % NaCl, w/v in distilled water).
3. 70 % ethanol.
4. Vehicle: 1 % Monastral Blue inert tracer in saline solution.
5. LPS: Dissolve 1 mg in 1 ml of vehicle. Aliquot small amounts (ca 15–20 μl) into Eppendorf® tubes. Keep the tubes on ice if LPS is going to be used fresh. For long storage, use screw cap vials; glass vials are preferred. Store the aliquots at −20 °C and avoid repeated thawing/freezing cycles.

3 Methods

The procedure described below is carried out in accordance with the guidelines of the European Union Council (86/609/EU) for the use of laboratory animals. Before starting with the surgery, check that all reagents and materials are ready.

1. Prepare a couple of Eppendorf® tubes containing distilled water and 70 % ethanol, respectively, for washing the syringe.
2. Thaw an aliquot of LPS; keep it on ice.
3. Prepare all the surgical material listed above, disinfecting them with 70 % ethanol.
4. Clean the Hamilton syringe barrel inner space with distilled water and 70 % ethanol (*see* **Note 1**).
5. Place the Hamilton syringe in the syringe holder (*see* **Note 2**).
6. Place a clean cloth over the stereotaxic apparatus platform where the animal will be placed (*see* **Note 3**).
7. Weigh and anesthetize the animal. Different injectable anesthetics can be used as already mentioned above (*see* **Note 4**).
8. Place the animal in a cage until it is deeply asleep (*see* **Note 5**).
9. Take an aliquot of LPS solution and fill the Hamilton syringe avoiding bubbles.
10. To decrease the chance of infection at the incision, shave the fur on the top of the head with an electric shaver.
11. Place the animal in the stereotaxic apparatus on horizontal position. Put the ears and the tooth bar, immobilizing the skull to avoid movements of the head (*see* **Note 6**) (*see* Figs. 2 and 3). Cover the body with the cloth (*see* **Note 7**).
12. Wipe the shaved zone with a sterile cotton pad soaked in 70 % ethanol. Let it dry.
13. Make a midline incision on the skin with the scalpel, starting between the eyes; about 2 cm in length should be enough.
14. Keep the skin open with two metal clips. Scratch and remove the membrane above the bone; clean the bleeding with a sterile cotton pad and keep the bone dry; the bregma is easily located (*see* Fig. 4a).
15. Place the needle on the bregma (*see* **Note 8**).
16. Read the three coordinates (AP, LM, and DV) on the manipulator (in mm) and write them down; this is our starting point. Move the needle up a few millimeters.
17. Set the AP and LM coordinates of the injection point by adding the previously calculated coordinates to the manipulator. Mark the point on the skull (*see* **Note 9**).

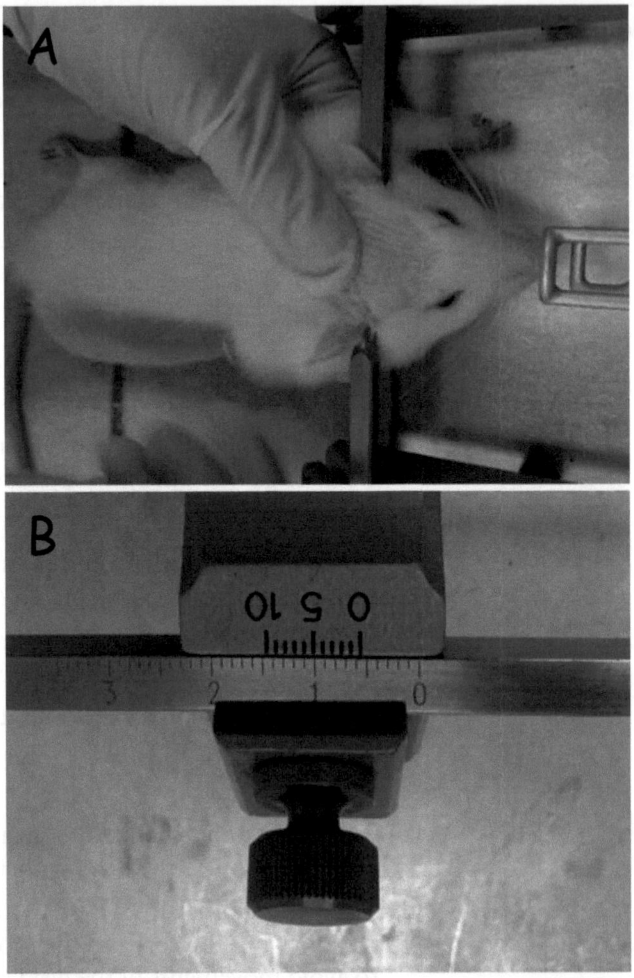

Fig. 2 (**a**) Placing ear bars. (**b**) Adequate position of the bar with respect to the bar holder

18. Retract the arm with the needle and drill a small hole in the skull bone above the mark (*see* **Note 10**; *see* Fig. 4b).
19. Put the arm in position and check that the needle enters (just the tip) through the drill hole centered. If the needle tip is close to the borders, make the hole a bit bigger but do not change needle coordinates (*see* **Note 11**).
20. Slowly, set the DV coordinate of the injection point; this inserts the needle tip into the brain till the injection point. Wait 1 min with needle in place.
21. Start releasing the LPS solution (2 μl) at an approximate flow rate of 0.5 μl/min (*see* **Note 12**).
22. Leave the needle in place for 5 min to avoid reflux along the injection track (*see* **Note 13**).

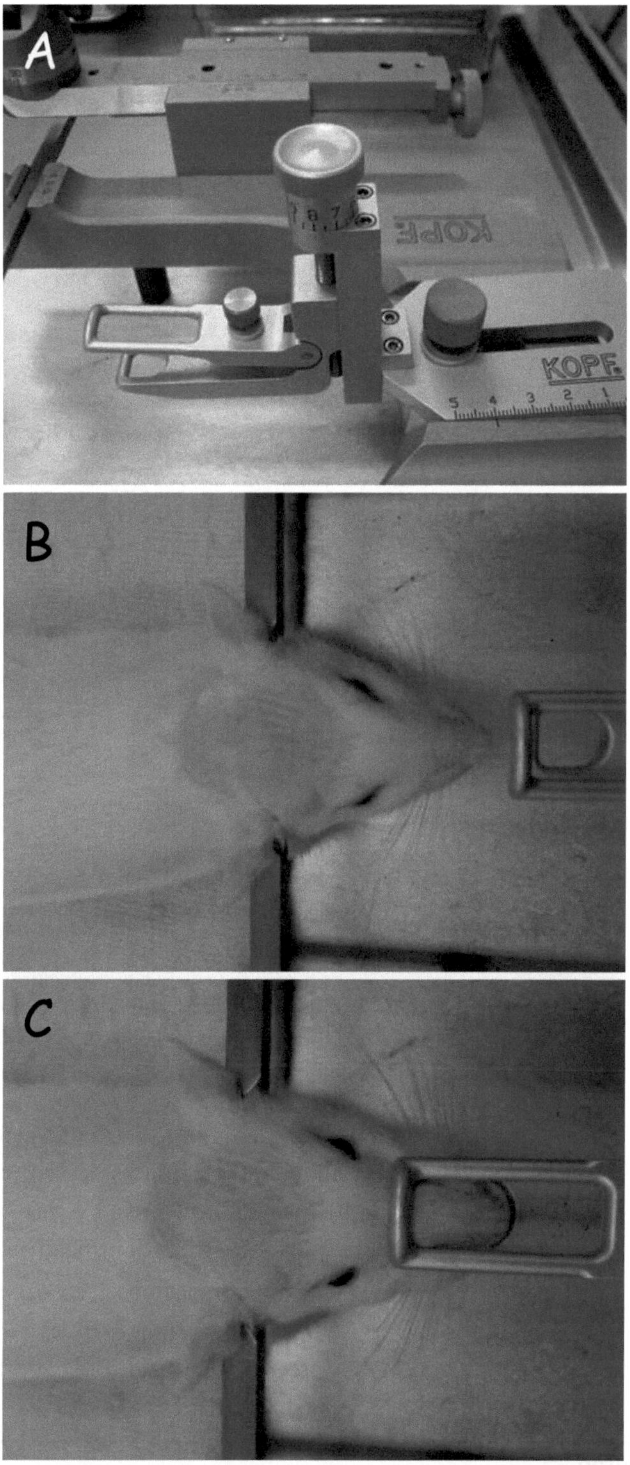

Fig. 3 (**a**) Tooth bar. The lower part hooks the teeth and the upper part pushes the rat nose down to the required flat position of the skull. Rat before (**b**) and after (**c**) fixing the head with the tooth bar

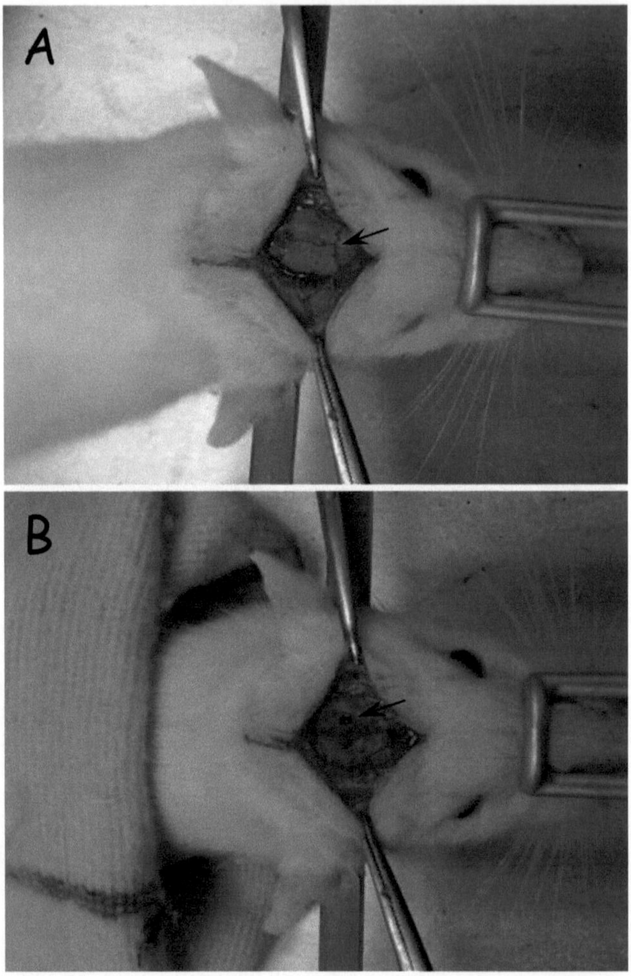

Fig. 4 (**a**) Rat skull after cutting the skin and removing bone membrane. The *arrow* shows the bregma point. (**b**) Drill hole (*arrow*) at the projection point of the SN on the skull

23. Take the needle out very slowly. Retract the arm, empty the syringe, and clean it as described above.
24. Close the wound area by suturing, apply disinfectant, and mark the rat for further identification.
25. Place the animal in a clean cage until it regains consciousness and recovers (**Notes** 7 and **14**).
26. Disinfect the surgical tools and the table.

4 Notes

1. Fill and empty the syringe several times with distilled water, then with 70 % ethanol, and again with distilled water. Cleanliness of the needle is very important to prevent infection and to check if the syringe is occluded.

2. Make sure that the needle is not bended at all; needle orientation must be absolutely vertical with respect to the apparatus platform to avoid injection out of the desired structure.
3. Animals can experiment some hypothermia under anesthesia.
4. Inhaled anesthetics can also be used. In such cases, the animal is usually introduced in a chamber where the gas is released until it is completely asleep. Gases commonly used are either isoflurane or halothane. A stereotaxic adapter is needed to influx anesthesic gas during the entire surgical procedure. In such a case, surgery should be performed in a ventilated bench or fume hood to prevent gas inhalation.
5. Look for reflexes by pinching the tail or paws to ensure that the animal is anesthetized.
6. The procedure is delicate and requires time until the necessary ease is got; you may experience some difficulty the first few times you do this if you have no previous experience. With practice, placing the rat in the apparatus takes less than 1 min. It is essential that the head is positioned correctly so that the plane containing bregma and lambda points is flat.

   Ear bars should not be pushed with strength to avoid unnecessary damage to the skull and brain; pressing with one finger is enough. Ear bars include a graduated scale; when the rat is placed in the right position, the number 1 on the bars should coincide approximately with the 0.5 that is in the bar holder (Fig. 2b). When the bars enter the earhole in the skull, a soft click can be heard; this is normally accompanied by eye blinking of the rat. At this moment, the up-down movement of the head is still free around the interaural line.

   Next, fix the teeth of the animal on the tooth bar to prevent the head moving up and down. This operation is much easier than placing the ear bars. The tooth bar can be moved freely backward and forward; softly, push the bar inside the rat mouth, push the rat nose down, and pull the tooth bar out of the rat's mouth. If there is resistance to traction, the bar is in the correct position and has hooked the incisors. Be sure to tighten the screw that secures the position of the bar to the U frame. Clip the nose of the rat by lowering the metal piece softly (Fig. 3c).
7. Cover the rat with a cloth to keep it warm; body temperature decreases because of anesthetics. Alternatively, you may use a thermal blanket lying below the animal.
8. It is very important not to damage the needle tip, just touch the skull with the very tip, and no pressure is required at all. When using a beveled needle, be much careful not to blunt the tip.
9. This allows finding the vertical projection of the desired structure on the skull surface. Move the needle down with much care, looking for the touching point on the skull. Move the

needle up and mark the point with a pencil (a waterproof ink pen is also fine); check that the needle touches the skull just in the marked point.

The stereotaxic atlas let us to determine the coordinates of the desired brain structure. Coordinates are expressed in millimeters relative to bregma. These are the coordinates of some structures we use frequently.

SN pars compacta: AP = −5.5, LM = ±1.8, DV = −8.3;

Striatum: AP = +0.5, LM = ±3.0, DV = −6.4;

Hippocampus: AP = −3.6, LM = ±2.0, DV = −3.4.

Negative values have different meanings depending on the coordinate; for the AP, negative means posterior; for the LM, it means right hemisphere; and for DV, down (the only way to reach the brain, actually).

10. Adjust drill speed to your convenience; low to medium speeds can provide more confidence to beginners. No pressure is required, just hold the drill and push down very softly with your fingertips. Be very careful in this step; drill tip should never touch the brain when drilling. In fact, meninges under the bone should not be broken by the drill. Once the hole is open in the bone, punch the meninges with a sterile needle. There might be some bleeding now, especially when drilling near the midline. Clean the blood with a sterile cotton pad. Do not continue with the injection procedure if there is bleeding through the hole to avoid putting blood into the brain with the needle.
11. Depending on the shape of the drill tip, hole walls may not be straight vertical, with the lower part of the hole narrower than the upper. Due to the needle flexibility, if the needle touches the lower hole border, its tip will end far from the desired injection place although the needle is apparently straight over the hole.
12. Do this very slowly; it helps turning the plunger softly (as if screwed) instead of pushing it down.
13. Interestingly, other substances may need longer times.
14. Alternatively, use a heat lamp.

References

1. McGeer PL, Itagaki S, Boyes BE et al (1988) Reactive microglia are positive for HLA-DR in the substantia nigra of Parkinson's and Alzheimer's disease brains. Neurology 38:1285–1291
2. McGeer PL, McGeer EG (2008) Glial reactions in Parkinson's disease. Mov Disord 23:474–483
3. Klegeris A, McGeer EG, McGeer PL (2007) Therapeutic approaches to inflammation in neurodegenerative disease. Curr Opin Neurol 20:351–357
4. Rogers J, Civin WH, Styren SD et al (1992) Immune-related mechanisms of Alzheimer's disease pathogenesis. In: Khachatunan ZS, Blass JP (eds) Alzheimer's disease, new treatment

strategies. Marcel Dekker, New York, NY, pp 147–163
5. Compston A (1993) Inflammation and the brain. Mol Chem Neuropathol 19:47–64
6. Herrera AJ, Castaño A, Venero JL et al (2000) The single intranigral injection of LPS as a new model for studying the selective effects of inflammatory reactions on dopaminergic system. Neurobiol Dis 7:429–447
7. Burrell R (1990) Immunomodulation by bacterial endotoxin. Crit Rev Microbiol 17: 189–208
8. Bourdiol F, Toulmond S, Serrano A et al (1991) Increase in ω3 (peripheral type benzodiazepine) binding sites in the rat cortex and striatum after local injection of interleukin-1, tumour necrosis factor-α and lipopolysaccharide. Brain Res 543:194–200
9. Andersson PB, Perry VH, Gordon S (1992) The acute inflammatory response to lipopolysaccharide in CNS parenchyma differs from that in other body tissues. Neuroscience 48:169–186
10. Montero-Menei CN, Sindji L, Pouplard-Barthelaix A et al (1994) Lipopolysaccharide intracerebral administration induces minimal inflammatory reaction in rat brain. Brain Res 653:101–111
11. Montero-Menei CN, Sindji L, Garcion E et al (1996) Early events of the inflammatory reaction induced in rat brain by lipopolysaccharide intracerebral injection: relative contribution of peripheral monocytes and activated microglia. Brain Res 724:55–66
12. Szczepanik AM, Fishkin RJ, Rush DK et al (1996) Effects of chronic intrahippocampal infusion of lipopolysaccharide in the rat. Neuroscience 70:57–65
13. Piani D, Frei K, Do KQ et al (1991) Murine brain macrophages induce NMDA receptor mediated neurotoxicity in vitro by secreting glutamate. Neurosci Lett 133:159–162
14. Chao CC, Hu S, Molitor TW et al (1992) Activated microglia mediate neuronal cell injury via a nitric oxide mechanism. J Immunol 149:2736–2741
15. Banati RB, Gehrmann J, Schubert P et al (1993) Cytotoxicity of microglia. Glia 7: 111–118
16. Giulian D (1993) Reactive glia as rivals in regulating neuronal survival. Glia 7:102–110
17. Giulian D, Vaca K, Corpuz M (1993) Brain glia release factors with opposing actions upon neuronal survival. J Neurosci 13:29–37
18. Lees GJ (1993) The possible contribution of microglia and macrophages to delayed neuronal death after ischemia. J Neurol Sci 114: 119–122
19. Paxinos G, Watson C (1986) The rat brain in stereotaxic coordinates, 2nd edn. Academic, London

Chapter 27

Analyses of Microglia Effector Function Using CX3CR1-GFP Knock-In Mice

Jenny A. Garcia, Sandra M. Cardona, and Astrid E. Cardona

Abstract

The generation of bone marrow radiation chimeric mice is a beneficial tool to utilize when studying inflammation of the central nervous system (CNS). It is widely accepted that blood-derived progenitors are capable of populating the CNS during chronic diseases and severe injuries; however, they are neither consistent nor efficient in doing so. The lack of the appropriate recruitment could explain delays in recovery and repair after an increase of toxic proteins in chronic neurodegenerative diseases. With the ingenious development of bone marrow chimeric mice, some of these concerns can be addressed and allow us to hypothesize about further implications and possible mechanisms that may lead to medicinal applications. Bone marrow chimeric mice are often used to distinguish the intrinsic versus extrinsic effects of specific mutations. In our case, chimeras help us to better understand the role of CX3CR1 in microglia and peripheral myeloid cells. To detect cell autonomous effects on myeloid cell differentiation, CX3CR1-deficient mice are used as donors and wild-type mice are used as recipients. In order to detect effects on the "immune cell environment," wild-type donors are used for the transfer into $Cx_3cr1^{-/-}$ recipients. The resulting chimeric mice can then be used for the analysis of microglial motility, regulation of neuroinflammation, and persistence. This technique can be applied to a broad spectrum of research ranging from neurodegenerative diseases to viral and parasitic pathogenicity and everything in between. This protocol describes the approach to generate chimeric mice and analyze the role of CX3CR1 in CNS inflammation in bone marrow radiation chimeras.

Key words CX3CR1, Bone marrow, Chimeras, Microglia, Trafficking, Chemokines, Radiation

1 Introduction

Microglia are specialized, surveillant phagocytic cells [1–3] that colonize the CNS during embryonic development and are best known for their ability to scan the environment with their extensive, motile processes. The knowledge of responses and pathogenic functions of microglia can be useful in developing inhibitors and therapeutic targets for disease modulation [3]. In the CNS, microglia are the only cells that express CX3CR1, also known as the fractalkine receptor [4]. It has been shown via in vivo models of autoimmune diseases

Bertrand Joseph and José Luis Venero (eds.), *Microglia: Methods and Protocols*, Methods in Molecular Biology, vol. 1041,
DOI 10.1007/978-1-62703-520-0_27,

that in the absence of CX3CR1, dysregulation of microglial responses occurs leading to neurotoxicity. CX3CR1 deficiency has been associated with neuronal cell death subsequent to lipopolysaccharide (LPS) challenge [4]. To monitor the role of CX3CR1 in the CNS, we utilized mice in which the *Cx3cr1* gene was disrupted by the insertion of DNA encoding green fluorescent protein (GFP), allowing for the direct visualization of microglial cells [4–8]. The generation of bone marrow chimeric mice allows us to investigate the effects of infiltration and adaptive immunity triggered by extraneous bone marrow cells on their endogenous counterparts. Considerable speculation of the chimeric model is focused on the ability of radiation to break down the blood–brain barrier (BBB). This issue can be addressed by shielding the brain from irradiation while knocking out the rest of the immune system. The only downfall is that the efficiency of chimerism is sacrificed due to the fact that skull bone marrow can contribute to hematopoiesis [9]. To further investigate roles of adaptive immunity when exploring the pathogenicity of multiple sclerosis (MS), we utilize the animal model experimental autoimmune encephalomyelitis in conjunction with the chimeric model [2, 10]. In such an experiment, C57BL6 recipient mice are challenged with MOG_{35-55} peptide 6 weeks post-chimerism via active or passive immunization. A variety of combinations can be investigated when using this approach. For instance, CD45.1 or Ly5.1 donor cells deficient of CX3CR1 can be tracked in recipient CD45.2 or Ly5.2 WT mice via flow cytometry analysis by the detection of the CD45.1 marker in addition to the microglia expressing GFP. Alternatively, CX3CR1-deficient mice can serve as recipients for CD45.2 *Cx3cr1* WT donor cells. Overall, this protocol can be applied to any genetic models of interest with potential implications in translational and medicinal research in the development of systems to moderate microglial function.

2 Materials

Prepare solutions and handle all tools under sterile conditions to prevent any source of contamination.

2.1 Radiation

1. RS 2000 Rad Source Biological Irradiator with radiation output of 160 kV, 25 mA, 0.3 mm Cu filter (Rad Source Technologies, Alpharetta, GA, USA) or any X-ray machine or cesium source of gamma irradiation available to your institution.
2. 1/16″ lead helmets (New Shield Inc., Pomona, NY, USA) (*see* **Note 1**).
3. Anesthesia cocktail: Mix Ketaset and Xylazine in 1× HBSS for a final concentration of 10 mg/ml Ketaset and 1 mg/ml Xylazine, and inject 100 mg/kg Ketaset and 10 mg/kg Xylazine.

2.2 Bone Marrow Isolation

1. Mice: B6.SJL-*Ptprc*^{a} *Pep3*^{b}/BoyJ (CD45.1 WT) #002014, B6.129P (Cg)-*Ptprc*^{a} *Cx3cr1*^{tm1Litt}/LittJ (CD45.1 KO), and C57BL/6J (CD45.2 WT) #000664 (The Jackson Laboratory, Bar Harbor, MA, USA).
2. 6-well plate (BD Falcon).
3. KimWipes or autoclaved paper towels.
4. 70 % ethanol.
5. 70 μM cell strainer (Fisher Brand #22363548 Fisher Scientific).
6. 2–200 ml beakers for holding tools in 1× PBS and 1 sterile with 70 % ethanol for submerging mouse to disinfect skin.
7. Sterile 1× PBS.
8. Absorbent pads.
9. Paper towels.
10. Plastic bags to collect carcass.
11. Scissors.
12. Forceps.
13. 10, 50 ml polypropylene conical tubes.
14. 1.5 ml graduated microcentrifuge tubes.
15. 3 ml Luer-Lock tip syringes (BD #309657).
16. 0.4 % trypan blue diluted to 0.05 % in HBSS (Gibco).
17. Iscove's/10 % FBS/0.5 % gentamicin reagent solution (fetal bovine serum – heat inactivated) (Sigma-Aldrich).
18. Needles: PrecisionGlide 25G $^{5/8}$ and 23G ¾ (BD #305122 and #305143, respectively).
19. Hemocytometer.
20. Bright-field microscope.
21. Carbon dioxide asphyxiation chamber.

2.3 Injection of Bone Marrow Cells

1. Iscove's Modified Dulbecco's Medium supplemented with L-glutamine, 25 mM HEPES (Gibco #12440), and 50 μg/ml gentamicin reagent solution (Gibco).
2. 1 ml tuberculin slip tip syringe with PrecisionGlide 27G ½ needle (BD #309623).
3. 1.5 ml graduated microcentrifuge tubes.
4. Isoflurane and vaporizer with oxygen chamber.

2.4 Flow Cytometry Analysis of Peripheral Blood

1. Double-distilled sterile water.
2. 1×HBSS free of calcium, magnesium chloride, and magnesium sulfate (Gibco).
3. 10 mM HEPES.

4. 10× HBSS.
5. 15 ml conical tubes.
6. 1.5 ml microcentrifuge tubes.
7. Sterile 5 mm Goldenrod Animal Lancet.
8. FC block: rat anti-mouse CD16/CD32 clone 2.4G2 in cell staining buffer (optimal concentration should be determined by the investigator) (BD Pharmingen).
9. Cell staining buffer (BioLegend).
10. Fluorescent antibodies: CD45.1 anti-mouse PE-Cy7 clone A20 in cell staining buffer (eBioscience) and CD45.2 anti-mouse APC clone 104 in cell staining buffer (eBioscience) (Note: optimal concentration should be determined by the investigator).
11. Prepare Sorenson's buffer by adding 3.59 g sodium phosphate monobasic and 24.7 g sodium phosphate dibasic heptahydrate to 400 ml ddH2O. Ensure that pH is between 7.0 and 7.3. Adjust final volume to 500 ml with ddH2O and store at room temperature.
12. 4 % paraformaldehyde in phosphate buffer (Sigma-Aldrich): First, prepare 8 % PFA by dissolving 40 g paraformaldehyde to 400 ml ddH2O and stir for 15 min at 55–65 °C. Add 100 μl 5 M NaOH for 100 ml of 8 % PFA solution. Stir at 30 °C and add ddH2O for a final volume of 500 ml. Add 200 ml of Sorenson's buffer to 500 ml of 8 % PFA and 300 ml ddH2O. Filter, sterilize, and store at −20 °C.
13. Heparin sodium salt 5,000 U/ml (Sigma-Aldrich).

2.5 Flow Cytometry Analysis of Microglia

1. Isotonic Percoll (GE Healthcare) buffered with 10× HBSS. Prepare 100 % ISP by adding 45 ml of isotonic Percoll to 5 ml 10× HBSS. Prepare 70 % Percoll by adding 21 ml of 100 % ISP to 9 ml 1× HBSS. This is enough for 9 samples. Scale up volumes as needed.
2. RPMI 1640 supplemented with L-glutamine without phenol red (Gibco).
3. Dounce homogenizers for 7 and 15 ml volume with A (loose) and B (tight) pestles (Pyrex, Wheaton).
4. S1 Pipet Filler (Thermo Scientific #9531) or Pipet-Aid.
5. 1× HBSS without calcium, magnesium chloride, and magnesium sulfate (Gibco) with 10 mM HEPES (Gibco).
6. FC block, antibodies, and reagents as in Subheading 2.4 in addition to any other antibodies of interest.

3 Methods

3.1 Radiation Conditioning of Recipient Mice

1. Subject the recipient mice (*see* **Note 1**) to radiation to deplete the immune system. This will also help prevent any rejection of injected cells. Determine if your experiment requires total body or brain-protected radiation.
2. For total body radiation, subject the recipient mice to 900 rads, a sublethal dose of radiation. Place all mice in an animal cage that does not contain any metal and place cage into radiation chamber with lid secured. Deliver the required amount of radiation and remove cage from radiation chamber.
3. If you decide to protect the head from radiation treatment, anesthetize the mice to ensure that they will not move around in the cage potentially causing their protective helmet to fall off (*see* **Note 2**). Once the mice have been anesthetized, roll them on to their back and place the helmet onto the face of the mouse. The idea is to protect the brain while keeping the thymus exposed to radiation treatment. Once the helmets are in place, replace the cage lid and place the cage into the radiation chamber. For protected head recipients, it is important to note that two consecutive doses of 600 rads are required.
4. To ensure a good percentage of chimerism, it is recommended to treat recipient mice with radiation the same day just prior to injecting bone marrow-isolated cells as seen in the next step.
5. If anesthesia was used, it is imperative to monitor the mice until they regain consciousness.

3.2 Isolation of Bone Marrow Cells

1. Euthanize donor mice by CO_2 asphyxiation and submerge them in 70 % ethanol to disinfect the fur and skin. Place disinfected mouse onto absorbent pads or paper towels to collect any bodily fluids or tissues that may be extracted.
2. Remove the skin from the lower limbs and cut away the muscles revealing the femur and tibia bones.
3. After removing as much muscle as possible, cut the entire leg off with scissors by cutting above the femoral head making sure to keep the femur intact (*see* **Note 3**).
4. Once the hind limb is removed from the body, you can carefully cut off the foot making sure to keep the tibia bones intact. Clean the bones with KimWipes or sterile paper towels of any excess muscle and collect bones in a 6-well plate with Iscove's medium containing 10 % FBS and 50 μg/ml gentamicin and keep on ice.
5. After collecting all hind limbs, separate the femur and tibia bones by cutting at the patella, area between the femur and tibia, making sure not to fracture the bones (*see* **Note 4**).

6. Using KimWipes or autoclaved paper towels, carefully remove any excess tissue from the bones until you can clearly see the epiphyses and entire shaft of the bone. This is important so that you can see when the red bone marrow flushes out.
7. Fill a 3 ml syringe with Iscove's/10 %FBS/50 μg/ml gentamicin. Cut off the epiphyses at both ends of the femur and use filled syringe with 25G 5/8 needle to flush out the bone marrow collecting the flow through in a 50 ml conical tube. While holding the bone tightly with forceps, insert the needle inside the bone in an up and down motion while pressing the plunger. The bone will turn white as the cells are flushed out. Flip the bone over and repeat flushing from the opposite end (*see* **Note 5**). Follow the same procedure for the tibia.
8. After you have collected all marrow cells, pour them through a 70 μM mesh filter and collect in a clean 50 ml conical tube. If there are clumps on the filter, dissociate them with a syringe plunger and wash with Iscove's/10 % FBS/50 μg/ml gentamicin to ensure that you recover as many cells as possible.
9. Centrifuge the cell suspension at $700 \times g$ for 7 min at 4 °C. Aspirate supernatant and suspend pellet in 1 ml Iscove's with 0.5 % gentamicin.
10. Count cells and adjust the final volume with Iscove's with 50 μg/ml gentamicin to prepare cells at $130\text{–}140 \times 10^6$ cells/ml.

3.3 Injection of Bone Marrow Cells

1. Anesthetize recipient mice with an isoflurane vaporizer by placing the mice in the gas-filled chamber.
2. Remove mouse from the chamber and inject 150–200 μl of your cell suspension to deliver $10\text{–}20 \times 10^6$ cells per mouse via retro-orbital injection using a 1 ml syringe with 27 G needle (*see* **Note 6**).
3. Monitor recipient mice until they are fully awake.
4. Treat recipient mice with 50 μg/ml gentamicin in drinking water for 2 weeks to avoid infection of immunocompromised mice.
5. Four weeks later, evaluate percentage of chimerism by flow cytometry analysis of peripheral blood.

3.4 Flow Cytometry Analysis of Peripheral Blood

1. Using a 5 mm lancet, draw 100–150 μl of blood from cheek pouch and collect in 60 μl heparin sodium salt solution (*see* **Note 7**). Be sure to mix well to prevent coagulation. Transfer blood to 15 ml conical tube and measure with a pipette to determine the volume of water (ddH_2O) needed to lyse red blood cells. Determine the appropriate amount of ddH_2O and 10× HBSS buffer needed to lyse your sample on a 20× final volume. For example, to lyse 0.1 ml of blood, you will need 1.7 ml of water and 0.2 ml of 10× HBSS, so the final volume of the lysed suspension is 2 ml (20× excess of the initial 0.1 ml of blood) and in 1× buffer.

2. Have the appropriate volumes of water, 10× HBSS and 1× HBSS/10 mM HEPES, measured out and ready as overtreatment with water can lead to lysis of white blood cells which are needed for staining. Add the correct amount of water, close cap tightly, and mix by inversion for 20 s. Immediately add 10× HBSS and invert tube twice to mix. Then add 5 ml 1× HBSS/10 mM HEPES to ensure complete buffering of your cell suspension.
3. Centrifuge for 7 min at 700 ×*g* at 4 °C. Aspirate supernatant and suspend pellet in 1 ml cell staining buffer and transfer cells to 1.5 ml microcentrifuge tube.
4. Centrifuge for 1 min at 8,000 ×*g* at room temperature. Aspirate supernatant and suspend in 50 μL of FC block (CD16/32) to prevent nonspecific binding of antibodies (*see* **Note 8**). Incubate on ice for 10 min.
5. Prepare antibody mix at appropriate dilutions for all samples. Be sure to choose antibodies that will allow you to stain for donor and recipient cells (i.e., if you are using CD45.1 donor mice and CD45.2 recipients, be sure that you are using antibodies that can identify those markers). Add 50 μl antibody mix to all samples and incubate for 30 min on ice protected from light.
6. Add 1 ml of cell staining buffer to wash any excess antibodies from the cell suspension and centrifuge at 8,000 ×*g* for 1 min at room temperature.
7. Aspirate supernatant and suspend pellet in 150 μL cell staining buffer and then transfer to equal volume of 4 % paraformaldehyde (PFA) making a final concentration of 2 % PFA. The final volume of the cells depends on the volume needed to acquire by your flow cytometer. Cells can be kept at 4 °C protected from light until ready to acquire, but no longer than 36 h.
8. Analyze results by arranging markers for recipient versus donor mice on the *X* and *Y* axis respectively. The resulting populations will show the percentage of chimerism (Fig. 1).

3.5 Flow Cytometric Analysis of Microglia

1. Anesthetize mouse with Ketaset/Xylazine cocktail at 100 mg/kg Ketaset and 10 mg/kg Xylazine and perfuse with 1× HBSS by cutting the thoracic cavity open and cut the right atrium of the heart allowing blood to flow out. Inject 20 ml of ice cold 1× HBSS into the left ventricle to completely remove blood from tissues.
2. Remove the brain from the skull and obtain the spinal cord by flushing with 1× HBSS (*see* **Note 9**). Transfer tissues to 6-well plate containing RPMI and keep on ice until all tissues are collected.
3. Fill Dounce homogenizers with 3.5 ml RPMI without phenol red, label them accordingly, and keep on ice. It is possible to homogenize up to two brains in one 7 ml homogenizer. When combining more than two brains, use a 15 ml homogenizer.

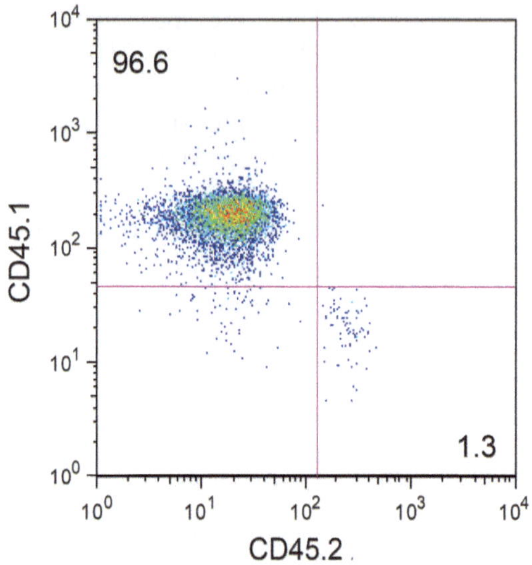

Fig. 1 Peripheral blood analysis from a CD45.2 WT mouse reconstituted with CD45.1 KO bone marrow cells. CD45.1 cells are stained with anti-mouse CD45.1 phycoerythrin-Cy7 (PE-Cy7). CD45.2 (recipient) cells are stained with anti-mouse CD45.2 allophycocyanin (APC). Data shows 96.6 % CD45.1 reconstitution of the CD45.2 recipient indicating a successful chimerism

4. Transfer tissues to their respective homogenizer and grind the tissue using the "A" or "loose" plunger until the suspension has reached confluency and then use the "B" or "tight" plunger. Once the suspension is homogeneous, transfer the suspension to a 15 ml conical tube and top off with RPMI to 7 ml. Continue this step with all samples and keep on ice before moving to the next step.
5. Add 3 ml of 100 % Percoll solution to the cell suspension and mix well to gain confluency. Slowly layer the homogenized suspension onto a new 15 ml conical tube containing 2 ml of 70 % Percoll. The idea is to set up a density gradient in which the brain mononuclear cells including microglia will layer atop the 70 % Percoll (*see* **Note 10**).
6. Centrifuge at 550 × *g* for 30 min at 18 °C.
7. Locate the interphase which contains the microglia just above the 2 ml graduation on the 15 ml tube. It is easily seen as a cloudy halo when held up to the light. Using a transfer pipette, carefully remove the interphase, approximately 2 ml, and transfer to a 15 ml tube containing 8 ml of 1× HBSS/10 mM HEPES to wash cells of Percoll.
8. Centrifuge for 7 min at 700 × *g* at 4 °C. Aspirate supernatant and suspend pellet in 1 ml cell staining buffer and transfer to 1.5 ml tubes.
9. Centrifuge cells for 1 min at 8,000 × *g*, aspirate supernatant, and suspend pellet in volume necessary for blocking step (*see* **Note 8**). Incubate cells in FC block solution for 10 min on ice.

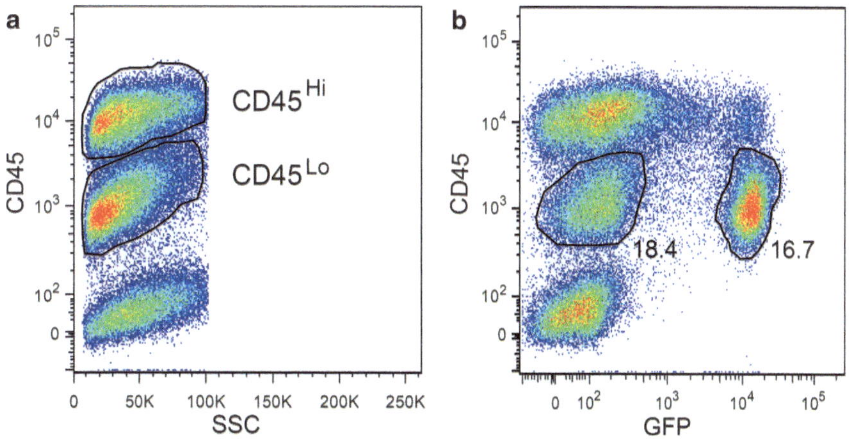

Fig. 2 (**a**) Cell homogenates were prepared from the brain of EAE-affected *Cx3cr1*$^{gfp/gfp}$ → wild-type chimeric mice and separated over density Percoll gradients. CD45 (APC) antibody was used to locate the CD45^{hi} population of hematogenous cells and CD45^{lo} or resident microglia. (**b**) From the same population of cells seen in (**a**), we can determine that 18.7 % of the population is GFP positive (FITC) indicating donor microglia. The other population (18.4 %) is FITC negative representative of resident microglia. Additional markers can be incorporated to determine differences in effector function between the different experimental groups

10. Without washing, add equal volume of antibody mix to the cells and incubate for 30 min on ice protected from light.
11. Wash cells of any excess antibody by adding 1 ml cell staining buffer to each tube and centrifuge for 1 min at 8,000 × *g*.
12. Aspirate supernatant and resuspend pellet in cell staining buffer and then transfer to equal volume of 4 % paraformaldehyde (PFA) for a final concentration of 2 % PFA. The final volume of the cells depends on the volume needed to acquire by your flow cytometer. Cells can be kept at 4 °C protected from light until ready to acquire.
13. To analyze microglia, you must first distinguish your donor cells from your recipients based on the expression of CD45 and the presence or absence of GFP (*see* **Note 11**). With CD45 on the *Y* axis and SSC on the *X* axis, locate the microglial population expressed as CD45^{lo}. This population denotes both donor and recipient microglia. By changing the *X* axis to FITC to observe the presence of GFP, it is now possible to differentiate donor from recipient microglia. (Note: other antibodies can be included to compare effector function or proliferation) (Fig. 2).

4 Notes

1. Select donor and recipient mice properly, ensuring the differentiation of peripheral cells from injected cells by using different strains of mice that you can identify by flow cytometry

staining. For example, in the figures shown, we used CD45.1 KO donor mice and CD45.2 WT recipients. Anti-CD45.1- and anti-CD45.2-specific antibodies are then used to determine the percentage of chimerism of the recipient mice. Our *Cx3cr1*$^{-/-}$ mice contain a knock-in of GFP within CX3CR1 making it possible to determine KO from WT cells.

2. To protect the mouse brain from radiation, you must first anesthetize it using 100 mg/kg Ketaset/10 mg/kg Xylazine. For this procedure, inject 100 μl of the solution for every 10 g of body weight via intraperitoneal (i.p.) route. Secure the helmet as discussed in the procedure and treat with correct dosage of radiation. Be sure to monitor the mice until they regain righting reflex and normal activity which takes approximately 1–2 h after injection of cocktail.
3. It can be helpful to cut into the spine just above the femoral head when detaching the limb. This will ensure that you will not break the femoral bone resulting in the loss of cells. After removing the skin and muscle, locate the femoral epiphysis and cut into the spine just above it. Once you have removed the limb, you can cut off any excess tissue and spinal bones with greater ease.
4. You can separate the bones by bending the knee backwards with a twisting motion until the bones detach. This aids in keeping the femur and tibia bones intact.
5. As you flush out cells from either end of the bone shaft, you might find a population of cells located in the middle portion of the shaft. Once you see that the epiphyseal end of the bone is white, you can trim the bone to better reach the middle of the shaft. Take care when cutting the bone as splitting can occur.
6. Retro-orbital injections take some practice. It is recommended to practice by injecting 150 μl of saline or PBS into other mice prior to your experiment. After you have anesthetized the mice as listed in the procedure, roll the mouse on its side. Secure the head with your index finger behind the skull and thumb under the chin. Insert the needle at a 45° angle into the conjunctiva, inner corner of the eye, until you feel a soft spot about a quarter inch deep. Begin to inject saline/cells. If you see the fluid coming out of the eye, you are not deep enough. Insert the needle deeper and continue injecting fluid. After injecting the correct amount of saline/cells, remove the needle and blot any excess fluid from the eye. The mice should wake within 2 min. Be sure to monitor the mice until they regain consciousness.
7. First, locate the submandibular vein. This is usually found behind the cheek below the ear and above the jaw bone. Make a deep puncture with an animal lancet and collect up to 200 μl of blood in a tube with 60 μl heparin. Mix blood with heparin instantly to prevent coagulation.

8. It is recommended to suspend sample pellets in 50 μl of FC block solution for incubation. After blocking the cells, add 50 μl of the antibody mix to the cells in FC block for a final volume of 100 μl.
9. To flush out the spinal cord, you will need to remove the entire spinal cord from the carcass. With a 10 ml syringe and blunt 16 gauge needle, find the spinal cord within the column; hold the column tight while you press the plunger with great force. It is important to hold tight to the spinal column as the suction is required to expel the spinal cord. The spinal cord should shoot out in one piece.
10. If cells are pipetted too fast or into the 70 % Percoll instead of on top, the cells will be lost indefinitely. It is helpful to use the gravity or lowest setting on the pipette if possible.
11. As previously stated, we use CD45.1 donor mice and use the CD45.1 anti-mouse PE-Cy7 to identify those cells. We use CD45.2 anti-mouse APC to label our recipient endogenous cells. The *Cx₃cr1*$^{-/-}$ mice contain GFP which allows us to visualize donor microglia. The CD45^{lo} GFP- population represents the resident microglia. The CD45^{lo} GFP+ population of cells is representative of CD45.1 donor microglia. CD45^{hi} populations are indicative of infiltrating monocytes and can also be segregated by the presence or absence of GFP.

Acknowledgments

This work was supported by NIH SCIGM095426 and National Multiple Sclerosis Society TA-3021-A-1 grant to AEC.

References

1. Soulet D, Rivest S (2008) Bone-marrow-derived microglia: myth or reality? Curr Opin Pharmacol 8:508–518
2. Mizutani M, Pino PA, Saederup N et al (2012) The fractalkine receptor but not CCR2 is present on microglia from embryonic development throughout adulthood. J Immunol 188:29–36
3. Prinz M, Priller J, Sisodia SS et al (2011) Heterogeneity of CNS myeloid cells and their roles in neurodegeneration. Nat Neurosci 14:1227–1235
4. Cardona A, Pioro EP, Sasse ME et al (2006) Control of microglial neurotoxicity by the fractalkine receptor. Nat Neurosci 9:917–924
5. Jung S, Aliberti J, Graemmel P et al (2000) Analysis of fractalkine receptor CX(3)CR1 function by targeted deletion and green fluorescent protein reporter gene insertion. Mol Cell Biol 20:4106–4114
6. Geissmann F, Gordon S, Hume DA et al (2010) Unravelling mononuclear phagocyte heterogeneity. Nat Rev Immunol 10:453–460
7. Auffray C, Fogg DK, Narni-Mancinelli E et al (2009) CX3CR1+ CD115+ CD135+ common macrophage/DC precursors and the role of CX3CR1 in their response to inflammation. J Exp Med 206:595–606
8. Cardona AE, Sasse ME, Mizutani M et al (2008) Scavenging roles of chemokine receptors: chemokine receptor deficiency is associated with increased levels of ligand in circulation and tissues. Blood 112(2):256–263
9. Prinz M, Mildner A (2011) Microglia in the CNS: immigrants from another world. Glia 59:177–187
10. Ransohoff RM, Cardona AE (2010) Microglia: myeloid cells of the central nervous system parenchyma. Nature 468:253–262

Chapter 28

In Vivo Two-Photon Microscopy of Microglia

Satoru Kondo and Shigeo Okabe

Abstract

In vivo imaging with two-photon microscopy is becoming an indispensable technique to investigate cellular and subcellular phenomenon in living tissues including the central nervous system. This microscopy enables to image dynamics of molecules, morphology, and excitability with minimal invasion to tissues. Microglia are residual immune-responsive cells in the central nervous system and show highly dynamic response to the environmental alterations. Diverse roles of microglial functions in the intact and pathological brain are still largely unknown. In this chapter we describe the detailed method to image the dynamics of microglia in the mouse brain in vivo.

Key words Microglia, Two-photon microscopy, In vivo imaging, Cell dynamics, Mouse

1 Introduction

Among in vivo imaging techniques, two-photon microscopy is the only method so far to image biological phenomenon with both spatially and temporally high resolution. This method was first applied for brain and now has been used for various tissue observations. The central nervous system (CNS) is mainly composed of neurons and glia whose cell diameter is only around 10 μm with fine subcellular structures. In the brain, two-photon microscopy has revealed the neuronal dynamics of structure by observing the morphology of fluorescently labeled neurons and excitability by measuring the calcium dynamics with calcium indicator. Historically, glia have been long regarded as static cells that only support the structure of CNS. However, recent findings suggest that glia are rather highly dynamic cells and indispensable for the neuronal functions. Microglia are one of glial subtypes in the CNS. They are quiescent cells with fine ramified processes in the non-pathological state but once insult occurs, they rapidly become activated (processes are retracted and thicken, become amoeboid) and migrate to the site of injury [1]. Microglia are electrically non-excitable but

Bertrand Joseph and José Luis Venero (eds.), *Microglia: Methods and Protocols*, Methods in Molecular Biology, vol. 1041, DOI 10.1007/978-1-62703-520-0_28, © Springer Science+Business Media New York 2013

structurally highly dynamic. Several transgenic mouse lines that express fluorescent proteins specifically in microglia are available [2, 3]. By utilizing such mice, microglial dynamics in the intact or pathological state can be investigated [4–8]. These studies revealed that even in a non-pathological state, their fine processes are highly motile and continuously survey brain parenchyma. This protocol describes the imaging of microglia in the living mouse brain using two-photon microscopy.

2 Materials

2.1 Reagents

1. Mice: Iba1-EGFP transgenic mice or Cx3cr1-EGFP knock-in mice are useful to image structural dynamics of microglia in vivo. Both mice express EGFP in microglia, and fine processes can be imaged. Iba1-EGFP transgenic mice were generated by Hirasawa et al. [2]. Iba1 (ionized calcium-binding adaptor molecule 1) is specifically expressed in microglia and macrophages, and EGFP is expressed under the regulation of this gene promoter. The availability of the mice should be addressed to the authors. CX3CR1 (CX3C chemokine receptor 1) is a receptor protein for chemokine and expressed in microglia, monocyte, and other immune-related cells. Cx3cr1-EGFP mice were generated by Jung et al. [3] and are available from The Jackson Laboratory (stock # 005582).
2. Anesthetics: Ketamine hydrochloride salt (50 mg/ml). Xylazine hydrochloride (20 mg/ml). Ketamine/xylazine mixture is prepared at the concentration of 10 mg/ml of ketamine and 1 mg/ml of xylazine diluted in saline. Lidocaine hydrochloride (2 % xylocaine jelly, 2 % xylocaine solution).
3. 0.9 % Saline: Saline is consisted of 0.9 % NaCl. Sterilized solution is available from pharmacy or can be prepared in laboratory by autoclaving.
4. Disinfectant (0.1 % Gentacin ointment).
5. Iodine tincture (povidone-iodine solution).
6. Ophthalmic ointment (Mycochlorin).
7. Absorber: High absorbable cotton paper.
8. Absorbable gelatin sponge.
9. Cotton buds.
10. Adhesive: Jelly-type instant adhesive.
11. Hair removal cream (Nair, USA).

2.2 Equipments and Tools

1. Two-photon excitation microscopy: Commercially built microscopes are available from some companies (e.g., FV1000MP,

Olympus, Japan; A1RMP, Nikon, Japan; LSM7MP, Zeiss, Germany). These microscopes are fully equipped, ready to use, and almost maintenance-free. Half custom-made microscopy can be constructed by purchasing system or kit from some companies (e.g., Thorlabs, USA; Prairie Technologies, USA; Sutter Instruments, USA; Femtonics, Hungary; and Scientifica, UK). For the construction of full custom-made microscopes, refer some literatures [9].

2. Low-magnified, high numerical aperture, water-immersion objective lens (e.g., XlPlan N 25× 1.05 NA, Olympus, Japan; CFI Apo LWD 25× 1.1 NA, Nikon, Japan).
3. Drill and bur: High-speed (ca 30,000 rpm) drill. Micro drill bur (φ0.7 mm, Fine Science Tools, USA). It is better to choose micro drill bur made from sterilizable stainless steel.
4. Scissors (BH-12S, Napox, Poland).
5. Forceps (Dumont #5 and #7, Fine Science Tools, USA).
6. Needle holder (Olsen-Hegar with scissors, Fine Science Tools, USA).
7. Microsurgical blade.
8. Scalpel.
9. Surgical needle with suture.
10. Head plate: Custom-made. 60 mm (L) × 10 mm (W) × 0.5 mm (H) stainless steel (Fig. 1).
11. Head-holding stage: Custom-assembled. All parts are available from some companies (e.g., Breadbord MB4, Clamping Fork CF175, Pedestal Post Holder PH1E, Stainless Steel Post TR1.5, Thorlabs, USA) (Fig. 1).
12. Balance: This is for weighing mouse. Readability of 0.1 g is preferable.
13. Cold light source: High luminance LED is preferable to use.
14. CCD or Digital camera: Either black/white or color CCD camera can be used. Blood vessels can be distinguished easily with color CCD camera. Compact digital camera for photography can be alternatively used. Attachment parts to mount on various microscopes is available from some companies (e.g., Microscope Network Co., Japan).
15. Stereomicroscope: Microscope with large field of view and high magnification (ca 40×) is recommended (S6, Leica, Germany).
16. Heat pad: Temperature-controlled heat pad with rectal probe (DC Temperature Control System Cat# 40-90-8, FHC, USA). Pet heat pad (Flexiguard, Petnap Ltd., UK).
17. Shaver for small animal.

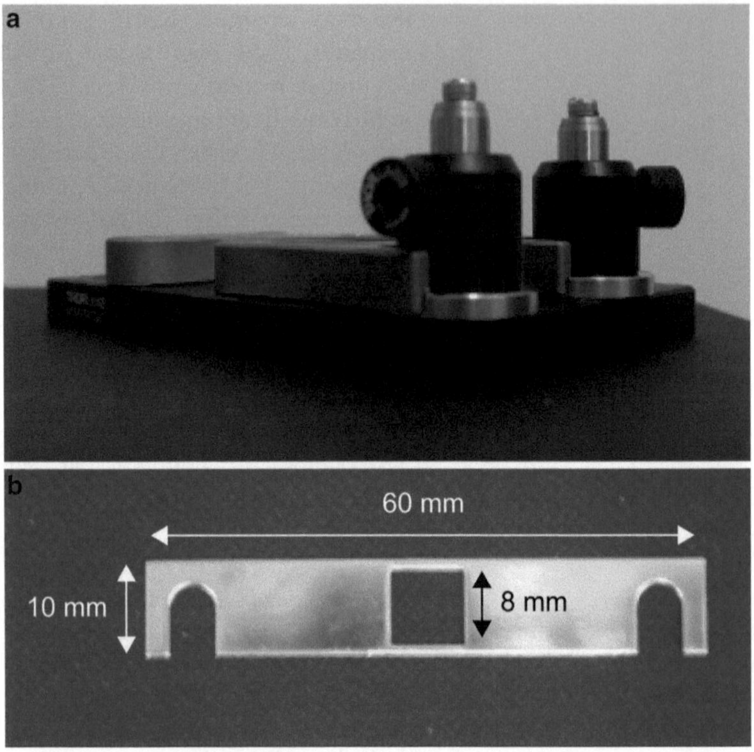

Fig. 1 Experimental equipments to hold the head of mouse. (**a**) Head-holding stage is shown. Height of the pole is adjustable by the screw on the side, depending on the size of mouse. Head plate is fixed on top of the poles with screw. (**b**) Head plate is shown with its sizes. This is made of stainless steel and can be reused repetitively. *Square hole* in the middle is for the observation window

3 Methods

To perform in vivo imaging of fluorescently labeled cells in the brain, skull that hinders the penetration of excitation and emission light should be treated to increase the transmission of light. To this end, there are two methods to prepare the observation window in the skull [10, 11]. One is called open-skull and other is thinned-skull window. Open-skull window is formed by a craniotomy, in which a part of skull is removed and replaced with cover glass. Thinned-skull window is prepared by scraping a part of skull with microsurgical blade to the thickness of 10–20 μm but leaves the skull intact. Distortion of light is smaller in open-skull window and better images can be obtained from deeper depth of tissue than thinned-skull window. However, it is reported that open-skull window activates microglia and astrocytes by inflammatory processes [9], yet this is still in debate [11]. Thinned-skull window can avoid activation of microglia and astrocytes, but optical accessibility is limited. Due to the larger light scattering and aberration by the remained skull, maximal depth of imaging of fine structure is limited to 100–200 μm from the cortical surface. Despite of some limitations for the observation, transcranial imaging might be preferable for the studies of microglial dynamics that is sensitive to minimal invasion to the brain tissue.

Microglia is present not only in the brain but in the spinal cord as well. Different technique is necessary for the spinal cord imaging [12].

3.1 Attachment of Head Plate to the Mouse Skull

1. Sterilize all the surgical tools and materials before surgery (*see* **Note 1**).
2. Place the mouse in a small box and weigh it. Administer ketamine/xylazine anesthesia based on the body weight (*see* **Notes 2** and **3**).
3. Transfer mouse on the temperature-controlled heat pad (*see* **Notes 4** and **5**). Protect mouse's eyes with ophthalmic ointment.
4. Shave fur with shaver from the base of ears to eyes. Alternatively, hair remover can be used.
5. Disinfect the skull skin with 70 % ethanol and iodine tincture.
6. Apply local anesthesia (2 % lidocaine) into the skull skin.
7. Make incision at the midline from ears to eyes and retract skin left and right. Remove the periosteum of the skull and clean the surface (*see* **Notes 6** and 7).
8. Thoroughly dry up the skull surface (*see* **Note 8**). Mark the center of the area where images are going to be taken.
9. Put glue on both skull and head plate and then attach a head plate to the skull tightly being the marked spot in the center (*see* **Notes 9** and **10**).
10. Move the mouse to the head-holding stage and attach head plate to the head-holding stage tightly (*see* **Note 11**).

3.2 Thinned-Skull Surgery

Thinning process consists of two steps. First step is carried out by high-speed drill for wider area, and second step is done by microsurgical blade and narrows the thinning area. To avoid the mechanical damage to the brain parenchyma (to avoid the activation of microglia) during the thinning, the area for the optical access (thinned-skull window size) should be limited to 0.5 mm square (ca. 0.3 mm^2).

1. Determine the thinning area for the first step (*see* **Note 12**). The area of thinning is ca. 1.5 mm square (Fig. 2a).
2. Start thinning with high-speed drill (*see* **Note 13**). The speed of drill is 20,000 rpm keeping the drill bur angle at 30° (*see* **Note 14**).
3. The skull should be cooled frequently with cold saline to prevent the heating due to the frictional heat between bone and drill bur.

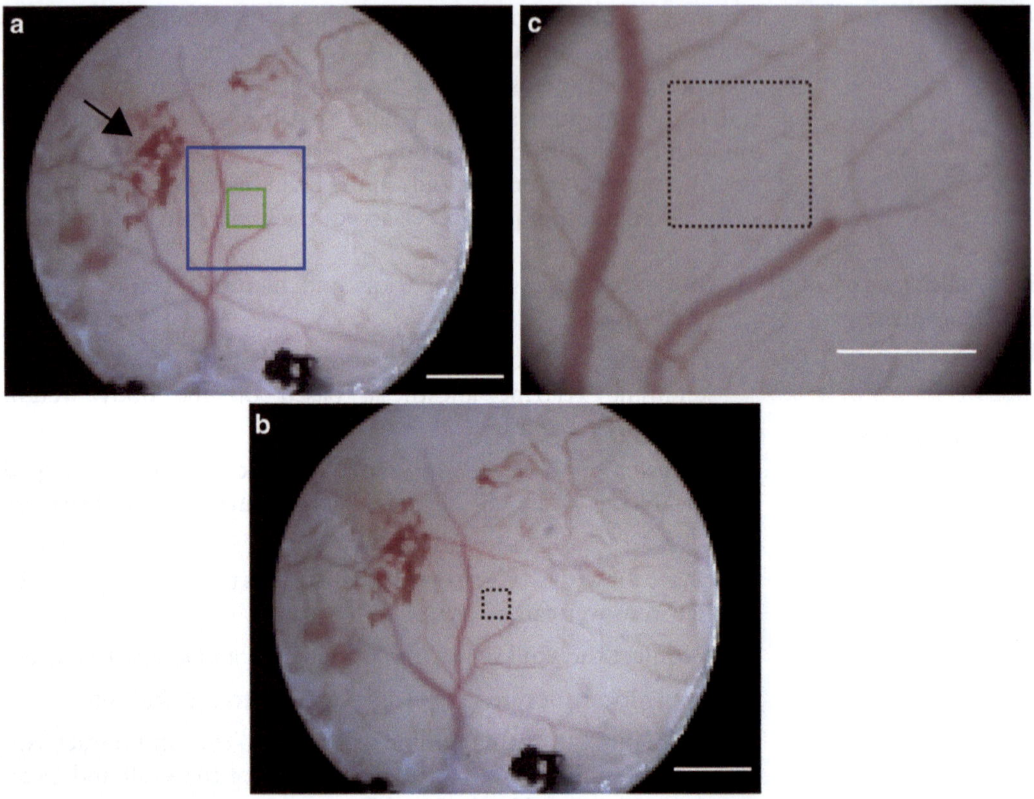

Fig. 2 Example photographs of the thinned-skull window. (**a**) A low-magnified photograph after the thinned-skull preparation. The thinning areas for the high-speed drill (*blue*) and for the microsurgical blade (*green*) are shown. The sizes of area are 1.5 mm square and 0.5 mm square, respectively. *Black arrow* shows blood vessels in the skull. (**b**) After the experiment, the imaged area is marked (*dotted square*). (**c**) High-magnified photographs of imaged area. Imaged area is marked with *dotted square*. Vasculature pattern helps to find the same area for the next imaging session. Scale bars, 1 mm

4. The thinning is performed until the spongy bone is removed (*see* **Note 15**). When the thinning reached to the third layer (compact bone), stop the thinning and investigate the thickness (*see* **Note 16**). At this stage, the skull thickness is still around 100 μm.
5. Determine the final thinning area (0.5 mm square) (*see* **Note 17** and Fig. 2a).
6. The thinning is further continued with drill. If the bone becomes too soft to thin with drill, thinning is switched to pursue with microsurgical blade by hand. The thinning should be done with greater caution. It is necessary to avoid pushing downward to the skull. Keep the angle of the microsurgical blade at 45° (*see* **Note 18**).
7. Stop thinning when thickness reached 10–20 μm (*see* **Note 19**).
8. Take image of cranial window to record the thinning area by CCD camera attached to the stereo microscope (*see* **Note 20** and Fig. 4a).

3.3 Imaging Microglia with Two-Photon Microscopy

Choice of objective lens is quite important to obtain the bright and high-resolution images with low excitation laser power. Higher laser power exposure may cause the microglial activation. Recently released objective lenses with low magnification and high numerical aperture are recommended to use. Olympus 25× NA1.05 or Nikon 25× NA1.10 can achieve better excitation of fluorophores and better collection of fluorescence in vivo.

Some methods to improve the excitation and detection efficiencies are recently reported, but most of them are still not installed in the commercially built microscopes (e.g., adaptive optics for wave-front correction or passive pulse splitters).

1. Transfer the mouse mounted on the head-holding stage under the microscope. Set and fix the position of mouse properly.
2. By using microscope's ocular and low-magnified objective lens (4×), adjust the focus on the skull surface by epi-illumination with cold light. Center the cranial window and take photograph by CCD camera attached on two-photon microscopy to record the vasculature pattern for the future relocation (Fig. 2a).
3. Change objective lens to higher magnification and immerse saline between objective lens and skull. Shield with blackout curtain or plate to prevent the light leakage.
4. By using microscope's ocular and epi-illumination with cold light, adjust the image area by comparing the previously taken low-magnified photograph (Fig. 2a). Vasculature pattern can be used as a landmark.
5. Switch to the LSM mode and start imaging. Set the HV value of PMT and laser intensity as appropriate (*see* **Note 21**).
6. If the focus is on top of the skull, initial image is the autofluorescence from skull. As the objective lens goes downward, subsequent image is the autofluorescence from thin layer of dura mater. Finally fluorescently labeled microglia will appear (*see* **Note 22**).
7. Acquire imaging data as required (*see* **Notes 23** and **24**) (Fig. 3).
8. At the end of each imaging session, move up the objectives to the surface and take photograph to record the vasculature pattern (Fig. 2c) and mark the position of imaged area on the low-magnified photograph of blood vessel pattern (*see* **Note 25** and Fig. 2b).
9. During the imaging, take care of the depth of the anesthesia and administer the additional anesthetics as necessary.
10. After all the imaging, detach the mouse from head holder and remove the head plate from the skull carefully. Clean the cranial

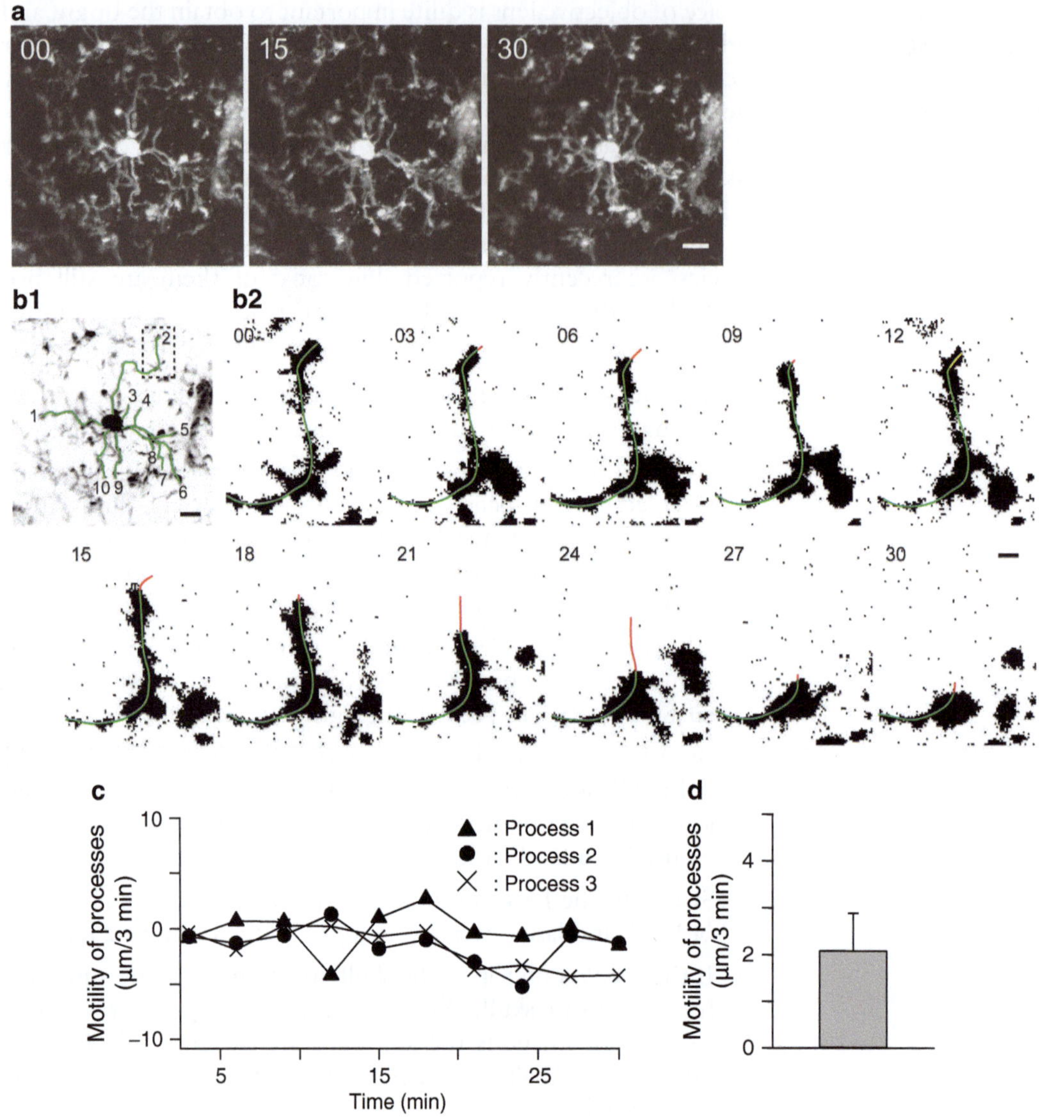

Fig. 3 An example of in vivo time-lapse imaging of microglial processes. (**a**) Morphology of microglia imaged by in vivo two-photon microscopy of Iba1-EGFP mice. Time-lapse imaging was performed, and images with 15 min intervals are shown. Scale bar, 10 μm. (**b**1) This image shows the binary image shown in A at time point of zero min, with an overlay of *green lines* (numbered from 1 to 10) on individual processes. The tips of these *green lines* were recorded, and their distances between adjacent time frames (3 min intervals) were calculated as an index of process motility shown in **c** and **d**. Scale bar, 10 μm. (**b**2) Binary images of a microglial process (*green line*) extracted from time-lapse sequences in the image area marked by a *dotted rectangle* in **b**1, illustrating the extent (*yellow line*) of rapid growth and shrinkage (*red line*). Numbers in the *upper left corner* of images indicate elapsed time in minute. Scale bar, 2 μm. (**c**) The motility of microglial processes. Extension and retraction of processes were plotted for three representative processes with time intervals of 3 min. Positive values indicate process extension, and negative values indicate retraction. (**d**) Summation of absolute distances of extension and retraction during the observation period of 30 min was calculated, and the average speeds per frame (3 min) were estimated from 10 processes

window with sterilized water and suture the skin. Put the disinfectant on the sutured skin and return the mouse into the cage. Keep the cage on the heat pad until awakening.

11. Clean the objective lens with appropriate cleaning solution and disinfect the area under the microscope with 70 % ethanol.

3.4 Repetitive Imaging

Bone has ability to grow again, and skull can gradually recover the thickness where thinned-skull operation was carried out. Therefore, skull rethinning is necessary, depending on the situation, before the repetitive imaging. The degree of bone regrowth depends on the time interval of imaging. Usually within 2–3 days after the last imaging, it is not necessary to do the rethinning. However, after around 4 days, connective tissue appears on the cranial window and regrowth of bones becomes evident after a week. Basically rethinning can be performed by the same procedure as the thinning process at the first time. Regrowth bone is something different from the original bone. It has inhomogeneous hardness and irregular structure. Because the difficulty of rethinning increases as the number of times of this process, maximum number of times that imaging can be performed is limited to 3–4.

Recently transparent cyanoacrylate cement can stabilize the thinned-skull window and prevent bone regrowth, achieving the repetitive observation without rethinning [13].

1. Place the mouse in a small box and weigh. Administer ketamine/xylazine anesthesia based on the body weight (*see* **Notes 2** and **3**).
2. Transfer mouse on the temperature-controlled heat pad (*see* **Notes 4** and **5**). Protect mouse's eyes with ophthalmic ointment.
3. Remove fur if necessary. Disinfect the skull skin with 70 % ethanol and iodine tincture.
4. Apply local anesthesia (2 % lidocaine) into the skull skin.
5. Make incision at the midline from ears to eyes and retract skin left and right.
6. Confirm the location of the observation window comparing the photograph that was taken at the previous imaging (Fig. 28.4a). Clean up the skull with cotton buds or soft paper not to damage the observation window (*see* **Note 26**). At this stage do not clean the observation window yet.
7. Put glue on both skull and head plate and then attach a head plate to the skull tightly being the observation window in the center (*see* **Notes 9** and **10**).
8. Move the mouse with heat pad to the head-holding stage and attach head plate to the head-holding stage tightly (*see* **Note 11**).
9. Start cleaning the observation window. Under the stereomicroscope investigate the condition of cranial window. If it is covered

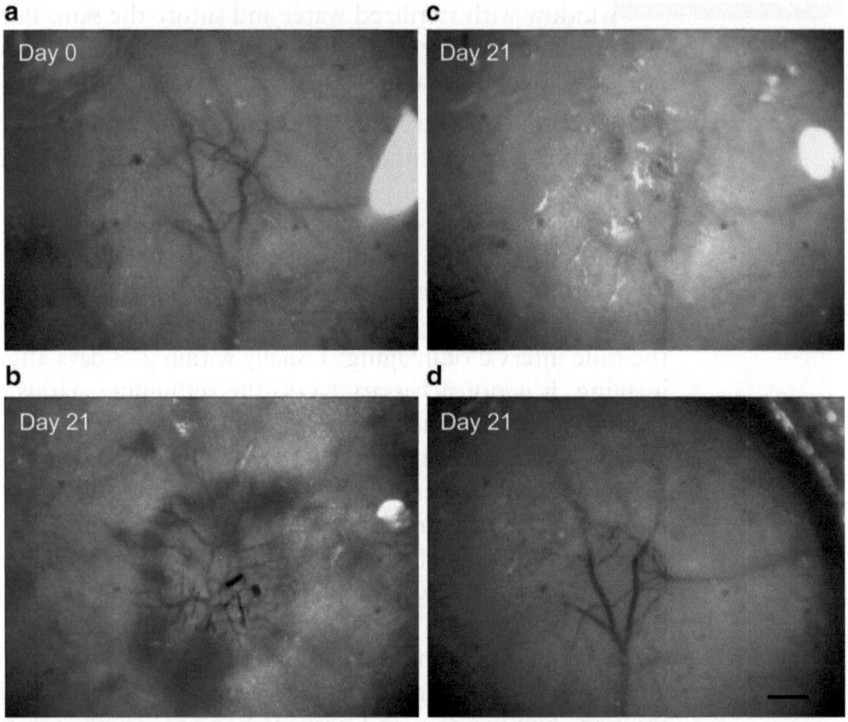

Fig. 4 Bone regrowth occurs after long-time intervals. (**a**) Photograph of thinned-skull window at day 0. High transparency of thinned-skull area is observed. (**b**, **c**, **d**) Photographs of thinned-skull window at day 21. (**b**) Thinned-skull area is covered with connective tissues. (**c**) Connective tissues are thoroughly removed. Transparency of thinned-skull window is lost, and bone regrowth evidently occurred. (**d**) Rethinning with high-speed drill and microsurgical blade recovered the high transparency of the observation window. Scale bar, 1 mm

only with a small piece of tissue, high transparency of thinned-skull window should be maintained beneath. Usually after the removal of the tissue observation can be done without the necessity of rethinning. If the cranial window is covered with dense connective tissue and blood vessels, bone regrowth could occur (Fig. 4b). Firstly, connective tissue should be thoroughly removed with scalpel or microsurgical blade. Untransparent regrowth bone will appear (Fig. 4c). Secondly, skull rethinning is curried out with high-speed drill and microsurgical blade as described in the Subheading 3.2. As mentioned earlier, the quality of regrowth bone is something different from the original bone. More attention should be paid to perform the skull rethinning than the first-time thinning.

10. Transfer mouse under the microscope and set the position of the mouse on the head-holding stage.
11. By using low-magnified objective lens (4×) and microscope's ocular, observe the skull surface by epi-illumination with cold

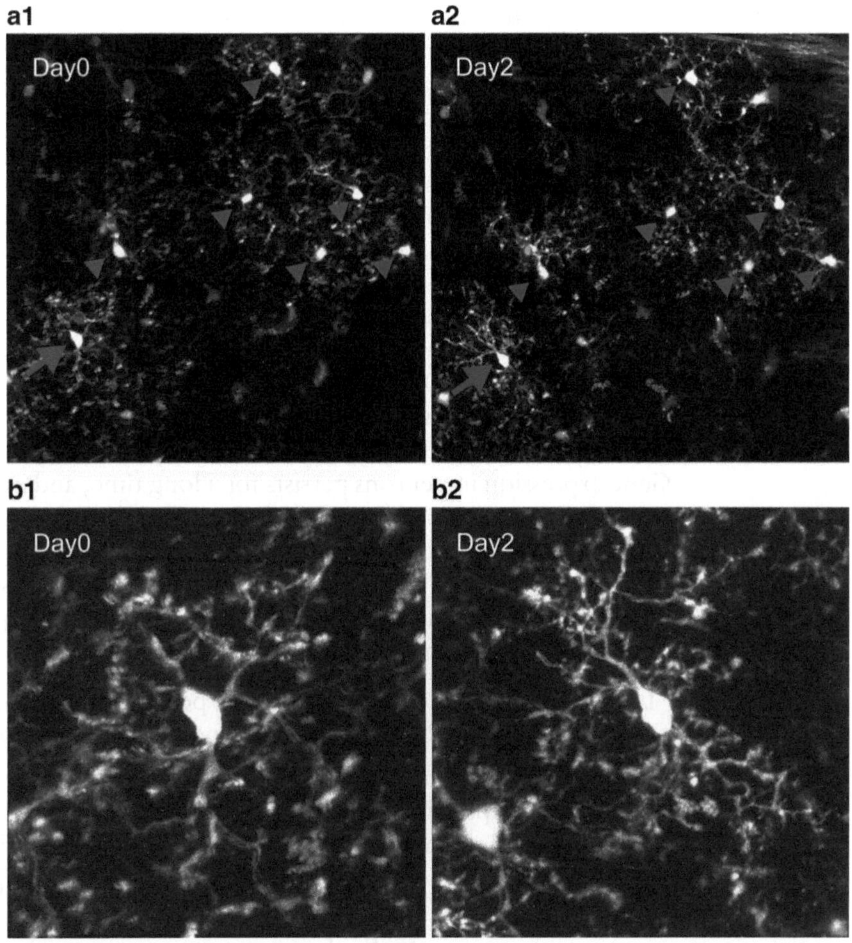

Fig. 5 An example of time-lapse imaging over different days. (**a**1, **a**2) Low-magnified images of Iba1-EGFP mouse. Same areas taken at two different days are shown. Several microglia can be seen, and cell bodies of microglia observed on both days are marked with *arrow* and *arrow heads*. Most of the microglia remained allocated at the same position during this time period (2 days). (**b**1, **b**2) Higher-magnified images of microglia from a1 and a2 (indicated by *arrow*) are shown. Although the position of cell bodies remained unchanged, extension patterns of their processes are largely altered as can be predicted from the short-term imaging (Fig. 3)

light. Compare observation window with the reference image that was taken previously and locate the area where imaging was performed previously (*see* **Note 27**).

12. Acquire imaging data as required (*see* **Note 28**) (Fig. 5).
13. After all the imaging, detach the mouse from head holder and remove the head plate from the skull carefully. Clean the cranial window with sterilized water and suture the skin. Put the disinfectant on the sutured skin and return the mouse into the cage. Keep the cage on the heat pad until awakening.
14. Clean the objective lens with appropriate cleaning solution and disinfect the area under the microscope with 70 % ethanol.

3.5 Simultaneous Imaging of Microglia and Neuron

Besides the immune function of microglia, increasing number of studies has demonstrated its interactions with neurons, especially at the synapses during physiological and pathological states. Simultaneous two-photon in vivo imaging of microglia and neurons could elucidate the roles of microglia in the synapses. To characterize the microglia-synapse interactions, double transgenic mice (Iba1-EGFP/Thy1-YFP or Cx3cr1-EGFP/Thy1-YFP), in which both microglia and layer V neurons are fluorescently labeled, are used [6, 7, 12]. Although these mice are quite useful for simultaneous imaging, emission wavelength of EGFP and YFP are very close to each other, and it is difficult to separate these signals with optical filter [14]. One of the alternative ways is to perform in utero electroporation of gene that codes red fluorescent protein to Iba1-EGFP or Cx3cr1-EGFP mouse embryo and label neurons. Gene expression in neurons persists for a long time, and imaging at later postnatal stage is possible. Red fluorescent protein is a better choice than the neuronal labeling. Basically there is no large difference for the imaging method as mentioned in the earlier sections.

Detailed description of in utero electroporation technique is out of the scope of this chapter. Please refer appropriate literatures [15]:

1. In utero electroporation of gene is performed to Iba1-EGFP or Cx3cr1-EGFP mouse embryos and raise mouse until appropriate age (*see* **Note 29**).
2. Surgical preparations and setup for observation are carried out by the same methods as described in the Subheadings 3.1–3.3.
3. Acquire imaging data as required (*see* **Note 30**) (Fig. 6).
4. Finish the imaging procedure as described in the Subheading 3.3.

4 Notes

1. All the surgical tools made of stainless steel should be prepared to sterilize in the dry heat oven. Sterilization for other surgical instruments and materials can be achieved by autoclave (high-pressure steam). Autoclaved materials can be dried up in the low-temperature oven kept at 50 centigrade.
2. The dose is 10 μl/g body weight (0.1 mg of ketamine and 0.01 mg of xylazine/g body weight). This mixture of anesthesia can be administered either intraperitoneally or intramuscularly.
3. Alternatively, isoflurane or other anesthesia can be chosen. In terms of the controllable and stable anesthesia, isoflurane is preferable, but from our experiences, motion movement of images is less in mouse anesthetized by ketamine/xylazine mixture.

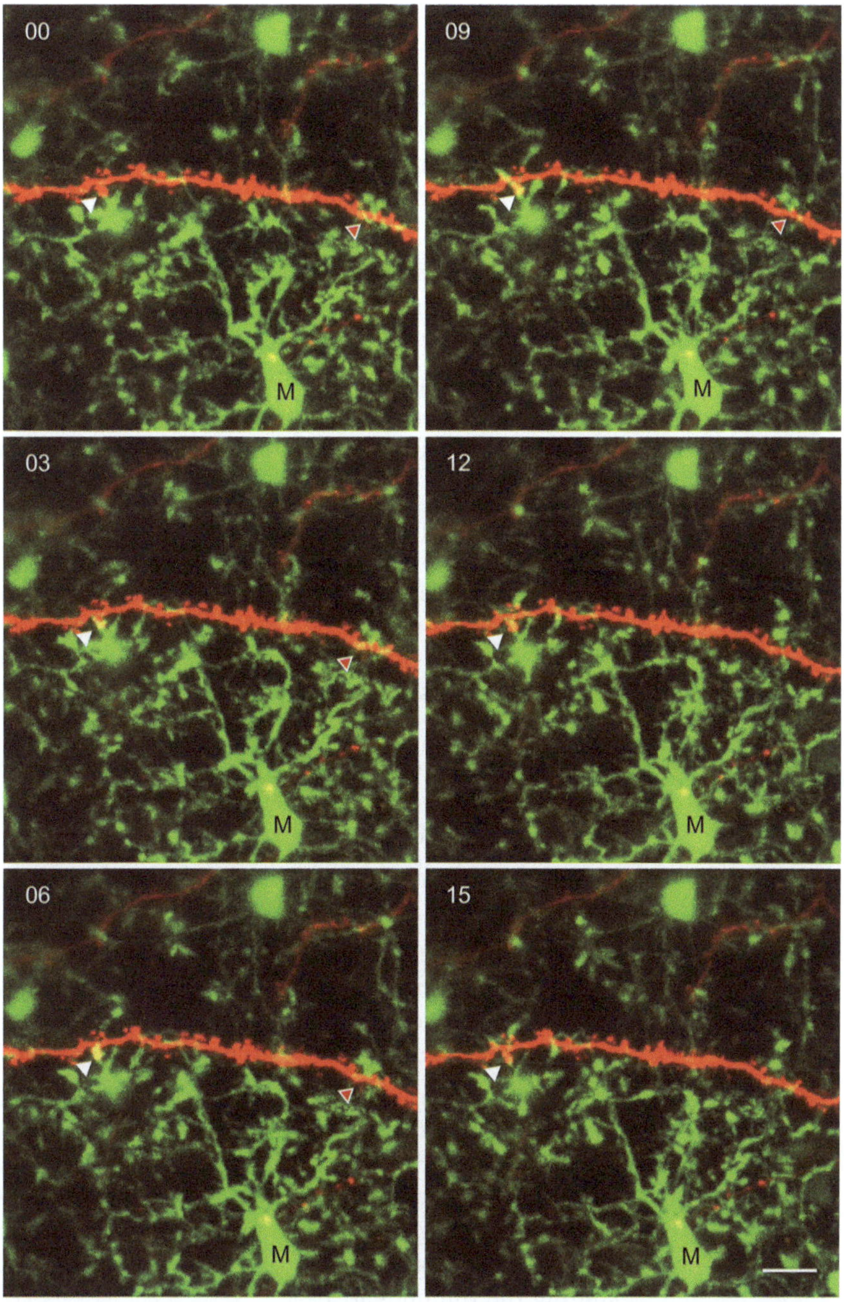

Fig. 6 Example images of simultaneous dual color imaging of microglia and neurons. Simultaneous dual color time-lapse imaging was performed with Iba1-GFP mouse, in which red fluorescent protein dsRed2 is expressed in neurons by in utero gene electroporation method. Cell body of microglia (*green*) is marked as M, and neuronal processes (*red*) are visible. Time-lapse images over 15 min show dendritic structures contacted by microglia (*yellow*, *white*, and *red arrowheads*). Numbers in the *upper left corner* of images indicate elapsed time in minute. Scale bar, 10 μm

4. In the case of isoflurane, mouse is placed on the heating pad attaching nose cone for maintaining anesthesia.
5. Confirm that mouse is deeply anesthetized by checking the absence of paw withdrawal from pinching the toe. If withdrawal reflex remains, adjust anesthetic level by additional administration of ketamine/xylazine or increasing the isoflurane concentration.
6. Sometimes bleeding from skull occurs during the cleaning, but it stops spontaneously within a short while. However if it prolongs, resolve the bleeding by using gel form.
7. To attach head plate, it is not necessary to remove cervical muscle for somatosensory or motor cortex imaging, but for other areas of the cortex, retract or remove cervical muscle as necessary.
8. This process is quite important. If the surface of skull is not exposed enough and cleaned, metal head plate will not be adhered firmly. Immediate adhesive glue is convenient to use. The head plate should be fixed tightly during surgery and imaging period yet be detached from skull after imaging. And also head plate is ideally adhered to the skull as immediate as possible.
9. It is important to check that the surface of the head plate and the skull is in parallel. In some area of cortex, the skull outline and cortical surface are not in parallel.
10. Usually wait for about 10 min until the head plate attaches to skull completely.
11. By narrowing the free space for the head of the mouse (the distance between the top of the head plate and the base of the head holder), spontaneous movements of mouse due to the breath or heartbeat is partially suppressed.
12. It is quite important to avoid large blood vessel within the observation window. Before starting the thinning process, confirm that there is no large blood vessel under the area that was marked in the previous section. Vasculatures on the cortical surface become visible by immersing the skull surface with saline. If any large vessel is visible, it is better to avoid it by shifting the thinning area.
13. The thinning should be performed uniformly and keep the skull surface flat.
14. Great caution should be applied not to push the skull during thinning.
15. The skull is consisted of three layers, two compact bones sandwich a spongy bone, and the total thickness is ca. 0.5 mm. As the thinning is going on, the transparency of the skull becomes higher, and blood vessels on the brain surface can be seen clearer.

It is possible to know which layer of the skull is currently thinned; within compact bone layer, drilling causes small bone powder whereas rough powder in spongy bone layer. Spongy bone layer often contains small canals for blood vessels and bone marrow (Fig. 2a), and bleeding may occur during the thinning but usually stops within a few minutes.

16. As the thinning proceeds, the bone becomes very soft not like bone but something like paper sheet. Drill bur should be moved as if it traces on a sheet of thin paper. The thickness of the third layer (compact bone) is ca. 100 μm. At this thickness, the blood vessel on the cortical surface can be seen without immersing the bone with saline.
17. Usually it is the center of the initial thinning area. If large blood vessel that may hinder the imaging appears during thinning, set the final thinning area properly.
18. It is easier to scalp when the bone is dried, but saline immersion should be done frequently to remove the heat, bone shaving, and assess thinning progress. The thinning should be done as flat as possible to decrease the light scattering and increase the resolution of image.
19. The thickness of the bone can be assessed by two-photon microscopy observation by measuring the autofluorescence of bone. If the thickness does not reach 10–20 μm, thinning process should be repeated.
20. This image is indispensable for the repeating imaging. If the imaging interval is more than a week, rethinning of observation window is necessary to do. This image helps to locate the thinning area precisely.
21. Initially laser intensity should begin from minimal; otherwise, microglia can be activated unexpectedly. As the laser power increases, autofluorescence of skull will be seen. The thickness of the bone can be measured by subtracting the depth at the start and the end point of autofluorescence. Dura mater has autofluorescence as well, but the appearance of autofluorescence of bone and dura mater is different. The autofluorescence of bone is relatively homogeneous but of dura mater looks fiber-like structure.
22. Due to the remained skull, light scattering and aberration are significant in thinned-skull preparation. The maximal depth of imaging is around 200 μm from the surface, and typically imaging depth is restricted to 100 μm for fine structure observation of microglial processes without any correction of aberration. It is better to avoid the recording site close to the large blood vessels, because aberration and also motion artifact is larger surround.

23. One of problems during imaging is motion artifact. This artifact problem is substantial to the three-dimensional stack image and time-lapse imaging. Motion artifact primarily results from respiration and heartbeat. Lateral image movement can be aligned offline by registration algorithm to a certain degree (e.g., Image J plugins), but axial movement remains difficult to align. To reduce the motion artifact, changing the posture of mouse or depth of anesthesia level is sometimes effective. Artificial respiration may help to diminish the motion. Respiration can be controlled with tightly attached mask or by tracheal intubation. These methods can be also applied for mouse, but it is difficult to maintain during a long time period imaging. It is recommended to check periodically if the lens is immersed enough. Deterioration of images during the imaging is often related with the inadequate or lack of immersion of the lens.
24. The resolution of images varied within the same window. It often happens despite of the careful flat thinning of skull. It is presumably related with the natural inhomogeneity in bone and unflatness of the opposite surface of skull.
25. This photograph is indispensable to relocate the same area at later time point.
26. The skull may be covered with connective tissues and sometimes contains newly invaded blood vessels. The thickness of connective tissue depends on the interval from the previous imaging. If intensive removal is necessary, use scalpel and forceps.
27. Usually branching and running pattern of blood vessels are almost unchanged. Therefore, vasculature pattern is a good landmark to find the previously imaged area.
28. Dimmer or more distorted images are sometimes observed at repetitive imaging. First point to suspect is the setting of microscopy. It is necessary to check the output laser power and excitation and collection light path. During the long-term repetitive imaging expression, the level of fluorescent protein may decrease. Bone regrowth on the opposite side may occur and increased inhomogeneities in bone resulted in the reduction of signal intensity. Further rethinning may improve the image quality. Increase of excitation power should be performed carefully not to give photodamage and activate microglia.
29. Among red fluorescent proteins, dsRed2 or tdTomato is useful for the simultaneous dual color imaging, because they can be excited at the same wavelength as GFP and the emission spectra of GFP and these red fluorescent proteins have very little overlap. Therefore, this combination would be excellent for separating each signal.

30. For the simultaneous dual color imaging, microscope is necessary to be equipped with two PMTs. The excitation wavelength for dual color imaging should be determined by the two-photon cross-section spectra. For the case for GFP and dsRed2, wavelength at 920 nm can be used for the simultaneous excitation with high efficiency but low water absorption [16]. Emission maxima of GFP are around 510 nm; hence, dsRed2 is around 580 nm upon simultaneous excitation. The appropriate filter set to separate these two signals could be 500–550 nm for GFP, 570–600 for dsRed2, and 560 nm for the dichroic mirror.

Acknowledgments

S.O. is supported by Grants-in-Aid for Scientific Research (18200025, 20019013, 21220008, and 22650070), Global COE Program (Integrative Life Science Based on the Study of Biosignaling Mechanisms), and Strategic Research Program for Brain Sciences from MEXT Japan and by Takeda Science Foundation.

References

1. Tremblay ME, Stevens B, Sierra A et al (2011) Mini-Symposium: The role of microglia in the healthy brain. J Neurosci 31:16064–16069
2. Hirasawa T, Ohsawa K, Imai Y et al (2005) Visualization of microglia in living tissues using Iba1-EGFP transgenic mice. J Neurosci Res 81:357–362
3. Jung S, Aliberti J, Graemmel P et al (2000) Analysis of fractalkine receptor CX3CR1 function by targeted deletion and green fluorescent protein reporter gene insertion. Mol Cell Biol 20:4106–4114
4. Nimmerjahn A, Kirchhoff F, Helmchen F (2005) Resting microglial cells are highly dynamic surveillants of brain parenchyma in vivo. Science 308:1314–1318
5. Davalos D, Grutzendler J, Yang G et al (2005) ATP mediates rapid microglial response to local brain injury in vivo. Nat Neurosci 8:752–758
6. Wake H, Moorhouse AJ, Jinno S et al (2009) Resting microglia directly monitor the functional state of synapses in vivo and determine the fate of ischemic terminals. J Neurosci 29:3974–3980
7. Tremblay ME, Lowery RL, Majewska AK (2010) Microglial interactions with synapses are modulated by visual experience. PLoS Biol 8:e1000527
8. Kondo S, Kohsaka S, Okabe S (2011) Long-lasting effect of a transient peripheral immune response on cortical spine dynamics revealed by two-photon microscopy in vivo. Mol Brain 4:27
9. Majewska A, Yiu G, Yuste R (2000) A custom-made two-photon microscope and deconvolution system. Eur J Physiol 441:398–408
10. Xu HT, Pan F, Yang G, Gan WB (2007) Choice of cranial window type for in vivo imaging affects dendritic spine turnover in the cortex. Nat Neurosci 10:549–551
11. Holtmaat A, Bonhoeffer T, Chow DK et al (2009) Long-term, high-resolution imaging in the mouse neocortex through a chronic cranial window. Nat Protocols 4:1128–1144
12. Farrar MJ, Bernstein IM, Schlafer DH et al (2012) Chronic in vivo imaging in the mouse spinal cord using an implanted chamber. Nat Methods 9:297–302
13. Drew PJ, Shih AY, Driscoll JD et al (2010) Chronic optical access through a polished and reinforced thinned skull. Nat Methods 7:981–984
14. Feng G, Mellor RH, Bernstein M et al (2000) Imaging neuronal subsets in transgenic mice expressing multiple spectral variants of GFP. Neuron 28:41–51
15. Saito T (2006) In vivo electroporation in the embryonic mouse central nervous system. Nat Protocols 1:1552–1558
16. Drovizhev M, Makarov NS, Tillo SE et al (2011) Two-photon absorption properties of fluorescent proteins. Nat Methods 8:393–399

Chapter 29

Use of Confocal Microscopy in the Study of Microglia in a Brain Metastasis Model

Manuel Sarmiento

Abstract

Confocal imaging of brain slices is a worthwhile analysis method to study the structure and function of resting and activated microglia with submicrometer resolution. This chapter will focus on acquisition of high-resolution confocal image stacks where we will discuss the technical aspects of confocal imaging in brain sections as well as some of the currently suitable fluorescent markers for this type of work.

Key words Confocal, Microglia, Brain, Cancer, Metastasis, Immunofluorescence

1 Introduction

The initial steps of brain metastasis are difficult to detect in vivo, and this is one of the primary reasons for poor prognosis in patients with this disease; the median survival of untreated patients is 1–2 months [1]. Confocal microscopy has become a widely used method in the study of brain diseases. This widespread adoption of confocal microscopy is the result of the advantages of the method, namely, eliminating out-of-focus light which produces sharp, high-contrast images of cells and subcellular structures even within thick sections, combined with the availability of increasingly powerful, sensitive, and user-friendly instruments from several manufacturers.

Metastasis is the movement or spreading of cancer cells from one organ or tissue to another. In the case of brain metastasis, the lymph nodes are not considered due to the absence of a lymphatic system within the central nervous system (CNS). The process of brain metastasis depends on the success of several steps, including cancer cell escape from the primary tumor, distribution and survival within the circulation, adhesion and penetration of the blood–brain barrier (BBB), and proliferation within the brain microenvironment. Importantly, the brain is the only site of tumor

Bertrand Joseph and José Luis Venero (eds.), *Microglia: Methods and Protocols*, Methods in Molecular Biology, vol. 1041,
DOI 10.1007/978-1-62703-520-0_29, © Springer Science+Business Media New York 2013

relapse in ~60 % of lung cancer patients, ~25 % of breast cancer patients, and ~55 % of melanoma patients [2] and is a frequent site of therapeutic failure.

The purpose of this chapter is to provide guidance in the use of confocal microscopy in the study of the behavior of microglia in one metastasis model. Microglia are the resident macrophages of the CNS [3]. Their roles in neurodegenerative and neuroinflammatory response have been intensely investigated, whereas their potential function in metastasis received almost no attention. Microglia have been described as enhancing invasion and colonization of brain tissue by breast cancer cells, serving both as active transporters and guiding rails to further invasion [4]. Recent insights into the mechanisms of cancer progression demonstrate the critical role of the tumor microenvironment in brain metastasis, tempting speculation with the fact that the resident macrophages/microglia may play an even more important role for colonization than the peripheral blood-derived macrophages. Other clinical reports state that once the tumor is in the CNS (i.e., glioma grade IV), macrophages/microglia support tumor growth by changing its microenvironment via neovascularization [5].

Several features of confocal scanning make it advantageous compared to conventional wide-field microscopy observations and measurements in brain sections. Probably the most important feature is that it allows the visualization of much smaller objects (e.g., microglia engulfing tumor cells). Moreover, it also makes possible the ability to correctly interpret the 3-dimensional organization of fine brain structure, for example, cell-to-cell contacts.

2 Materials

2.1 Tissue Preparation

1. *Heparinized sodium chloride*: Make up 50 mL of 0.9 % sodium chloride in distilled water and add 500 U of heparin (sodium heparin 125,000 U/5 mL).
2. PLP_{light}: Periodate-lysine-paraformaldehyde (PLP) containing only 0.025 % glutaraldehyde (*see* **Note 1**).
3. 0.1 M *phosphate buffer*: Mix 3.1 g of sodium dihydrogen orthophosphate in 500 mL distilled water. Mix 13 g of Sorenson's salt ($(Na_2HPO_4)2H_2O$) in 400 mL distilled water. Add together and make up to 1 L. If necessary adjust pH to 7.4.
4. *30% sucrose (in phosphate buffer)*: The brains will be cryoprotected in a high-concentration sucrose solution composed by 30 g of sucrose in 100 mL of phosphate buffer (*see* **Note 2**).
5. *Isopentane*: To speed the freezing process, the brains will be frozen in isopentane at –20 °C for at least 30 min.

2.2 Tissue Processing

Other chapters within this book describe stereotaxical procedures involved with in vivo models. Thus, this chapter aims to begin explanation of the methods used to process our samples once the tumor cells are already implanted in the brain parenchyma.

1. *PBS*: Dissolve 50 tablets of phosphate-buffered saline (Dulbecco A, Oxoid, UK) in 5 L of distilled water.
2. *Quench solution*: Add 2.5 mL of hydrogen peroxide 30 % (Sigma Aldrich, UK) to 250 mL of methanol (*see* **Note 3**).
3. *Blocking solution*: Block endogenous streptavidin and biotin using the streptavidin/biotin kit (Vector Laboratories, USA). After this, we need also to block sections with Tris–NaCl blocking buffer (TNB, Perkin-Elmer, USA).
4. Image-IT® FX signal enhancer (Invitrogen, USA) for 30 min. This reagent will improve the signal of the fluorophore.
5. *Primary antibody* against microglia (1:200 in TNB). Iba-1 (DAKO, USA. Ionized calcium binding adaptor molecule 1).
6. *Secondary antibody*: Biotinylated antibody against the species where the primary antibody was raised in (1:200 in TNB; Vector Laboratories Inc, USA).
7. *Streptavidin-HRP* (1:100 in TNB; Perkin-Elmer, USA).
8. *TSA-biotin* (Tyramide Signal Amplification-biotin) (1:100 in amplification buffer; Perkin-Elmer, USA) (*see* **Note 4**).
9. *Streptavidin-Cy3* fluorophore (1:200 in TNB; Invitrogen, USA).
10. *Vectashield* mounting medium (Vector Labs, USA) with DAPI.
11. *Cover slips.*

2.3 Imaging

1. *Confocal microscope* (LSM-710, Carl Zeiss Microimaging, *Jena, Germany*) (*see* Fig. 1).
2. *Oil immersion* (either oil or water immersion solution depending on the lens).
3. *Lens tissues* (*see* **Note 5**) (*see* Fig. 1a, b).

3 Methods

Fluorescence occurs when a molecule absorbs one or more photons to reach the excited state from which it can relax to the ground state; this process is accompanied by emission of a photon. Therefore, confocal imaging in brain sections requires that the cells or their elements must be tagged with some kind of fluorophore. The labelling can be achieved not only by immunofluorescence but also by other techniques such as viral gene expression systems or DNA transfection. In this case, our mouse mammary carcinoma cells (4T1) are tagged with a previously transfected GFP fluorophore [6].

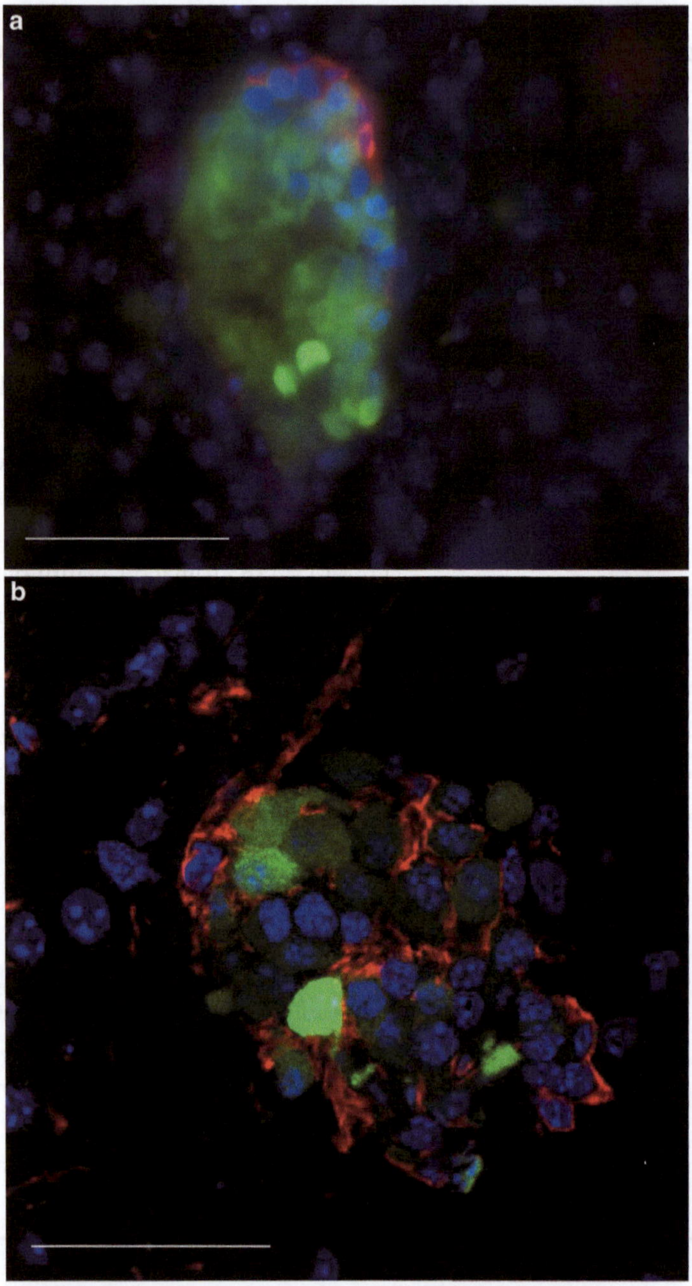

Fig. 1 Comparison of a tumor colony within the striatum of a mouse brain taken under two different microscopes. (**a**) Photomicrograph captured from an epi-fluorescence microscope. (**b**) Picture taken from a Zeiss confocal microscope. Both pictures were acquired from the same brain section. *Green*: tumor cells tagged with GFP, *Blue*: DAPI to stain nuclei, *Red*: Iba-1 antibody against microglia. Scale bars, 50 μm

Conventional confocal imaging of cells within brain sections is usually limited to the upper 60 μm of a section. The average penetration of commercial antibodies is around 20 μm; therefore, the thickness of the sections should be restricted to these measures.

3.1 Tissue Preparation

1. Inject anesthetic to the animal and start transcardiac perfusion. Inject into the left ventricle 50 mL of heparinized saline followed by 200 mL of PLP_{light} at ~40 mL/min. Ensure that the right atrium of the heart is cut to relieve pressure.
2. Harvest the brain and postfix it for 4 h at 4 °C in PLP_{light}. After this, transfer the brain into sucrose 30 % until sink, at 4 °C (after 24–36 h the brain will sink, indicating that cryoprotection is complete).
3. Place the brain in isopentane at –20 °C for 30 min. Once frozen, wrap it in tin foil and transfer to –20 °C for long-term storage.
4. The brains will be placed in a Leica cryostat at –20 °C temperature, and the thickness of the sections will be of 10 μm. Once cut, the sections will be placed on the slides (*see* **Note 6**).

3.2 Tissue Processing

1. Hydrate 10 μm sections for 5 min in PBS (*see* Subheading 3.1).
2. Quench the sections with 1 % hydrogen peroxide (*see* Subheading 2.2, **item 2**) for 15 min.
3. Streptavidin- and biotin-blocked (Vector Laboratories, USA; 15 min each; *see* Subheading 2.2, **item 3**).
4. 30 min incubation in Image-IT® enhancer (Subheading 2.2, **item 4**) plus 30 min in Tris–NaCl blocking buffer (TNB; *see* Subheading 2.2, **item 3**).
5. Incubate overnight with anti-Iba-1 antibody raised in rabbit (Wako, USA; *see* Subheading 2.2, **item 5**).
6. Rinse with PBS (three rinses, 10 min each).
7. Incubate with a biotinylated secondary antibody anti-rabbit (Subheading 2.2, **item 6**) for 1 h.
8. Rinse sections (three rinses, 10 min each).
9. Incubate with streptavidin-HRP (*see* Subheading 2.2, **item 7**) in TNB for 30 min.
10. Rinse with PBS (three times, 10 min each).
11. Incubate sections for 8 min in the dark with TSA-biotin (*see* Subheading 2.2, **item 8**) diluted in amplification buffer (*see* **Note 4**).
12. Rinse slides in PBS (three times, 10 min)
13. Incubate sections with a streptavidin-Cy3 fluorophore (1:200 in TNB; Invitrogen, USA) (*see* Subheading 2.2, **item 9**) for 30 min.
14. Cover-slip the slides using Vectashield mounting medium (*see* Subheading 2.2, **item 10**) with DAPI to stain nuclei. The slides are now ready to move to the microscope.

3.3 Examination Under Confocal Microscopy

Images can be acquired using an inverted confocal microscope (LSM-710, Carl Zeiss Microimaging, Jena, Germany) and analyzed using Image J (rsbweb.nih.gov/ij) and Zen (Carl Zeiss) software.

1. Place the slide in the sample holder. Start from a low magnification lens in order to find the tumor colony.
2. Apply immersion medium (oil or glycerol depending on which class of lens used) to the objective of the ×63. Focus on the region of interest and choose the channels to be used depending on the fluorophores used in your immunofluorescence. In this case we will use lasers at 405 nm (DAPI), 488 nm (GFP), and 568 nm (Cy3).
3. Adjust laser intensity. In order to avoid saturation of the images, each channel's laser intensity must be checked individually and adjust appropriately (*see* **Note 7**).
4. Set all the pinholes at ≤1 airy unit (AU). With a pinhole diameter <1 AU, resolution improves (better point separation), which is penalized by a drastic loss in energy (*see* **Note 8**).
5. Detection ranges were set to eliminate crosstalk between fluorophores: 409–485 nm for DAPI, 494–553 nm for Alexa Fluor, and 564–712 nm for Cy3. Detection windows can be tuned individually in order to avoid the overlap of the emission spectra of the fluorophores (*see* **Note 9**).
6. *Z*-stack: use the micrometer to focus at both levels of the sample (upper and lower). Save both coordinates in order to set the thickness of our range of detection (*see* **Note 10**).
7. Adjust the confocal parameters to obtain the better resolution image (*see* **Note 11**) (*see* Fig. 2).

4 Notes

1. One of the reasons to use PLP_{light} (a "soft" fixative) is the use of frozen sections during the processing of the samples. This kind of fixative helps an easier penetration of the antibody within the sample, allowing a more efficient binding between the Iba-1 antibody and its epitope.

   Recipe: Dissolve 3.4 g lysine HCl in 94 mL distilled water. Dissolve 0.45 g disodium hydrogen orthophosphate dehydrate (Sorenson's salt) in 25 mL distilled water. Add this to lysine solution in order to adjust the pH to 7.4. Discard any remaining phosphate solution. Dissolve 5 g of paraformaldehyde in 50 mL of distilled water at 60 °C. Add paraformaldehyde and slowly add a few drops of 1 M NaOH as a catalyst while stirring until the solution clears. Leave to cool down. Prepare 0.53 g of sodium metaperiodate. Mix the lysine solution, paraformaldehyde solution, and the sodium periodate together. Make up to 250 mL with 0.1 M phosphate buffer

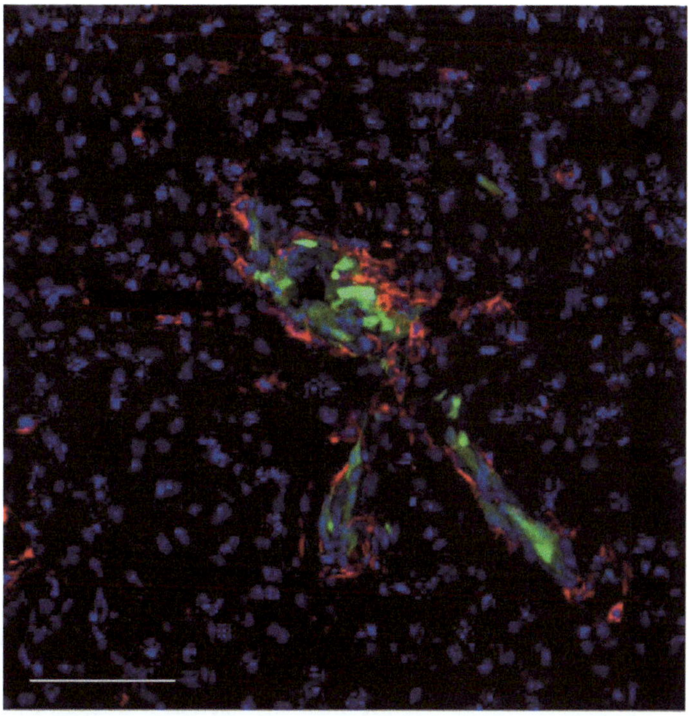

Fig. 2 Photomicrograph of a tumor within the striatum of a mouse 5 days after implantation. You can see the tumor cells growing along the vessels and microglial activation surrounding the boundaries of the colony. *Green*: tumor cells tagged with GFP, *Blue*: DAPI to stain nuclei, *Red*: Iba-1 against microglia. Scale bar, 100 μm

and adjust pH to 7.2 with NaOH. Finally, add 0.25 mL of glutaraldehyde 0.01 % for fixing tissue.

2. Depending on the accuracy of the transcardial perfusion, sometimes the brains take longer to sink when left in sucrose solution; if so, it is recommended to use sucrose gradient embedding. After the first 4 h in PLP_{light}, transfer the brain to 10 % until it sinks. Repeat this procedure with 20 % and finally 30 % solutions of sucrose. In this way we can ensure that the sucrose will perfuse inside the brain.
3. Some fluorophores are affected by the methanol quenching solution. In these cases, instead of methanol, use PBS.
4. The Perkin–Elmer kit provides the TNB blocking buffer and the amplification buffer into which the reagents will be diluted. It is possible to substitute these with PBS if necessary.
5. It is important to thoroughly clean the lens after using the immersion oil in order to avoid staining other lenses. The air lenses could be seriously affected if immersion agents are allowed to come into contact with them.
6. Once cut the brain, to avoid the detachment of the sections from the slide, it is important to leave them dry properly; in this case we left them overnight at room temperature in darkness to avoid photobleaching of the GFP present in the tumor cells.

7. We need to check that the images are not overexposed to the fluorophore; to do that, we must check that the image has no saturated voxels. The exact method for doing so varies between different manufacturer's software, but in principle there will be an option to highlight every saturated pixel. If necessary, we can reduce the gain/power of the laser to remove them.
8. A small pinhole diameter will increase the depth of focus, but reduce the light intensity received so that the projected image will be dimmer. For a proper confocal microscopy, the pinhole shouldn't be wider than 1 AU. However, it should be considered that it will depend on the signal level which noise source dominates. With high-amplitude signals (number of detected photons >10,000), laser noise is the dominating effect, whereas the quality of low signals (number of detected photons <1,000) is limited by the shot noise of the light. 1 Airy Unit is a good value to enable a confocal fluorescence XY-image to be obtained.
9. Even without a fluorescent dye, brain sections contain fluorescent molecules. Some of these molecules, such as tyrosine, tryptophan, and NADH, will only be fluorescent when excited with UV light (<400 nm). In confocal imaging, UV light is not normally used as it does not penetrate well into the tissue. The 488 nm light from an argon laser (the most commonly used wavelength in a single-photon system) excites other endogenous molecules, in particular flavins and porphyrins. It is important to note that most of these molecules emit in the green–yellow part of the spectrum (500–600 nm). Thus, it is important to take into account the autofluorescence problems of the fluorophores variants of green fluorescent protein (GFP) that are imaged using 458 or 488 nm excitation and so emit in the green part of the spectrum, where both absorption and emission of endogenous molecules are the strongest [7].

   Molecules that are mainly responsible for autofluorescence are mostly unexcitable by >600 nm light (e.g., 633 nm line of red diode lasers). Thus, the red part of the spectrum seems to be ideal for single-photon imaging of difficult and deep objects in brain sections where autofluorescence is intense (i.e., Cy3 or Cye5 dye) [7].
10. Confocal imaging gives us the option of either choosing a single plane of the section to image or to obtain the average of all the slices exposed (known as maximum intensity projection). Depending on the quality of our immunofluorescence, and our objectives, either thing is possible (*see* Fig. 3).
11. The system determines the settings for the Gain (Master) and digital offset to match the scan speed, pinhole size, and laser power. The higher digital offset, the lower background, but the intensity of the signal received will be also reduced (*see* Fig. 3a, b).

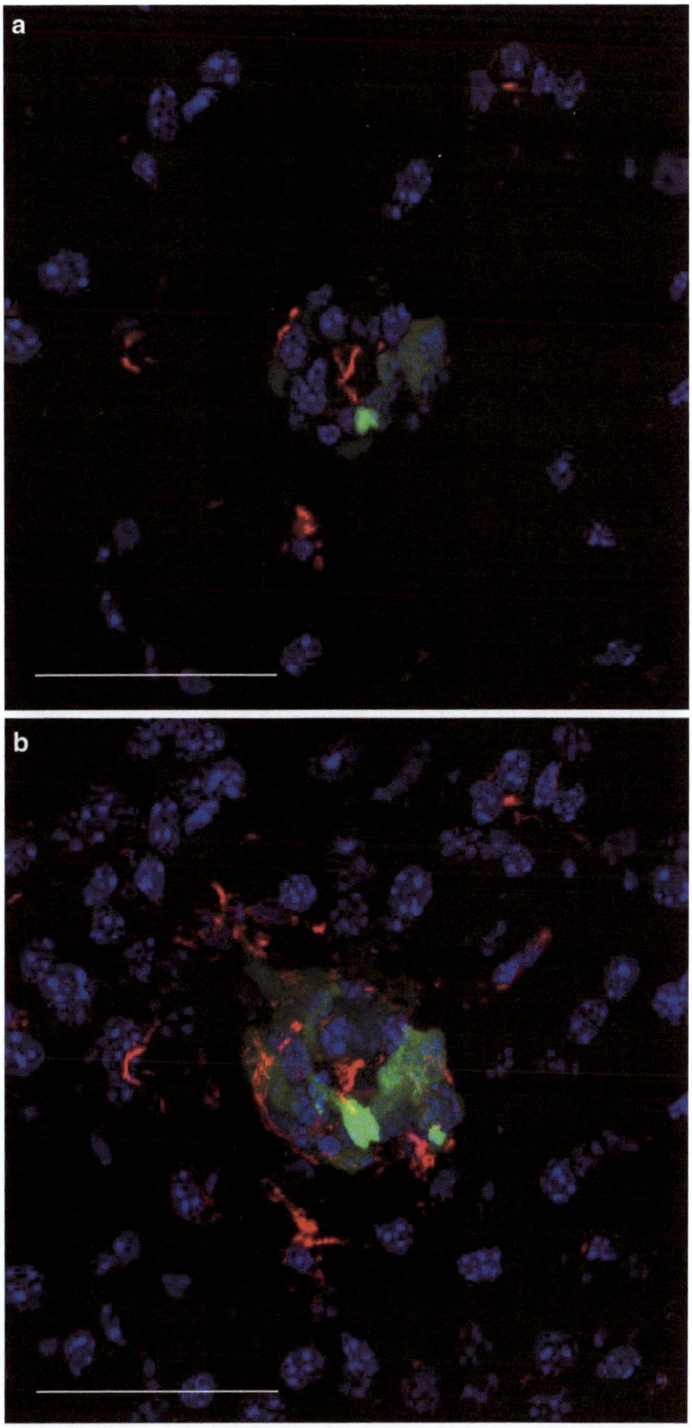

Fig. 3 These two photomicrographs are taken from a single tumor colony in a mouse 5 days after tumor implantation. (**a**) A single *X*–*Y* plane section of the slide. (**b**) A combined image using the average of intensity of 20 planes of the same slide (maximum intensity projection). Scales bars, 50 μm

References

1. Eichler AF, Loeffler JS (2007) Multidisciplinary management of brain metastases. Oncologist 12:884–898
2. Sampson JH, Carter JH Jr, Friedman AH et al (1998) Demographics, prognosis, and therapy in 702 patients with brain metastases from malignant melanoma. J Neurosurg 88:11–20
3. Hanisch UK, Kettenmann H (2007) Microglia: active sensor and versatile effector cells in the normal and pathologic brain. Nat Neurosci 10:1387–1394
4. Pukrop T, Dehghani F, Chuang HN et al (2010) Microglia promote colonization of brain tissue by breast cancer cells in a Wnt-dependent way. Glia 58:1477–1489
5. Komohara Y, Ohnishi K, Kuratsu J et al (2008) Possible involvement of the M2 anti-inflammatory macrophage phenotype in growth of human gliomas. J Pathol 216:15–24
6. Aslakson CJ, Miller FR (1992) Selective events in the metastatic process defined by analysis of the sequential dissemination of subpopulations of a mouse mammary tumor. Cancer Res 52:1399–1405
7. Kasparov S, Teschemacher AG, Paton JF (2002) Dynamic confocal imaging in acute brain slices and organotypic slice cultures using a spectral confocal microscope with single photon excitation. Exp Physiol 87:715–724

Index

Bertrand Joseph and José Luis Venero (eds.), *Microglia: Methods and Protocols*, Methods in Molecular Biology, vol. 1041, DOI 10.1007/978-1-62703-520-0, © Springer Science+Business Media New York 2013

M

N

O

P

R

S

T

V

W

MIX
Papier aus verantwortungsvollen Quellen
Paper from responsible sources
FSC® C105338

If you have any concerns about our products,
you can contact us on
ProductSafety@springernature.com

In case Publisher is established outside the EU,
the EU authorized representative is:
Springer Nature Customer Service Center GmbH
Europaplatz 3, 69115 Heidelberg, Germany

Printed by Libri Plureos GmbH
in Hamburg, Germany